In this unique volume, key researchers present newly emerging approaches to computer simulation models of large, forest landscapes. Over the past decade the field of landscape ecology has developed rapidly, focusing on the need to address ecological research and management at large spatial scales, and longer temporal domains. There is also great attention being focused on the use and management of forests throughout the world, particularly on issues such as long-term sustainability, ecosystem management, and biodiversity protection. These models have the potential to help answer research and management questions through simulation experiments that have not, in the past, considered spatial interactions among ecological processes and human activities.

Representing a rapidly emerging area in the field of landscape ecology, this volume will be of value to ecologists, forest and natural resource managers, as well as to wildlife biologists and conservationists.

DAVID J. MLADENOFF is an Associate Professor in the Department of Forest Ecology and Management at the University of Wisconsin–Madison.

WILLIAM L. BAKER is a Professor in the Department of Geography and Recreation at the University of Wyoming.

Spatial modeling of forest landscape change: approaches and applications

EDITED BY

DAVID J. MLADENOFF and WILLIAM L. BAKER
University of University of Wyoming
Wisconsin–Madison

PUBLISHED BY THE PRESS SYNDICATE OF THE UNIVERSITY OF CAMBRIDGE
The Pitt Building, Trumpington Street, Cambridge, United Kingdom

CAMBRIDGE UNIVERSITY PRESS
The Edinburgh Building, Cambridge CB2 2RU, UK   www.cup.cam.ac.uk
40 West 20th Street, New York, NY 10011-4211, USA   www.cup.org
10 Stamford Road, Oakleigh, Melbourne 3166, Australia
Ruiz de Alarcon 13, 28014 Madrid, Spain

© Cambridge University Press 1999

This book is in copyright. Subject to statutory exception and to the provisions of relevant collective licensing agreements, no reproduction of any part may take place without the written permission of Cambridge University Press.

First published 1999

Printed in the United Kingdom at the University Press, Cambridge

Typeset in Bembo 11/13pt [wv]

A catalogue record for this book is available from the British Library

Library of Congress Cataloguing in Publication data

Spatial modeling of forest landscape change:
approaches and applications/[editors, David J. Mladenoff, William L. Baker].
p.   cm.
Papers presented at symposium in Albuquerque, New Mexico, USA, 1997.
Includes index.
ISBN 0 521 63122 X (hardback)
1. Forest dynamics – Computer simulation – Congresses.   2. Landscape ecology – Computer stimulation – Congresses.   I. Mladenoff, David J.
II. Baker, William L. (William Lawrence)
QK938.F6S668 1999
577.3′01′13–dc21   98-50552 CIP

ISBN 0 521 63122 X hardback

Contents

List of contributors	*page* vii
Preface	xi

1 Development of forest and landscape modeling approaches
DAVID J. MLADENOFF and WILLIAM L. BAKER — 1

2 Modeling the competitive dynamics and distribution of tree species along moisture gradients
JOHN P. CASPERSEN, JOHN A. SILANDER, JR, CHARLES D. CANHAM and STEPHEN W. PACALA — 14

3 Spatial and temporal impacts of adjacent areas on the dynamics of species diversity in a primary forest
JIANGUO LIU, KALAN ICKES, PETER S. ASHTON, JAMES V. LAFRANKIE and N. MANOKARAN — 42

4 Scaling fine-scale processes to large-scale patterns using models derived from models: meta-models
DEAN L. URBAN, MIGUEL F. ACEVEDO and STEVEN L. GARMAN — 70

5 Simulating landscape vegetation dynamics of Bryce Canyon National Park with the vital attributes/fuzzy systems model VAFS/LANDSIM
DAVID W. ROBERTS and DAVID W. BETZ — 99

6 Design, behavior and application of LANDIS, an object-oriented model of forest landscape disturbance and succession
DAVID J. MLADENOFF and HONG S. HE — 125

7 Predicting forest fire effects at landscape scales
ROBERT H. GARDNER, WILLIAM H. ROMME and MONICA G. TURNER — 163

8 Mechanistic modeling of landscape fire patterns
   MARK A. FINNEY 186

9 Achieving sustainable forest structures on fire-prone landscapes while pursuing multiple goals
   JOHN SESSIONS, K. NORMAN JOHNSON, JERRY F. FRANKLIN and JOHN T. GABRIEL 210

10 Modeling the driving factors and ecological consequences of deforestation in the Brazilian Amazon
   VIRGINIA H. DALE and SCOTT M. PEARSON 256

11 Spatial simulation of the effects of human and natural disturbance regimes on landscape structure
   WILLIAM L. BAKER 277

12 HARVEST: linking timber harvesting strategies to landscape patterns
   ERIC J. GUSTAFSON and THOMAS R. CROW 309

13 Progress and future directions in spatial modeling of forest landscapes
   WILLIAM L. BAKER and DAVID J. MLADENOFF 333

Index 350

Colour plates between 162 and 163

Contributors

Miguel F. Acevedo, Department of Geography and Institute of Applied Sciences, University of North Texas, Denton, TX 76203, USA

Peter S. Ashton, Harvard Arboretum, Harvard University, Cambridge, MA 02318, USA pashton@oeb.harvard.edu

William L. Baker, Department of Geography and Recreation, University of Wyoming, Laramie, WY 82071, USA bakerwl@uwyo.edu

David W. Betz, US Forest Service, Targhee National Forest, PO Box 208, St Anthony, ID 83445, USA

Charles D. Canham, Institute of Ecosystem Studies, Millbrook, NY 2545, USA

John P. Caspersen, Department of Ecology and Evolutionary Biology, University of Connecticut, Storrs, CT 06269-3042, USA **Current address:** Department of Ecology and Evolutionary Biology, Princeton University, Princeton NJ 08544, USA jpc@eno.princeton.edu

Thomas R. Crow, US Forest Service, Forestry Sciences Laboratory, 5985 Highway K, Rhinelander, WI 54501, USA. Current address: School of Natural Resources and Environment, University of Michigan, Ann Arbor, MI, 48107, USA

Virginia H. Dale, Environmental Sciences Division, Oak Ridge National Laboratory, Oak Ridge, TN 37831-6036, USA vhd@ornl.gov

Mark A. Finney, Systems for Environmental Management, PO Box 8868, Missoula, MT 59808, USA mfinney@montana.com

Jerry F. Franklin, College of Forest Resources, Box 352100, University of Washington, Seattle, Washington 98195, USA

John T. Gabriel, Alsea Geospatial, Inc., Corvallis, Oregon 97333, USA

Robert H. Gardner, Appalachian Laboratory, University of Maryland Center for Environmental Science, Frostburg, MD 21532, USA gardner@al.umces.edu

Steven L. Garman, Department of Forest Science, Oregon State University, 3200 Jefferson Way, Corvallis, OR 97331, USA

Eric J. Gustafson, US Forest Service, Forestry Sciences Laboratory, 5985 Highway K, Rhinelander, WI 54501, USA egustafson/nc—rh@fs.fed.us

Hong S. He, Department of Forest Ecology & Management, University of Wisconsin–Madison, 1630 Linden Drive, Madison, WI 53706, USA

Kalan Ickes, Department of Biological Sciences, Louisiana State University, Baton Rouge, LA 70803-1715, USA

K. Norman Johnson, College of Forestry, Oregon State University, Corvallis, OR 97331–5706, USA

James V. LaFrankie, Center for Tropical Forest Sciences, c/o Nanyang Technological University, National Institute of Education, Singapore 1025

Jianguo Liu, Department of Fisheries and Wildlife, Michigan State University, East Lansing, MI 48824, USA jliu@perm3.fw.msu.edu

N. Manokaran, Forest Research Institute Malaysia, Kepong, 52109 Kuala Lumpur, Malaysia

David J. Mladenoff, Department of Forest Ecology and Management, University of Wisconsin–Madison, 1630 Linden Drive, Madison, WI 53706, USA djmladen-@facstaff.wisc.edu

Stephen W. Pacala, Department Ecology and Evolutionary Biology, Princeton NJ 08544, USA

Scott M. Pearson, Dept. of Biology, Mars Hill College, Mars Hill, NC 28754, USA

David W. Roberts, Department of Forest Resources and Ecology Center, Utah State University, Logan, Utah 84322-5215, USA dvrbts@nr.usu.edu

William H. Romme, Department of Biology, Fort Lewis College, Durango, CO 81301, USA

John Sessions, College of Forestry, Oregon State University, Corvallis, OR 97331-5706, USA john.sessions@cof.orst.edu

John A. Silander, Jr, Department of Ecology and Evolutionary Biology, University of Connecticut, Storrs, CT 06269-3042, USA

Monica G. Turner, Department of Zoology, University of Wisconsin–Madison, Madison, WI 53706, USA

Dean L. Urban, Nicholas School of the Environment, Duke University, Durham, NC 27708-0328, USA deanu@pinus.env.duke.edu

Preface

Our purpose in writing this book is to assemble a representative group of approaches being taken in spatial modeling of forest landscapes. This is a new, rapidly growing area of landscape ecology, that also draws on forestry and geographic information science. Research in this area produces contributions to fundamental ecological science as well as to applied science and landscape management. This new interface of sciences and management is currently an exciting area of work.

The genesis of this book grew out of our own work on models and applications in forest landscape ecology, observing the diversity of approaches being taken by colleagues, and frequent discussions on the need for a book such as this. Along with many other colleagues, including those in this volume, we perceived a need for such a book as a benchmark in a rapidly changing field, for scientists, teachers, and graduate students, as well as a useful reference for landscape managers.

Most of the chapters in this book are expanded versions of presentations in a symposium we organized with the same title as part of the 1997 annual meeting of the Ecological Society of America (ESA), in Albuquerque, New Mexico, USA. We are particularly grateful to the ESA Vegetation Ecology Section, and the North American Section of the International Association for Vegetation Science (NA-IAVS), for sponsoring and supporting the symposium.

Any edited volume is composed of a set of contributions that evolves from what fits the focus of the volume, the contributions desired, work known to the organizers, and contributors who can commit time to participate. It would be possible to compile a volume still larger than this, that more broadly covers landscape modeling, or forest models and planning systems. The scope of this book is based on our perceived need for a limited focus. Within the scope of the book, we attempted to bring together contributors working with approaches and in geographic locations that were as diverse as possible, or could be found through colleagues. We are extremely pleased with the diversity and quality of researchers represented in this volume. It could only be improved if we could have found additional representatives of the field beyond North America. A few colleagues were not able to contribute, owing to conflicts with other commitments. As we moved ahead with preparing the book, we learned of others within the field of forest landscape modeling. Although we could not include all, we take this as an indication of a field that continues to be healthy and growing, in both research and application.

In addition to a round of internal chapter reviews within the group of authors, we were fortunate to have outside reviews by several colleagues. We are indebted to Dennis Boychuk, Frank Davis, Michael Huston, Timo Kuuluvainen, William Laurence, Ajith Perera, Stephen Shifley, Volker Radeloff, Jan van Wagtendonk, and David Wallin. We are grateful for assistance from Ted Sickley.

We have established a World Wide Web site to accompany this book (*http://forestlandscape.wisc.edu/book/*), and to help keep interested users up to date. Included are test versions of some of the models and data included here, teaching exercises, and contacts for the authors. There are also links to individual authors' web sites, where additional information can be gotten on recent research and publications. We also provide links to other, related sites of interest. Suggestions of additional, related links are welcome.

David J. Mladenoff
William L. Baker

1

Development of forest and landscape modeling approaches

David J. Mladenoff and William L. Baker

Ecological basis of forest models

The forest landscape models encompassed by this book derive from several sources. Most largely descend from ecological models developed over approximately the past 30 years. They owe their conceptual basis to developments in concepts and theories of ecological succession, disturbance, and equilibrium and non-equilibrium ecological systems, including populations, communities, ecosystems, and landscapes. Clearly, these conceptual roots extend further back within ecology than the briefer, 30-year period of computer modeling. Most of the approaches presented here combine aspects of all of the above. Some of the models in this volume (most notably Sessions *et al.*, Chapter 9) also derive more broadly from the field of geographic information systems (GIS) and forest planning, management, and decision-making software.

The modeling approaches here also are generally both empirical and mechanistic, and many have practical application in analyzing change and management on real landscapes. Development and use of these forest landscape models liberally cross the lines of fundamental and applied research. The impetus for many of these modeling approaches clearly originates within the needs of resource management, and the growing interest in managing larger landscapes. Meeting these management needs often requires tools that can help assess the effects of different management scenarios. These scenarios often require decision-making horizons spanning broad temporal and spatial scales. Often, simulation models are the only way to assess alternatives that cannot be tested under such real-world conditions.

The models in this book also can be considered to lie within an area of ecology itself, namely landscape ecology. Landscape ecology, particularly in North America, derives its theoretical development most strongly from ecosystem and community ecology of the last four decades, and its applied aspects from environmental management (Golley, 1993). Spatial processes and interactions, large land areas, and heterogeneity are explicit added focuses of landscape ecology (Pickett and

Cadenasso, 1995; Turner, 1989). Both ecology and environmental or ecosystem management can trace their own growth to a common origin in the social and environmental climate of the 1960s, with the growth of publicly funded research and environmental concern and regulation (McIntosh, 1985).

Several concepts in ecology are, in particular, fundamental for the approaches described in this book. These are concepts of vegetation change based on disturbance and succession, and the non-equilibrium nature of vegetation and ecosystems, e.g., contributions and literature reviewed in Pickett and White (1985). Another important concept is that of scale, and especially the view of the importance of spatially explicit dynamics, operating at varied spatial and temporal scales, in vegetation change (Levin and Paine, 1974; Grubb, 1977; Loucks et al., 1985). These concepts, along with advances in technical tools such as the computational capability and low cost of new computers, GIS software, and remote sensing, provide the foundation for the spatial modeling approaches presented here.

Theory and concepts of succession, disturbance, and equilibrium

There have been many excellent reviews of the development of successional theory and concepts and this development will not be described in detail again here (Connell and Slatyer, 1977; Drury and Nisbet, 1973; West et al., 1981; McIntosh, 1985; Glenn-Lewin et al., 1992). However, a synopsis of historical context and development of vegetation change concepts in ecology will be provided. These concepts are ultimately embedded in the modeling approaches described in this book. Also, a general genealogy of forest models that lead to the models presented later will be outlined.

Systematic study of succession developed in the 1890s and early 1900s, especially with the work of Cowles (1899, 1911) and Cooper (1926). Detailed theory and description of mechanisms were comprehensively developed by Clements (1916). Clements was instrumental in articulating concepts and mechanisms of succession as an orderly, deterministic progression from beginning points of vegetation composition to a climatically determined climax, that represented a stable equilibrium (Clements, 1916). Most importantly, the plant community was seen as responding as a single organism itself through time, and the inevitable climax, once reached, exhibited long-term stability. The intuitive appeal and heuristic usefulness of such a theory is in its order and predictability – features of great utility in developing a predictive science of ecology, and ultimately, useful predictive models. There is some risk in portraying a caricature, or in over-simplifying, the rich and detailed work of Clements. Nonetheless, it is this simplified view that became dogma with ecology (McIntosh, 1985).

Alternative models were developed early as well, by Gleason (1926) in the US, Ramensky in Russia (1924; cited in McIntosh, 1985), and Tansley in England (1935). Gleason especially criticized Clements and showed that succession had a much larger stochastic component, and that species respond individualistically to the environment. Species composition is due in part to chance events resulting in an assemblage of species with similar environmental responses, but not due to deterministic change of the community as a whole (Gleason, 1927). Nevertheless, the intuitive appeal of the Clementsian approach largely prevailed in ecology until after the 1940s. In the 1950s more extensive, quantitative approaches validated Gleason's view, in locations of both steep environmental gradients (Great Smoky Mountains: Whittaker, 1956) and moderate gradients (Wisconsin: Curtis and McIntosh, 1951; Curtis, 1959).

Even after the 1950s, the Clementsian view persisted, particularly within the developing field of systems ecology (Odum, 1969, 1971; Margalef, 1963). For several decades a notable rift existed in ecology exemplified on the one hand by the Clementsian-derived approach of systems ecology and, on the other hand, by the Gleasonian-based community and population – evolutionary approach (McIntosh, 1980; Peet and Christensen, 1980). More recent approaches have recognized a more integrated view of succession, and include plant demographics (Huston and Smith, 1987; Glenn-Lewin et al., 1992).

Somewhat more slowly to develop than succession theory itself was the other half of the vegetation change equation – the importance of disturbance. This development required evolution in the concepts of succession and climax, and the implied stable equilibrium state of the climax, that had left little role for the perceived, rare disturbance events. Non-equilibrium notions of vegetation actually were articulated relatively early, and recognized the importance of observational scale, but were even more slowly accepted than were challenges to the Clementsian succession paradigm (Watt, 1947; Whittaker, 1953). More recent concepts of disturbance explicitly recognize the importance of stochasticity and variability in disturbance severity, spatial and temporal scale, and the integration of changing ecosystem processes, such as resource availability, with individual species responses to disturbance (Pickett and White, 1985). The recognition that these processes operate at varied spatial and temporal scales, that succession and disturbances can recur in ecological systems, and that these interactions have consequences for varied trajectories of vegetation change, set the stage for modeling forest landscapes in realistic ways.

Conceptual models of forest change

Several approaches developed to apply these changing successional concepts to predictive models, albeit non-spatial, during the 1970s. Those most developed were two related approaches, Markov transition models and vital-attribute models. These approaches exemplify more formal mathematical and rule-based formulations of succession theory, are useful heuristically, and began application of succession theory to studies of ecological change and management.

Stationary Markov models (Feller, 1968) have been used to characterize successional changes over time based on observed transitions in forests (Stephens and Waggoner, 1970; Waggoner and Stephens, 1970). Markov models are a mathematical approach utilizing a matrix of empirically determined transition probabilities to predict tree-species replacement and, therefore, composition over time. Shortcomings of this approach have been described many times, and include the problems that (i) successional change may not be stationary, i.e., the probabilities may not be constant over time; (ii) they are typically first-order models, meaning that only the current state is considered in determining the transition; and (iii) they are non-spatial, in that adjacency relationships are not considered in the transition probabilities (van Hulst, 1979; Binkley, 1980). Modifications of this approach exist, described as semi-Markov models, that address some of the above problems (Ginsberg, 1971).

Computer models of forest change

During the 1970s several approaches to converting succession theory and knowledge of forest stand dynamics appeared that we see as antecedents to many of the modeling methods in this book. These models were formulated at individual-tree, forest gap, and stand scales.

Relatively simple rule-based transition models, related to Markov models, were developed in Australia (Noble and Slatyer, 1980) and in the western US (Cattelino *et al.*, 1979). These transition models have greater flexibility than the simple Markov models, and began to incorporate various pathways due to disturbance and site characteristics, rather than converging to a single steady state. State changes were based on combinations of species life history characteristics or 'vital attributes' (Cattelino *et al.*, 1979) which respond to disturbances. The models proved to be ecologically useful, and were applied within the FORPLAN forest planning system of the US Forest Service (Potter *et al.*, 1979). These were some of the first models to be implemented on computers and used extensively in forest management for large areas, such as a National Forest, although, again, they were not explicitly spatial.

One of the most interesting and detailed of the early forest models was the FOREST model of Ek and Monserud (1974), developed for northern Wisconsin, USA. The model simulated reproduction, growth, and mortality of all trees in a forest stand. Each tree was spatially located, and adjacent effects of crown shading and seed dispersal were included. The model challenged the computational capabilities of computers at the time, included great detail, and also had extensive input parameter requirements. In many ways the FOREST model was a conceptually and mechanistically superior approach, especially for the 1970s. But the computational requirements and input needs limited its use.

Slightly earlier, Botkin *et al.* (1972) developed JABOWA, the first of the forest gap simulators, for forests of the Northeastern US. This model and its many descendants incorporated the ecological realism of simulating detailed succession at the forest gap level. The JABOWA model of Botkin *et al.* spawned numerous modifications and model versions, particularly by Shugart and colleagues (Shugart, 1984, 1998). Gap models have been adapted worldwide, and have made important contributions to understanding forest-stand dynamics in nearly every type of forest ecosystem. Useful reviews of these implementations and the gap model evolution are also numerous (Shugart, 1984, Urban and Shugart, 1992).

Applications of gap models are often assumed to portray change at the single-tree gap scale, and results interpreted accordingly. However, this was not typically the scale at which most versions operated. The original JABOWA model simulated a 0.01 ha (10 m × 10 m), tree-size plot. The FOREST model and its descendants simulate a plot size significantly larger than this, typically 0.08–0.1 ha (Shugart, 1984). The computational problems inherent in the FOREST model were somewhat avoided in this conceptualization. An entire forest stand simulation could be approximated, assuming homogeneity, by averaging multiple single plot runs. Later versions of the gap model added detail by a return to the single-tree scale, and with spatial implementations, simulating a network of plots in a stand-sized grid of 4–9 ha (Smith and Urban, 1988; Sarkar et al., 1996). Since the original gap models were single-plot, and non-spatial, they typically did not include disturbances such as fire (but see Kercher and Axelrod, 1984). Gap models have been usefully applied to many problems since early in their development, including management (Aber et al., 1982) and climate change (Solomon, 1986; Pastor and Post, 1988; Shugart, 1998).

Most models at this time were addressing within-stand change. A few approaches attempted to directly address larger-scale, landscape or regional forest change, and stand out for the early period of their development, as the FOREST model does at the stand level. Models were developed for regions of the US under the US IBP Program (Loucks et al., 1981). A differential equation model that simulated regional forest change across the North central US Lake States was developed by Shugart et al. (1973). The model was non-spatial, and did not include disturbance. Change was similar to the Markov tree models described above, but in this case transitions were between major forest cover classes. The third model, for the Georgia Piedmont, is particularly interesting in that it included land-use cover-classes as well as natural forest changes (Johnson and Sharpe, 1976). This model was not only a deterministic successional model, but included possible alterations on the landscape by logging, fire, and grazing. The model output described percentage changes in the cover classes for the region under different change rates for the disturbance parameters (Johnson and Sharpe, 1976).

Recently, Pacala et al. (1993) have developed a spatially explicit, stand-level model (SORTIE) based on new and more detailed field data and calibrations of seed dispersal, recruitment, growth, and mortality. In this approach, individual trees are each located in a stand. In this respect and in its greater data requirements, it owes much to the approach in the FOREST model developed by Ek and Monserud (1974). In its mechanisms and array of submodels, SORTIE derives its structure from the JABOWA-FORET gap models (Pacala et al., 1993). The advantages of SORTIE are that it incorporates stronger field data-derived relationships in its submodels than did the previous gap models. Its disadvantages are that, like FOREST in the past, simulating large areas with a spatially explicit, single-tree model has great computational cost. Typically, stands of <10 ha can be simulated with reasonable processing time (Pacala et al., 1993). The model also has its power built on the need for extensive new data collection for input parameter calibration, data not yet available for most locations outside of its original development locale of southern New England.

Forestry growth and yield models

Forestry models under this group differ in their purposes and derivation from the more ecological models described above. The development of this field has been reviewed several times (Munro, 1974; Loucks et al., 1981; Dale et al., 1985; Parks and Alig, 1988). These models derived from forest growth and yield data and predictive equations at first concerned with predicting aggregate stand growth increment for potential harvest, such as STEMS (Belcher et al., 1982). Later individual-tree simulation models bridged this field with the ecological models. The FOREST model is within this overlapping group and is appropriately described above as well.

Forest management and planning models

Models under this general area actually begin to separate from the more biologically based succession and forest growth models, and overlap with more general management planning, land-use planning, and decision-support software systems. There has often been a close link between the growth and yield models and these planning models (Iverson and Alston, 1986). Some distinction of scale can be made, differentiating larger scale, strategic planning or regional timber-supply models, and smaller scale, growth and yield and tactical decision models. One of the best known models is the FORPLAN model of the US Forest Service (Iverson and Alston, 1986). Models of this type are often complex, and have been criticized for past versions that lacked ecological dynamics and variability, or spatial considerations, which were then not included in resulting projections and plans (Johnson, 1992). This is a complex and growing area, and despite shortcomings, has been important in extending the development and use of computers models for large-scale planning, and assessing management consequences on large forest ownerships.

These models that developed within the forest management field also tended to diverge from other forest models after the 1970s, similar to the split between the forestry and ecological models. Within this management and planning field, GIS applications developed that were more decision-making software than models in a strict sense, and they were not ecological. These applications did not typically consider, for example, natural disturbance rates or variability, or spatial interactions in their planning algorithms (Johnson and Scheurman, 1977; Hoganson and Burke, 1997).

Disturbance models

The final group of models that form part of the basis for the models in this book are disturbance models. Much of the information and initial modeling efforts within this group were empirically based fire spread models, developed to understand fire behavior for suppression purposes (Van Wagner, 1969; Rothermel, 1972; Gardner *et al.*, Chapter 7). This information was later extended to forest management models (Kessell, 1976). Subsequent ecological research on forest disturbance, including fire (Heinselman, 1973, 1981; Van Wagner, 1978; Johnson, 1992) and later windthrow (Canham and Loucks, 1984; Runkle, 1982; Frelich and Lorimer, 1991), also laid important groundwork for later integration into ecological models.

Development of landscape modeling approaches

The spatial landscape models in this book have some of their roots in the forest ecology and modeling discussed in the last two sections, but technological developments in the 1980s shifted both these areas of research toward spatial phenomena and large land areas. Landscape ecology, the study of ecological phenomena on large land areas (the scale of kilometers), has roots much earlier in the twentieth century, but bloomed in the 1980s. While satellite imagery and geographical information systems (GIS) were available earlier, Landsat Thematic Mapper (TM) data at 30-m resolution and small workstations with GIS software made spatial analyses of large land areas more feasible in the 1980s.

When models of landscape change were reviewed in the late-1980s (Baker, 1989), there were relatively few examples of spatial models, in part because the impacts of this technological revolution were just unfolding. In the middle of the 1980s, common models of landscape change were dominated by distributional approaches. Distributional landscape models focus on ". . . the distribution of land area among classes of landscape phenomena . . ." (Baker, 1989 p.113); an example is Shugart *et al.*'s (1973) model of succession among 15 forest types in the western Great Lakes, based on forest cover data. While the Shugart *et al.* model used differential equations, also popular were difference-equation (matrix) models using Markov, semi-Markov, and projection approaches; an example is the Rejmanek *et al.* (1987) model of vegetation dynamics in the Mississippi delta region.

However, in the middle 1980s, there were several threads of development leading to spatially explicit landscape models. In spatially explicit models, the behavior of an individual cell or pixel cannot be predicted without knowing its location relative to other cells. The earliest spatially explicit landscape model may be the gradient fire model of Kessell (e.g., 1979), which uses spatially estimated vegetation and fuels data to simulate spatial fire patterns and post-fire succession. The Kessell model was deeply rooted in forest ecology and management traditions, building on gradient analysis and incorporating existing disturbance models discussed in the last section.

In addition to the Kessell model, in the last half of the 1980s spatially explicit models of shifting cultivation and secondary forest succession (Wilkie and Finn, 1988), and coastal marshland changes (Browder *et al.*, 1985; Sklar *et al.*, 1985) appeared. These models were less rooted in the forest ecology and modeling literature, but instead began to apply emerging landscape ecology ideas in a spatial modeling framework.

Another thread of development was from mathematical and physical theory about properties of arrays of cells. One aspect of this thread is cellular automata models, whose roots are in information theory of the 1960s. A cellular automaton in its simplest form is a grid-cell model where complex dynamics arise from simple neighborhood interaction rules (Wolfram, 1984). A neighborhood-based transition model of landscape changes in Georgia appeared in the late 1980s (Turner, 1988). A second aspect of this thread is percolation modeling, derived from fractal theory and spatial properties of arrays. In the late 1980s percolation models were used to analyze the spread of disturbances across a landscape (Turner et al., 1989).

Over the last decade these threads of development from forest ecology, disturbance ecology, landscape ecology, and the mathematics of cellular automata and percolation arrays have interacted to lead us to the present state of spatial modeling, demonstrated in this book. Models emphasizing forest ecology and modeling of small land areas, such as gap models, are now reaching development in a spatially explicit form on large land areas, represented by the ZELIG version FACET (Urban et al., Chapter 4). Models that have roots in the cellular automata framework, such as METAFOR (Urban et al., Chapter 4) and DISPATCH (Baker, Chapter 11) continue to be useful for modeling contagious processes such as disturbance spread, but have been broadened to include more complex neighborhood dynamics (e.g., FORMOSAIC: Liu et al., Chapter 3). Indeed, grid-cell models such as DISPATCH (Baker, Chapter 11) and HARVEST (Gustafson and Crow, Chapter 12) may have roots in cellular automata, but include processes operating at scales other than the immediate neighborhood.

This trend toward multi-scale models is paralleled by a trend toward multi-process spatial models, exemplified in LANDIS (Mladenoff and He, Chapter 6), LANDSIM (Roberts and Betz, Chapter 5), FORMOSAIC (Liu et al., Chapter 3) and DELTA (Dale and Pearson, Chapter 10), but present in many of the other models, even those with narrower purposes. For example, the fire spread model, *FARSITE* (Finney, Chapter 8), uses grid-cell input data, but a vector-format to model the spreading fire front and has exogenous climate drivers that control fire spread, as well as a spotting routine that leapfrogs local dynamics of fire spread. The Sessions et al. (Chapter 9) SAFE FORESTS model focuses on fire dynamics and timber harvesting, but with other constraints. These multi-scale, multi-process models take cellular or vector neighborhood models, that recall the automata approach, and marry them with successional models, dispersal models, movement models, and disturbance models that interact in a spatially explicit format that is not simply neighborhood-based. Bottom-up models, such as cellular automata, demonstrated that complex dynamics can arise from simple neighborhood interactions, but these new models include important non-neighborhood processes that modify and control local neighborhood dynamics. Underlying this shift toward multi-scale, multi-process models is a changing world view away from a simple choice between top-down or bottom-up approaches (Baker and Mladenoff, Chapter 13).

Current challenges in forest landscape modeling

The forest landscape models presented in the rest of this book share a common, but varied, lineage with the historical work described in this chapter. To some degree, most are based on either ecological disturbance and succession principles from ecology and forestry, or earlier landscape transition models that were non-spatial.

How much multi-scale, multi-process detail is needed to model a forest landscape in a spatially explicit way? In the last chapter of this book an attempt will be made to synthesize how the models in this book answer this question. However, here at the beginning, only some of the relevant parts of this query will be posed. How much of the detail of individual-tree and population processes inside forest stands is necessary, and what are the essential processes? Are within-stand processes alone, if replicated across the landscape, sufficient to capture the essential dynamics of the landscape? How should stand-level processes vary in different environments? What processes inside a stand lead to natural (e.g., fire) and human (e.g., logging) disturbances, and how should these be modeled? Forest fragmentation is an increasing global phenomenon. Do our models include the essential changes that accompany fragmentation? Some landscape models include spatial dispersal and disturbance spread, but are other spatial processes needed? Regional and global processes are of increasing importance, as are local and global socioeconomic forces, and it may be asked how well our models represent these processes. Do people play a role in the dynamics of our models? Modelers to some extent answer these questions by including richness of detail in processes they feel are important, and by de-emphasizing or excluding processes that they feel are less important. However, what is included in each model is also a function of the purpose of the model.

In reading about the models in the following chapters, consider their genealogy, their purpose, and how these questions are answered. Viewed from two decades ago, the achievements represented here were almost inconceivable; a decade ago there was only a hint of the emerging directions. It is hoped you will see that these models are headed somewhere that will be important to us, even though it clearly will not be a single, grand, final model. Perhaps a diversity of spatial models of landscapes can extend our vision in different ways, for various purposes and scales. It has always been found hardest to call up the long-term vision and sweeping gaze needed to see the landscape for the trees.

References

Aber, J. D., Melillo, J. M. and Federer, D. A. (1982). Predicting the effects of rotation length, harvest intensity, and fertilization on fiber yield from northern hardwood forests in New England. *Forest Science*, **28**, 31–45.

Baker, W. L. (1989). A review of models of landscape change. *Landscape Ecology*, **2**, 111–33.

Belcher, D. M., Holdaway, M. R. and Brand, G. A. (1982). *A description of STEMS –*

the stand and tree evaluation and modeling system. USDA Forest Service, General Technical Report NC-79. St. Paul, MN.

Binkley, C. S. (1980). Is succession in hardwood forests a stationary Markov process? *Forest Science*, **26**, 566–70.

Botkin, D. B., Janak, J. F. and Wallis, J. R. (1972). Some ecological consequences of a computer model of forest growth. *Journal of Ecology*, **60**, 849–73.

Browder, J. A., Bartley, H. A. and Davis, K. S. (1985). A probabilistic model of the relationship between marshland–water interface and marsh disintegration. *Ecological Modelling*, **29**, 245–60.

Canham, C. D. and Loucks, O. L. (1984). Catastrophic windthrow in the presettlement forests of Wisconsin. *Ecology*, **65**, 803–9.

Cattelino, P. J., Noble, I. R., Slatyer, R. O. and Kessell, S. R. (1979). Predicting the multiple pathways of plant succession. *Environmental Management*, **3**, 41–50.

Clements, F. E. (1916). *Plant succession: an analysis of the development of vegetation.* Carnegie Institute Publication 242. Washington, DC.

Connell, J. H. and Slatyer, R. O. (1977). Mechanisms of succession in natural communities and their role in community stability and organization. *American Naturalist*, **111**, 1119–44.

Cooper, W. S. (1926). The fundamentals of vegetation change. *Ecology*, **7**, 391–413.

Cowles, H. C. (1899). The ecological relations of the vegetation on the sand dunes of Lake Michigan. *Botanical Gazette*, **27**, 95–117, 167–202, 281–308, 361–9.

Cowles, H. C. (1911). The causes of vegetation cycles. *Botanical Gazette*, **51**, 161–83.

Curtis, J. T. and McIntosh, R. P. (1951). An upland continuum in the prairie–forest border region of Wisconsin. *Ecology*, **32**, 476–96.

Curtis, J. T. (1959). *The vegetation of Wisconsin*. Madison, WI: University of Wisconsin Press.

Dale, V. H., Doyle, T. W. and Shugart, H. H. (1985). A comparison of tree growth models. *Ecological Modelling*, **29**, 145–69.

Drury, W. H. and Nisbet, I. C. T. (1973). Succession. *Journal of the Arnold Arboretum*, **54**, 331–68.

Ek, A. R. and Monserud, R. A. (1974). *FOREST: A computer model for the growth and reproduction of mixed species forest stands.* Research Report A2635. College of Agricultural and Life Sciences. University of Wisconsin–Madison. Madison, WI.

Feller, W. (1968). *An introduction to probability theory and its applications.* Vol. 1 3rd. edn. New York, NY: Wiley.

Frelich, L. E. and Lorimer, C. G. (1991). Natural disturbance regimes in hemlock–hardwood forests of the upper Great Lakes region. *Ecological Monographs*, **61**, 159–62.

Ginsberg, R. B. (1971). Semi-Markov processes and mobility. *Journal of the Mathematical Society*, **1**, 233–62.

Gleason, H. A. (1926). The individualistic concept of the plant association. *Bulletin of the Torrey Botanical Club*, **53**, 1–20.

Gleason, H. A. (1927). Further views of the succession concept. *Ecology*, **8**, 299–326.

Glenn-Lewin, D. C., Peet, R. K. and Veblen, T. T. (eds.). (1992). *Plant succession: Theory and prediction.* London, UK: Chapman and Hall.

Golley, F. B. (1993). Development of landscape ecology and its relation to environmental management. In *Eastside forest ecosystem health assessment.* Volume II.

Ecosystem management: principles and applications, ed. M. E. Jensen and P. S. Bourgeron, pp. 37–44. USDA Forest Service, Missoula, MT, USA.

Grubb, P. J. (1977). The maintenance of species richness in plant communities: The importance of the regeneration niche. *Biological Reviews of the Cambridge Philisophical Society*, **52**, 107–45.

Heinselman, M. L. (1973). Fire in the virgin forests of the Boundary Waters Canoe Area, Minnesota. *Quaternary Research*, **3**, 329–82.

Heinselman, M. L. (1981). Fire and succession in the conifer forests of northern North America. In *Forest succession: concepts and application*. ed. D. C. West, H. H. Shugart, and D. B. Botkin. New York, NY: Springer-Verlag.

Hoganson, H. M. and Burk, T. E. (1997). Models as tools for forest management planning. *Commonwealth Forestry Review*, **76**, 11–17.

Huston, M. A. and Smith, T. M. (1987). Plant succession: Life history and competition. *American Naturalist*, **130**, 168–98.

Iverson, D. C. and Alston, R. M. (1986). *The genesis of FORPLAN: A historical and analytical review of Forest Service planning models*. USDA Forest Service, General Technical Report INT-214.

Johnson, E. A. (1992). *Fire and vegetation dynamics: Studies from the North American boreal forest*. Cambridge, UK: Cambridge University Press.

Johnson, K. N. and Scheurman, H. L. (1977). Techniques prescribing optimal timber harvest and investment under different objectives – discussion and synthesis. *Forest Science Monograph* 18. Washington, DC: Society of American Foresters.

Johnson, K. N. (1992). Consideration of watersheds in long-term forest planning models: The case of FORPLAN and it use on the national forests. In *Watershed management: Balancing sustainability and environmental change*, ed. R. J. Naiman, pp. 347–60. New York, NY: Springer-Verlag.

Johnson, W. C. and Sharpe, D. M. (1976). Forest dynamics in the northern Georgia piedmont. *Forest Science*, **22**, 307–22.

Kercher, J. A. and Axelrod, M. C. (1984). A process model of fire ecology and succession in a mixed-conifer forest. *Ecology*, **65**, 1725–42.

Kessell, S. R. (1976). Gradient modeling: A new approach to fire modeling and wilderness resource management. *Environmental Management*, **1**, 39–48.

Kessell, S. R. (1979). *Gradient modeling: resource and fire management*. Springer-Verlag, New York.

Levin, S. A. and Paine, R. T. (1974). Disturbance, patch formation, and community structure. *Proceedings of the National Academy of Sciences, USA*, **71**, 2744–7.

Loucks, O. L., Ek, A. R., Johnson, W. C. and Monserud, R. A. (1981). Growth, aging, and succession. In *Dynamic properties of forest ecosystems*, ed. D. E. Reichle, pp. 37–86, Cambridge, UK: Cambridge University Press.

Loucks, O. L., Plumb-Mentjes, M. L. and Rogers, D. (1985). Gap processes and large-scale disturbances in sand prairies. In *The ecology of natural disturbance and patch dynamics*, ed. S. T. A. Pickett and P. S. White, pp. 71–83, Orlando, FL: Academic Press.

Margalef, R. (1963). On certain unifying principles in ecology. *American Naturalist*, **97**, 357–74.

McIntosh, R. P. (1980). The background and some current problems of theoretical ecology. *Synthese*, **43**, 195–255.

McIntosh, R. P. (1985). *The background of ecology: Concept and theory*. Cambridge, UK: Cambridge University Press.

Munro, D. D. (1974). Forest growth models. A prognosis. In *Growth models for tree and stand simulation*, ed. J. Fries, pp. 7–21. Department of Forest Yield Research, Royal College of Forestry, Stockholm, Sweden.

Noble, I. R. and Slatyer, R. O. (1980). The use of vital attributes to predict successional changes in plant communities subject to recurrent disturbances. *Vegetatio*, **43**, 5–21.

Odum, E. P. (1969). The strategy of ecosystem development. *Science*, **164**, 1289–93.

Odum, E. P. (1971). *Fundamentals of ecology*. 3rd edn. Philadelphia, PA: Saunders.

Pacala, S. W., Canham, C. D. and Silander, J. A. Jr. (1993). Forest models defined by field measurements: I. The design of a northeastern forest simulator. *Canadian Journal of Forest Research*, **23**, 1980–8.

Parks, P. J. and Alig, R. J. (1988). Land base models for forest resource supply analysis: A critical review. *Canadian Journal of Forest Research*, **18**, 965–73.

Pastor, J. and Post, W. M. (1988). Response of northern forests to CO_2-induced climate change. *Nature*, **334**, 55–8.

Peet, R. K. and Christensen, N. L. (1980). Succession: A population process. *Vegetatio*, **43**, 131–40.

Pickett, S. T. A. and White, P. S. (eds.). (1985). *The ecology of natural disturbance and patch dynamics*. Orlando, FL: Academic Press.

Pickett, S. T. A. and Cadenasso, M. L. (1995). Landscape ecology: Spatial heterogeneity in ecological systems. *Science*, **269**, 331–4.

Potter, M. W., Kessell, S. R. and Cattelino, P. J. (1979). FORPLAN: A forest planning language and simulator. *Environmental Management*, **3**, 59–72.

Ramensky, L. G. (1924). Basic regularities of vegetation covers and their study. (In Russian). *Vestnik Optynogo dela Stedne-Chernoz. Ob., Voronezh*, 37–73.

Rejmanek, M., Sasser, C. E. and Gosselink, J. G. (1987). Modeling of vegetation dynamics in the Mississippi River deltaic plain. *Vegetatio*, **69**, 133–40.

Rothermel, R. C. (1972). *A mathematical model for predicting fire spread in wild land fuels*. USDA Forest Service, Intermountain Forest and Range Experiment Station, Research Paper INT-115. Ogden, UT.

Runkle, J. R. (1982). Patterns of disturbance in some old-growth mesic forests of eastern North America, *Ecology*, **63**, 1533–46.

Sarkar, S., Cohen, Y. and Pastor, J. (1996). Mathematical formulation and parallel implementation of a spatially explicit ecosystem control model. In *Conference proceedings, Grand challenges in computer simulations*, Society for computer simulations, New Orleans: LA.

Shugart, H. H., Jr., Crow, T. R. and Hett, J. M. (1973). Forest succession models: A rationale and methodology for modeling forest succession over large regions. *Forest Science*, **19**, 203–12.

Shugart, H. H. (1984). *A theory of forest dynamics: The ecological implications of forest succession models*. New York, NY: Springer-Verlag.

Shugart, H. H. (1998). *Terrestrial ecosystems in changing environments*. Cambridge, UK: Cambridge University Press.

Sklar, F. H., Costanza, R. and Day, J. W. Jr. (1985). Dynamic spatial simulation modeling of coastal wetland habitat succession. *Ecological Modelling*, **29**, 261–81.

Smith, T. M. and Urban, D. L. (1988). Scale and the resolution of forest structural pattern. *Vegetatio*, **74**, 143–50.

Solomon, A. M. (1986). Transient response of forests to CO_2-induced climate change: Simulation experiments in eastern North America. *Oecologia*, **68**, 567–79.

Stephens, G. R. and Waggoner, P. E. (1970). The forests anticipated from 40 years of natural transition in mixed hardwoods. *Bulletin of the Connecticut Agricultural Experiment Station*. New Haven, CT.

Tansley, A. G. (1935). The use and abuse of vegetational concepts and terms. *Ecology*, **16**, 284–307.

Turner, M. G. (1988). A spatial simulation model of land use changes in a Piedmont county in Georgia. *Applied Mathematics and Computation*, **27**, 39–51.

Turner, M. G. (1989). Landscape ecology: the effect of pattern on process. *Annual Review of Ecology and Systematics*, **20**, 171–97.

Turner, M. G., Gardner, R. H., Dale, V. H. and O'Neill, R. V. (1989). Predicting the spread of disturbance across heterogeneous landscapes. *Oikos*, **55**, 121–9.

Urban, D. L. and Shugart, H. H. (1992). Individual-based models of forest succession. In *Plant succession: Theory and prediction*, ed. D. C. Glenn-Lewin, R. K. Peet, and T. T. Veblen, pp. 249–93. London, UK: Chapman and Hall.

Van Hulst, R. (1979). On the dynamics of vegetation: Markov chains as models of succession. *Vegetatio*, **40**, 3–14.

Van Wagner, C. E. (1969). A simple fire growth model. *Forestry Chronicle*, **45**, 103–4.

Van Wagner, C. E. (1978). Age class distribution and the forest fire cycle. *Canadian Journal of Forest Research*, **8**, 220–7.

Waggoner, P. E. and Stephens, G. R. (1970). Transition probabilities for a forest. *Nature*, **225**, 1160–1.

Watt, A. S. (1947). Pattern and process in the plant community. *Journal of Ecology*, **35**, 1–22.

West, D. C., Shugart, H. H., and Botkin, D. B. (eds.). (1981). *Forest succession: Concepts and application*. New York, NY: Springer-Verlag.

Whittaker, R. H. (1953). A consideration of climax theory: The climax as a population and pattern. *Ecological Monographs*, **23**, 41–78.

Whittaker, R. H. (1956). The vegetation of the Great Smoky Mountains. *Ecological Monographs*, **26**, 1–80.

Wilkie, D. S. and Finn, J. T. (1988). A spatial model of land use and forest regeneration in the Ituri forest of northeastern Zaire. *Ecological Modelling*, **41**, 307–23.

Wolfram, S. (1984). Cellular automata as models of complexity. *Nature*, **311**, 419–24.

2

Modeling the competitive dynamics and distribution of tree species along moisture gradients

John P. Caspersen, John A. Silander, Jr, Charles D. Canham and Stephen W. Pacala

Introduction

Ecology seeks to elucidate the mechanisms underlying patterns in the distribution and abundance of species. Models play an increasingly important role in examining the relationship between pattern and process because the broadest patterns in nature are not amenable to direct experimental study. This trend is particularly evident in the proliferation of landscape models (Baker, 1989).

Landscape ecology is increasingly reliant on models because they provide the means to explore how large-scale patterns emerge from complex dynamics at smaller scales. Yet, landscape models must be empirically grounded in order to yield true understanding of the mechanisms that govern landscape patterns. In this chapter both the design and field calibration of an individual-based model of forest landscape dynamics are described. The model was developed to address how landscape-scale variation in soil moisture influences the competitive dynamics and distribution of tree species.

Variation in species composition along soil moisture gradients is one of the most striking and well-studied patterns of forest landscapes (Whittaker, 1967; Oliver and Larson, 1996). A long tradition of gradient analysis has employed moisture gradients to demonstrate that environmental heterogeneity is one of the primary determinants of landscape pattern. Yet the underlying processes that give rise to gradient patterns have rarely been examined. Smith and Huston (1989) were among the first to examine the relationship between pattern and process using an individual-based model to simulate how landscape variation in soil moisture influences the dynamics and distribution of tree species. This chapter seeks to demonstrate how calibrating such a model with field data yields new insights into the underlying causes of landscape patterns and dynamics. The development of gradient models is first reviewed to illustrate how these models have led to the development of the model presented in this chapter.

Gradient models

Gradient analysis and niche models

Gradient analysis was developed in the 1950s and 1960s to characterize community assembly patterns by correlating the distribution and abundance of species with environmental factors that vary across landscape gradients (Curtis and McIntosh, 1951; Whittaker, 1956, 1967). Early gradient studies revealed continuous variation in species composition along all but the steepest environmental gradients, confirming Gleason's individualistic hypothesis of community assembly (Curtis and McIntosh, 1951; Whittaker, 1956). Gradient analysis was later refined to fit explicit functions to species' distribution patterns and evaluate whether they conform to patterns predicted by niche models of resource partitioning (Gauch and Whittaker, 1972).

Niche theory posits that species exhibit symmetric or Gaussian distribution patterns and that species are regularly segregated along environmental gradients (Gauch and Whittaker, 1972). However, gradient analyses conducted in a wide variety of environments revealed that a majority of species distribution patterns did not conform to the regularly spaced, symmetric curves predicted by resource partitioning (Austin, 1985). Rather, the distributions of most species are skewed toward suboptimal conditions, suggesting that competitive displacement, rather than resource partitioning, is the dominant process governing the distribution and abundance of species along gradients.

Resource competition models

Before the widespread use of gradient analysis, the experimental studies of Ellenberg (1954) had demonstrated that the resource requirements of species cannot necessarily be inferred from their observed distribution patterns. Ellenberg showed that grass species generally exhibit peak abundance under mesic conditions when grown in monoculture, but that peak abundance was skewed toward extreme conditions when grown in competition (Ellenberg, 1954). In the quarter century following the publication of Ellenberg's pioneering competition experiments, plant community ecologists invested considerable effort in measuring species responses to environmental gradients and relating species resource requirements to the results of competition experiments (Parrish and Bazzaz, 1976; Pickett and Bazzaz, 1978*a*, *b*). However, few attempts were made to elucidate the mechanisms by which plants compete for resources.

In the 1980s, Tilman developed and tested non-spatial models of resource competition to examine whether the outcome of competition is determined by the ability of species to deplete limiting resources (Tilman, 1982). His studies demonstrated that variation in the supply rate of limiting resources can give rise to species replacement patterns because constraints on the acquisition of multiple limiting resources impose trade-offs on species' ability to compete along resource supply gradients (Tilman, 1987). Tilman has forcefully argued that such competitive trade-offs cause a broad range of vegetation patterns and that mechanistic (resource-based) models of competition must be developed to understand vegetation patterns (Tilman, 1990).

While non-spatial competition models have been extremely useful in understanding resource competition, they omit important processes that govern the interplay of population dynamics and resource availability. Non-spatial models assume that the spatial scale of competition and dispersal is large enough to prevent disturbance and demographic stochasticity from generating spatial heterogeneity in plant density and resource availability. Yet, studies have shown that biotic heterogeneity has a significant effect on the dynamics and composition of plant communities (Reynolds et al., 1997). Theorists have only just begun to develop models that are both spatial and analytically tractable, in order to study the consequences of biotically generated spatial structure (Tilman and Kareiva, 1998).

Individual-based models

Forest ecologists have long recognized the importance of the biotic heterogeneity generated by disturbance (Watt, 1925, 1947). Indeed, gap dynamics has been an organizing concept in forest ecology since Watt first articulated the importance of disturbance and regeneration in plant communities (Pickett and White, 1985). Thus, forest ecologists were among the first to develop spatially explicit computer models to simulate the competitive interactions among individual plants (Botkin et al., 1972a, b; Shugart et al., 1973; Botkin, 1993).

Individual-based models have typically been designed to simulate the dynamics of succession in a homogeneous environment. However, individual-based models have also been employed to examine how species competitive interactions vary across environmental gradients (Botkin et al., 1972; Pastor and Post, 1986; Solomon, 1986; Tilman, 1988; Smith and Huston, 1989). Like non-spatial competition models, some of these models have been used to examine how constraints on the acquisition and use of multiple limiting resources impose trade-offs on species' ability to compete along resource gradients (Tilman, 1988; Smith and Huston, 1989). Smith and Huston (1989) simulated species zonation along soil moisture gradients with an individual-based model in which species performance was governed by a trade-off between the ability to grow in low light and the ability to grow in low soil moisture conditions. This trade-off between shade tolerance and drought tolerance is premised on leaf-level constraints on balancing carbon gain and water loss, and whole-plant allocation constraints on balancing carbon gain and water supply.

In this chapter an individual-based model is presented that was calibrated with field data to address whether such a trade-off between shade tolerance and drought tolerance governs the distribution and abundance of species along landscape soil moisture gradients. The analysis presented employs the same conceptual framework developed by Smith and Huston (1989) to simulate the competitive dynamics and distribution of species along landscape moisture gradients. However, an attempt will be made to evaluate whether landscape patterns can be traced to aspects of whole-plant performance that can be measured in the field.

Model design and calibration

The model presented in this chapter is a simple extension of the SORTIE model of forest stand dynamics (Pacala *et al.*, 1996). In order to model landscape dynamics using SORTIE, species responses to landscape-scale variation in soil moisture have been incorporated. Nevertheless, the structure of the landscape model remains largely the same as in the original version of SORTIE. Thus, for the sake of clarity, SORTIE is first described as originally designed and calibrated and the changes required to incorporate landscape-scale variation in soil moisture are subsequently described. The data and methods are then briefly described for assessing how species performance varies with soil moisture. A full description of the structure, dynamics, parameter estimation and error analysis of SORTIE can be found in Pacala *et al.* (1996).

The SORTIE algorithm is written in C and various versions run on UNIX workstations and personal computers. The model contains a record of every individual tree's diameter, species identity, and x- and y-coordinates. SORTIE simulates the dynamics of succession by predicting the fate of each individual using submodels of tree growth, mortality, recruitment and resource depletion (Fig. 2.1). The submodels were designed simultaneously with the field methods and maximum likelihood estimators necessary to estimate their parameters directly from field measurements. Each of the submodels was calibrated in the transition oak-northern hardwood forests of northeastern Connecticut.

Resource and performance submodels of SORTIE

Light submodel

The light submodel calculates the light available to an individual as a function of the species identity and location of neighboring trees as well as the diurnal and seasonal movement of the sun. The model is defined by four attributes of the neighborhood and radiation regime: (i) species-specific equations relating tree height, crown diameter, and crown depth to stem diameter, (ii) species-specific light extinction coefficients, (iii) the diurnal and seasonal movements of the sun and (iv) the mix of diffuse and beam radiation. These attributes are used to compute the distribution of canopy openness above any given individual as a function of the density and species identity of neighboring trees:

$$\text{Openness} = e^{-\Sigma E_i(\text{\# crowns of species } i \text{ intercepted})} \tag{1}$$

where E_i is the light extinction coefficient for species i. Canopy openness is then combined with the distribution of sky brightness (from attributes (iii) and (iv)) to

Fig. 2.1. Species-specific sub-models of SORTIE (Equations 1–4 in the text). Clockwise from upper right: light extinction as a function of solar zenith and azimuth, crown geometry and light attenuation; radial growth as a function of light availability; mortality as a function of growth; recruitment as a function of distance and adult size. Species codes: red maple (RM), sugar maple (SM), American beech (Be), white ash (WA), white pine (WP), red oak (RO), eastern hemlock (He), yellow birch (YB), black cherry (BC).

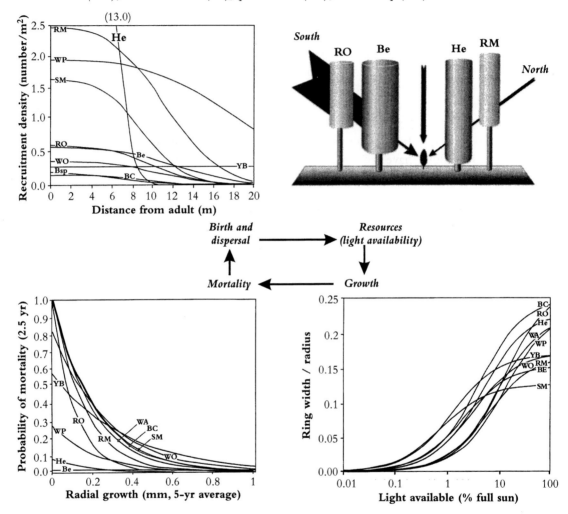

compute the Gap Light Index (GLI), a measure of whole-season photosynthetically active radiation in units of percent of full sun (Canham *et al.*, 1993).

Growth submodel

The growth submodel consists of species-specific equations that predict radial growth of saplings from stem radius and GLI:

$$\text{Annual radial increment} = \text{Radius} \left[(G_1 \times \text{GLI}) / ((G_1/G_2) + \text{GLI}) \right] \quad (2)$$

where G_1 is the asymptotic growth rate at high light and G_2 is the slope at zero light. In accordance with the Constant Area Increment Law (Phipps, 1967), the basal area increment of canopy trees cannot exceed a constant maximum rate, G_3.

Mortality submodel

The sapling mortality submodel consists of species-specific equations that predict the individual's probability of mortality as a function of recent growth:

$$\text{Probability of mortality} = M_1\, e^{-M_2 g} \qquad (3)$$

where g is radial growth, M_1 is the probability of mortality at zero growth and M_2 is the rate at which the probability of mortality decays with growth (Kobe et al., 1995). This model specifies the probability of mortality over a $2\frac{1}{2}$-year-period as a function of average annual growth during the previous 5 years. The mortality submodel also includes purely random disturbance whereby all individuals, including adults, have a constant probability of dying from density independent factors. For model runs described here, this background annual mortality rate was set to 0.01. This yields an average gap-to-gap interval of 100 years that is representative of gap-phase disturbance regimes in late-successional stands (Runkle, 1985).

Recruitment submodel

The recruitment submodel consists of species-specific equations that predict the number and spatial locations of seedlings produced by maternal trees as a function of the location and diameter of maternal trees (Ribbens et al., 1994). The number of seedling recruits per unit area is a function of the number of recruits produced by maternal trees and the proportion of those recruits that disperse to a given distance from the maternal tree:

$$\text{Seedling density} = R_2\, (\text{diameter}/100)^2 \times [\exp(-R_1 \times \text{distance}^3)/n] \qquad (4)$$

where R_2 is the number of 5-year-old recruits produced by an adult of 100 cm in diameter. The spatial dispersal of recruits is determined by the function in brackets, a radially symmetric probability density centered on the maternal tree. R_1 is the species-specific decay parameter and n is a normalizer that ensures the area of the density function equals one.

A run of the stand-level model is initiated with a random distribution of individuals of specified number, size, and species identity. At the beginning of each 5-year iteration, SORTIE uses the light model to calculate a GLI for each sapling and then computes growth rate from GLI and diameter. Competition occurs whenever individuals are shaded by taller neighbors. SORTIE then uses the estimated growth rate to calculate the probability of mortality using the mortality models and pseudo-random coin tosses. Finally, SORTIE uses the recruitment submodels to determine the number and spatial positions of all recruits produced by every tree. By repeated iterations of the model, SORTIE forecasts the long-term changes in the abundance and spatial distribution of all tree species populations in a stand.

The importance of spatial processes

SORTIE simulations have shown that the relative scale of dispersal and competition has profound effects on both the structure and function of forest ecosystems. First, short dispersal combined with demographic stochasticity causes significant clustering and species segregation to develop at scales of 25 m after 500 years of succession. As a consequence, a mosaic of monodominant stands develops late in the course of succession even though the environment is homogeneous and the spatial distribution distributions are initially random.

Second, the relative scale of clustering and competition is also a critical determinant of the total biomass accumulation and rate of turnover in community composition during the course of succession. Pacala and Deutschman (1995) developed a non-spatial version of SORTIE which omits spatial structure by assuming that the density in each plant's neighborhood equals the mean density of the stand as a whole. This non-spatial version of SORTIE predicts a two-fold reduction in basal area and a two-fold increase in species turnover as compared to the spatial version of SORTIE. The basal area is predicted to be higher in the spatial version of the model because clustering serves to decrease the intensity of competition in the stand as a whole. Species turnover occurs at half the rate in the spatial version of the model because clustering allows shade-intolerant species to persist in low density areas later into the course of succession.

Implementation at the landscape scale

In order to implement SORTIE at the landscape scale, topographic variation in soil moisture has been focused on because it is one of the principal determinant of landscape variation in forest community composition in the region (Damman and Kershner, 1977; Whitney, 1991). Many other factors are likely to affect species performance at landscape scales, but topographic variation in soil moisture has been chosen to focus on because the purpose of our model is to determine whether a simple suite of mechanisms is sufficient to produce patterns observed in natural landscapes. Moreover, competition for water is not included in the model because field studies demonstrate that the tree species in our study sites do not differ in their ability to deplete soil moisture (C. Canham, unpublished data). Thus, implementing the model at the landscape scale simply requires specifying how soil moisture varies as a function of landscape (topographic) position and how species performance varies as a function of water as well as light availability. First, the changes required to specify soil moisture as a function of landscape (topographic) position and the resulting spatial structure of the model are discussed. The field calibration and form of the submodels quantifying species performance as a function of water as well as light availability are subsequently described.

In the landscape version of SORTIE, variation in soil moisture is specified as a continuous function of the x- and y-coordinates of a simulated landscape. Just as each tree occupies a spatial position defined by its x- and y-coordinates, each tree experiences a soil moisture defined as a function of its x- and y-coordinates. Introducing abiotic heterogeneity introduces fixed spatial structure that influences the performance and competitive interactions of species. As a consequence, the trajectory of succession can vary with soil moisture across a heterogeneous landscape.

For the purposes of this chapter, landscape variation in soil moisture is specified as a linear gradient. It has been decided to collapse landscape variation in soil moisture onto a single dimension because it is easier to visualize species distribution and abundance patterns in one dimension as they have been traditionally illustrated in niche models and gradient analyses: the results can be readily generalized to a landscape in which soil moisture varies in two dimensions. Thus, the soil moisture experienced by an individual tree is simply defined as a linear function of its x-coordinate and expressed in percent volumetric content.

The range of variation in soil moisture and the distance over which it varies are chosen to be consistent with scale and magnitude of landscape variation in soil moisture observed in the field. The scale of topographic variation in soil moisture varies considerably in the uplands of northwestern Connecticut where SORTIE was calibrated. Both extremes of saturated soil and shallow ridgetop soil can be observed along slopes as short as a three to four hundred meters and a small watershed encompassing those extremes covers as little as 50 hectares or 0.5 km^2. A long hillslope slope gradient, however, may extend up to a few thousand meters. The landscape-level simulations presented in this chapter span a gradient that varies from 15–42% volumetric content over a distance of 1 km, the scale of a small watershed gradient in Connecticut. The plot size is 1000 meters by 1000 meters, or 1 km^2. Runs covering 1 km^2 and 500 years average approximately 5 hours on a UNIX workstation.

Field calibration

The primary study sites used to calibrate SORTIE are located in the gneiss and schist uplands in the vicinity of Great Mountain Forest (GMF) in northwestern Connecticut. The upland landscape in the vicinity of GMF is topographically heterogeneous, with elevations ranging from 250–700 m, encompassing a wide range of soil moisture conditions. Associated with landscape variation in soil moisture is variation in forest community composition (Damman and Kershner, 1977): xeric ridgetops are dominated by oak forests, mesic midslopes and valley bottoms above 250 m by transition oak – northern hardwood forests, and hydric wetland sites by stands of *Acer rubrum* and *Tsuga canadensis*. *Tsuga canadensis* also forms isolated stands on xeric ridgetops and steep slopes which gives rise to a bimodal distribution that is observed throughout much of its range, particularly in New England (Kessel, 1979). Forests in the vicinity of GMF are predominantly 80–120 years old.

To examine how landscape variation in soil moisture influences species distribution patterns, field studies were conducted to assess how the growth and mortality of the major tree species vary across landscape gradients of water availability. The seven species studied include *Tsuga canadensis* (eastern hemlock, He), *Fagus grandifolia* (beech, Be), *Acer saccharum* (sugar maple, SM), *Acer rubrum* (red maple, RM), *Quercus rubra* (red oak, RO), *Fraxinus americana* (white ash, WA), and *Pinus strobus* (eastern white pine, WP).

In the summer of 1996 several sites per species were located that encompassed a wide and continuous range of variation in soil moisture, excluding wetland or hydric soils. For the focal species at each site a random sample of the stems of live and recently dead saplings was collected and the proportion of live and recently dead saplings in the site estimated. Recently dead saplings included individuals whose leaf and bud retention and twig suppleness indicated that they had died within 2.5 years (Kobe et al., 1995). To ensure that the growth of live saplings was examined across a full range of orthogonal variation in light and water availability, several additional sites were located for each species that encompassed an even broader range of conditions, including full sun and seasonally saturated soils.

Using a digital ring analyzer the annual radial growth rate of the live saplings was measured in each of the previous 5 years. For the recently dead saplings the annual radial growth rate was measured in each of the 5 years preceding death. After harvesting each sapling light and water availability were also measured. To measure light availability, hemispherical photographs of the canopy above each sapling were taken and an index computed of whole season light availability from the digitized photograph. Soil moisture (% volumetric content) was measured to a depth of 15 cm at the base of live and dead sapling on two dates in July and August 1996 using time domain reflectometry (TDR). For some saplings soil moisture was measured on only a single date. While soil moisture is temporally variable, the range of variation along landscape gradients is large enough that even a single measurement is sufficient to quantify landscape-scale variation. In general, even simple topographic indices are adequate to accurately capture the range of variation in soil moisture in heterogeneous landscapes (Barling et al., 1994).

Despite the wide range of variation in soil moisture, regression analysis of growth in relation to both light and soil moisture reveals that the radial growth of saplings is not significantly correlated with soil moisture for any of the seven species (J. P. Caspersen and R. K. Kobe, unpublished data). Figure 2.2 illustrates that the 5-year average growth rate of live sugar maple saplings (circles) is uncorrelated with soil moisture, as is annual growth in any of the 5 years preceding harvest. In contrast, the probability of mortality increases dramatically with decreasing water availability. Under mesic conditions only those saplings that suffer suppressed growth are subject to mortality (squares), while saplings die at increasingly higher growth rates as soil moisture decreases (Fig. 2.2). Insofar as growth may vary with soil moisture, one would expect growth to vary in concert with mortality across the range of conditions sampled. Yet, Fig. 2.2 clearly illustrates that the effect of soil moisture on sapling growth is negligible relative to its effect on mortality.

To quantify the relationship between sapling mortality and soil moisture we modified the maximum likelihood methods used to estimate the probability of mortality as a function of growth (Eq. 3 in Kobe et al., 1995). The modified

Fig. 2.2. 5-year average radial growth rate and soil moisture of live and dead sugar maple saplings.

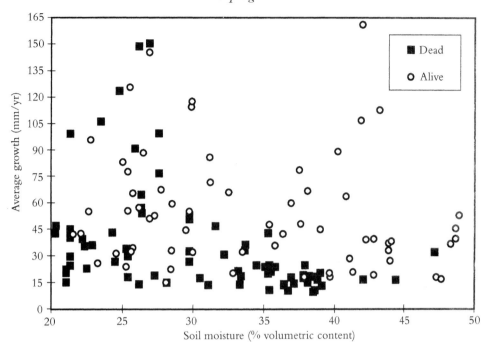

maximum likelihood estimator is based on the conditional probability distribution governing the distribution of growth rates and soil moisture among the live and dead individuals (Fig. 2.3) and the binomial distribution of live and dead individuals at each site. The method works by estimating the mortality function that best reshapes the bivariate distribution of growth and soil moisture prior to mortality into the distribution of growth and soil moisture among dead individuals (Fig. 2.2).

The modified mortality function specifies the probability of mortality over a 2.5-year period as a function of the average soil moisture as well as the average recent growth:

$$m(g) = e^{-M_1 g - M_2 gw} \tag{5}$$

where g is the average growth in the previous 5 years, w is the average percent volumetric water content as measured by time domain reflectometry and M_1 and M_2 are species-specific parameters that define the rate of decay of the probability of sapling mortality with increasing growth and soil moisture.

As with any short-term field study, our sampling protocol provides only a snapshot of sapling mortality rates because it includes only those saplings that had died in the 2.5 years preceding harvest. Given that drought stress varies from year to year, mortality rates during any given 2.5-year-period may not be representative of long-term average mortality rates. However, the 1994–1996 period was neither

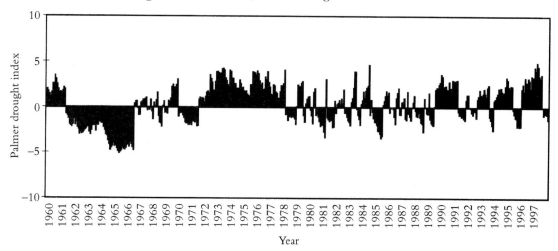

Fig. 2.3. Drought history of northwestern Connecticut. The Palmer drought severity index is derived from state climatic division monthly average temperature and precipitation (National Climatic Data Center). Mild drought = −1.0 to −2.0, moderate drought = −2.0 to −3.0, severe drought = −3.0 to −4.0, extreme drought < −4.0.

unusually dry or wet compared to the entire climate record for northwestern Connecticut, although it did include a moderate drought in 1995 (Fig. 2.3). Thus, the mean mortality rate during the 1994–1996 period represents as reasonable an estimate of mortality as could be obtained without employing long-term census data.

The estimated mortality functions reveal striking interspecific variation in the probability of sapling mortality with respect to both growth and soil moisture (Fig 2.4). Recall that growth itself varies as function of light but not water availability. Thus, the relationship between the probability of mortality and growth in part defines a species' shade tolerance: steeper slopes reflect greater shade tolerance. Both sugar maple and hemlock are shade tolerant under mesic conditions while red oak is not. For sugar maple, however, the slope of the relationship between growth and mortality flattens with decreasing soil moisture, indicating a trade-off between shade tolerance and drought tolerance. For hemlock, on the other hand, the probability of mortality does not vary with soil moisture. Thus, hemlock does not conform to the trade-off between shade- and drought-tolerance proposed by Smith and Huston (1989). Moreover, species differences in drought tolerance are determined largely by interspecific variation in mortality rather than growth.

The disparity between the mortality and growth responses suggests that water is only limiting to saplings during episodic drought. In a humid climate such as Connecticut's, the annual growth rate of saplings that survive episodic drought on xeric sites may not be measurably reduced because whole-season water balance remains favorable. Thus, the episodic nature of drought in humid climates likely explains why there was no significant relationship between soil moisture and growth across the range of conditions we studied. Because our results indicate that

Fig. 2.4. Probability of mortality as a function of growth and soil moisture (Eq. 5 in text) for (a) sugar maple, (b) hemlock and (c) red oak.

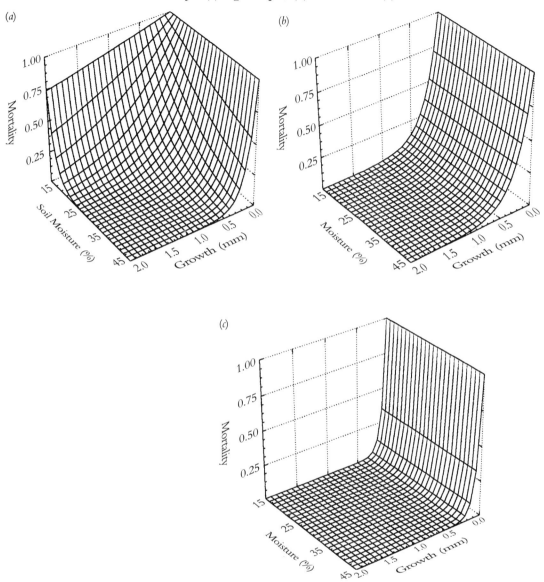

variation in annual radial growth is negligible relative to variation in mortality across the range of conditions sampled, only the effect of soil moisture on sapling mortality are included in the landscape model.

Hydric conditions

Assessing performance near the limit of a species distribution is hindered by the rarity of naturally occurring individuals. Under extreme conditions, mortality during establishment may be sufficiently high that few saplings naturally occur. In such cases, transplant experiments provide a means to assess how the limits of physiological tolerance affect species performance and distribution. A transplant experiment was conducted to assess the risk of mortality on seasonally saturated soils where considerable mortality occurs during establishment. Seasonally saturated soils are periodically inundated soils where the volumetric water content generally exceeds 60% and seedlings and saplings are occasionally subject to anoxia and frost heave. There is dramatic interspecific variation in mortality during establishment on saturated soils, with all species except white ash, red maple, and hemlock suffering more than 90% mortality on saturated soils. White ash exhibited the lowest mortality, 40%, and red maple and hemlock intermediate mortality rates, 56% and 73%, respectively. The probability of mortality during establishment on saturated soils is specified in the model as a species-specific constant M_4 (Table 2.1).

Performance metrics

During the course of succession, interspecific variation in performance is determined by the suite of parameters that define the various submodels. Thus, it is difficult to evaluate how a species will perform solely by examining individual parameter estimates. To clarify the competitive strategies of species we present measures of six aspects of species performance that, in addition to the probability of establishment on saturated soils (M_4), determine competitive dynamics under various conditions: the total amount of shade cast by a 30 cm diameter tree, time required to grow 3 m in full sun and at 1% full sun, mean dispersal distance, and 5-yr survivorship at low-light in mesic and xeric conditions (Table 2.1). Total shade was calculated as the absolute value of the difference between the spatial integral of ln(GLI) over the individual's shadow and the corresponding integral for full sun (GLI = 100).

Model dynamics and landscape pattern

Two sets of runs are presented to illustrate the dynamics of the model. First, stand-level runs are presented depicting the course of succession under mesic and xeric conditions. Second, landscape-level runs are presented depicting species' distribution and abundance along a continuous soil moisture gradient. Finally, the estimated establishment probabilities obtained from the transplant study are used to examine how species differences in establishment in seasonally saturated soils influence community dynamics.

Simulations

With the exception of the mortality parameters (Eq. 5), each set of runs was initiated with the same estimated parameter values used in the original version of SORTIE (Pacala *et al.*, 1996). Unestimated parameters, R_2, G_3 and initial diameter, were assigned based on extensive sensitivity analyses (Pacala *et al.*, 1996). Fecundity,

Table 2.1 Metrics summarizing interspecific variation in life-history traits (see explanation in text)

Species	Probability of establishment on saturated soils	Shade cast by 30 cm diameter tree	Time to 3 m height in full sun (yr)	Time to 3 m height in 1% sun (yr)	Mean dispersal distance (m)	5-yr survivorship of a 2 cm diameter sapling in 1% sun and 40% soil moisture	5-yr survivorship of a 2 cm diameter sapling in 5% sun and 20% soil moisture
RM	0.56	25.7	14.6	92.8	10.6	0.30	0.64
SM	0.90	27.0	15.5	75.3	4.1	0.96	0.50
Be	0.84	78.5	19.4	55.0	5.9	0.96	0.00
WA	0.40	19.5	11.9	100.6	16.3	0.39	0.77
WP	0.84	16.6	14.7	158.0	15.8	0.00	0.36
RO	0.91	19.5	11.9	125.4	8.7	0.10	0.89
He	0.73	46.0	15.5	55.0	5.9	0.99	1.00

Fig. 2.5. Species basal area during the 500 years of succession in (a) *mesic and* (b) *xeric conditions.*

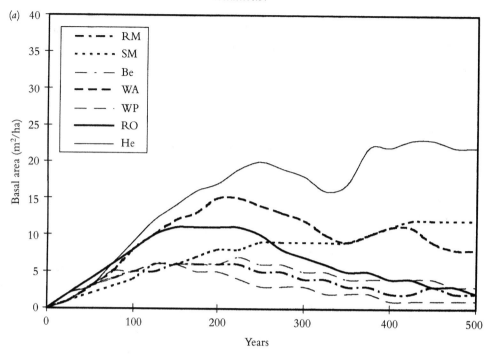

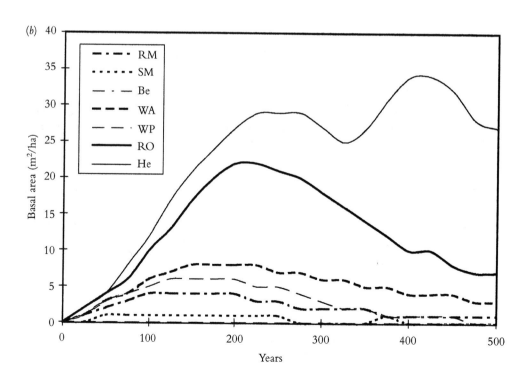

R_2, was set to 5 for all species, and individuals of dioeceous white ash were assigned a gender at random when born. The constant area increment of canopy trees, G_3, was assigned a value that corresponds to an annual radial increment of 1.5 mm for a tree 100 cm in diameter. New recruits were initiated with a diameter of 2 mm. Finally, each run was initiated with a random spatial distribution of 25 1 cm diameter saplings of each species per hectare.

Stand-level runs

The first set of runs described are stand-level runs that depict the course of succession under mesic and xeric conditions, 20% volumetric content and 35% volumetric content, respectively (Fig. 2.5). During the course of succession in mesic conditions, shade intolerant species increase in abundance initially then decline after reaching peak abundance between 100 and 200 years. Shade-tolerant species, including hemlock and sugar maple, continue to increase in abundance for 500 years (Fig. 2.5(a)). In xeric conditions, drought-intolerant species such as sugar maple, beech, red maple, and white ash suffer higher mortality. As a consequence, hemlock and red oak are dominant early in the course of succession until red oak declines and hemlock alone persists as the late-successional dominant (Fig. 2.5(b)).

Landscape-level runs

The landscape-level runs illustrate the distribution and abundance of species along a soil moisture gradient that varies from 15 to 42% volumetric content (Fig. 2.6). During the first 100 years of succession hemlock and red oak begin to dominate at the xeric end of the gradient because drought-intolerant species suffer higher mortality (Fig.2.6(a)). In contrast, the relative abundance of species remains relatively even in mesic conditions during the first 100 years of succession. By 250 years, the shade-intolerant species begin to decline under mesic conditions as well (Fig. 2.6(b)). As a result, the species that remain abundant exhibit skewed distribution patterns that reflect interspecific differences in shade tolerance and drought tolerance. By 500 years, hemlock alone is dominant across most of the gradient, except under mesic conditions where sugar maple continues to steadily increase (Fig 2.6(c)).

Seasonally saturated soils

The establishment probabilities obtained from the transplant experiment are incorporated into the model to examine how species differences in establishment influence community dynamics on seasonally saturated soils. White ash is predicted to be the dominant under hydric conditions, while hemlock and red maple are predicted to be subdominant (Fig. 2.7). Each of these three species are also predicted to dominate hydric soils in the absence of the other two (results not shown).

Causes of stand- and landscape-level predictions

The spatial and temporal patterns predicted by the model are governed by the suite of life-history traits summarized in Table 2.1. Several life-history traits, however,

30   J. P. Caspersen *et al.*

Fig. 2.6. Species basal area across a soil moisture gradient at (a) *150 years,* (b) *250 years, and* (c) *500 years.*

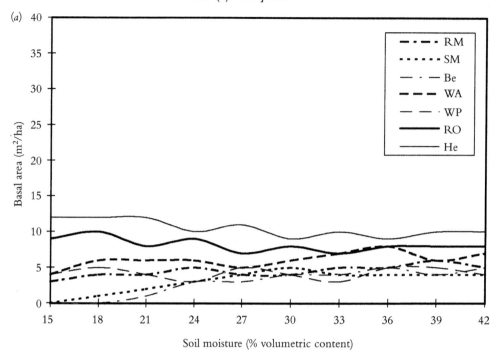

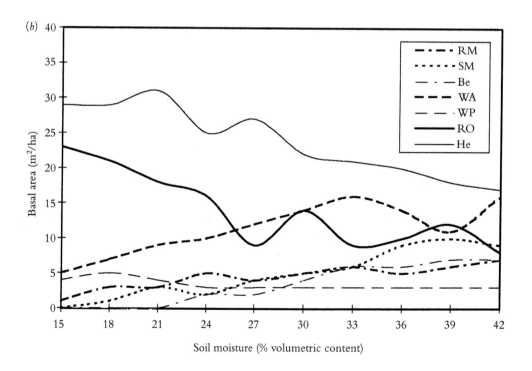

Modeling tree species along gradients

Fig. 2.6 (cont.)

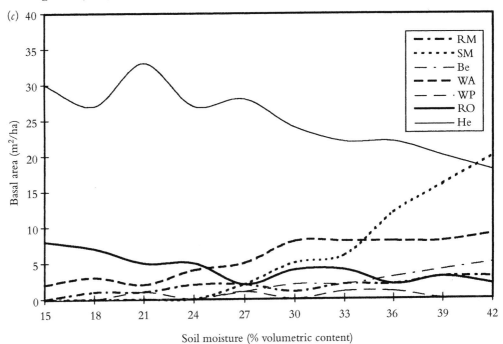

Fig. 2.7. Species basal area during the 500 years of succession in hydric conditions.

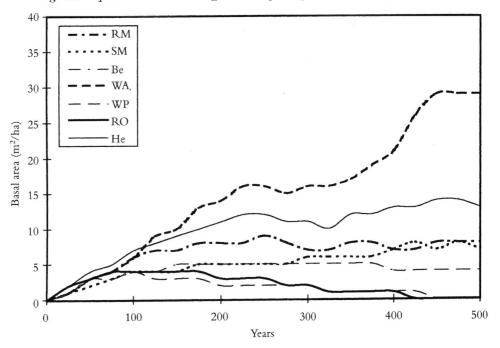

stand out as the critical determinants of the competitive dynamics and distribution of tree species along moisture gradients.

Under mesic conditions there is a prolonged trajectory of species turnover during the course of succession because there is a trade-off between high-light growth and low-light survivorship (Pacala et al., 1996). Thus, species which grow rapidly in high light increase initially, then are replaced by species that survive better in low light. In contrast, large interspecific differences in mortality cause an autosuccessional trajectory in xeric conditions. There is no appreciable turnover is species composition because species intolerant of the combined effect of drought and shade are suppressed early on by competition from hemlock and red oak.

The distribution of species along the soil moisture gradient reflects the interplay of competition and environmental heterogeneity. In the absence of competition, all of the species are able to survive and grow in all but the most xeric and hydric conditions (results not shown). Thus, species distribution patterns unfold during the course of succession as species respond differentially to joint limitation by light and water (Fig. 2.6(b)). The distribution of the drought intolerant species, particularly sugar maple and beech, are skewed towards mesic conditions where they suffer less drought-induced mortality. The distribution of the drought-tolerant species, red oak and hemlock, are skewed towards xeric conditions where they experience less competition. Hemlock, however, is abundant in both mesic and xeric conditions because it is both shade tolerant and drought tolerant.

Comparing predictions with observed patterns

To assess the landscape-level predictions of the model a GIS data set was assembled composed of a forest classification map and digitized soil survey maps. The forest classification map was derived from three merged Landsat TM images of northwestern Connecticut and adjacent states (Mickelson et al., 1998). The classification includes 20 forest classes (Fig. 2.8: for Fig. 2.8(b), see color section), most of which correspond to dominance or co-dominance by one or more of the seven species examined in the landscape model. The soil coverage was derived from digitized, geo-corrected soil polygon maps containing 74 soil types (Gonick and Shearin, 1970). The 74 soils types were aggregated into seven composite classes based on their associated soil moisture attributes: wetland (peat and muck soils), wet (poorly and very poorly drained soils), moist (somewhat poorly drained), intermediate (well and moderately well drained), dry (somewhat excessively well drained), dry and thin (thin well-drained and somewhat excessively well-drained soils with exposed bedrock), and very dry (excessively well drained). Of these, only acidic (non-calcareous) soils, typically above 250 m were considered.

To examine the association between forest cover and soil moisture, the frequency of occurrence of each of the forest classes on each soil class was calculated. The observed pixel distribution was then compared with what would be expected if a given forest cover class had been distributed at random across soil moisture classes. Goodness of fit G-tests on observed minus expected pixel frequencies were

Fig. 2.8. (a) *Location of study sites, GIS data coverages, and forest classification map (shown as cross-hatched area) in northwestern Connecticut and adjacent areas in New York and Massachusetts. (b: see colour section). Forest classification map (corresponding to cross-hatched region in (a)) derived from Landsat TM imagery. Legend shows the set of coverages, color coded, that were resolved in the image classification analysis (see Mickelson* et al. *in press for further details): RO – red oak* (Quercus rubra) *dominated stands; RO/Mx – red oak dominated, but with mixed hardwoods sub-dominant; RO/RM – red oak co-dominant with red maple* (Acer rubrum)*; RM–red maple dominated stands; SM – sugar maple* (Acer saccharum) *dominated stands; SM/RO/Mx – mixed sugar maple, red oak stands; WA/RM/Mx – mixed white ash* (Fraxinus americana) *and red maple stands; Bc/SM/Mx – black cherry* (Prunus serotina) *with sugar maple and mixed hardwoods; Be – american beech* (Fagus grandifolia) *dominated stands; NHd/Be/SM – northern hardwoods, dominated by beech and sugar maple; NHd/YB/RM/HE – mixed northern hardwoods dominated by yellow birch* (Betula alleghaniensis)*, red maple and eastern hemlock* (Tsuga canadensis)*; MxHd – mixed hardwoods with no specific dominant; Mx/Hd/WP – mixed hardwoods with eastern white pine* (Pinus strobus) *co-dominant; WP – white pine dominated stands; P/MxConif – red pine* (Pinus resinosa) *dominated stands; He/RM – hemlock and red maple dominated stands; He/MxHd – mixed hemlock and hardwood stands; He – hemlock dominated stands; Sp– spruce dominated stands, primarily* Picea mariana *or* P. rubens. *The designation of LU or HU after some of the forest stand types indicates either mountain laurel* (Kalmia latifolia) *dominated understory forest sub-type (LU) or a hemlock dominated understory sub-type (HU). Other landcover types indicated: PFO/PSS – palustrine forested or scrub shrub wetlands; open water (lakes, ponds, rivers, etc.); Ag – all agricultural lands (including croplands, pastures, etc.), undifferentiated; Urban/Imperv – urban areas and paved surfaces; Barren/Soil/Quarry – all surfaces with exposed soil or rock; Cloud/shadow – areas obscured by cloud and associated shadows in one or more TM scene; Non-classified – landcover was not classified for these sites.*

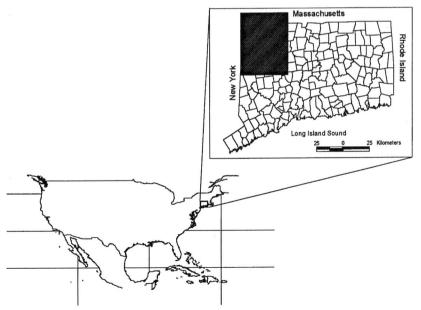

used to determine whether observed patterns in the distribution of forest cover types deviated significantly from random. Values of G were compared with critical values of the chi-square distribution (Sokal and Rohlf, 1995). To facilitate graphical comparisons, results are shown on a standardized percent basis of observed minus expected number of pixels for forest cover type (Fig. 2.9). All results were significant at the 0.005 level.

Most of the forest classes dominated by the species included in the landscape model exhibit skewed distribution patterns (Fig. 2.9(a)–(g)). The sugar maple, northern hardwoods (including beech and sugar maple), white ash, and red maple classes occurred more frequently than expected at random on intermediate to wet soils and less frequently than expected on dry soils. Red oak, on the other hand, occurred more frequently than expected on dry soils and less frequently than expected on intermediate to wet soils. Thus, the observed distribution of these species is consistent with the skewed distributions predicted by the model.

Hemlock exhibited a bimodal distribution, with a pronounced affinity for hydric soils and a less pronounced but significant affinity for xeric soils (Fig. 2.9(c)). This pattern agrees with the bimodal distribution pattern reported in a previous study of ecotypic variation in Hemlock (Kessel, 1979). The bimodal distribution is also consistent with our finding that hemlock is both shade tolerant and drought tolerant. However, hemlock was predicted to be dominant in mesic and xeric conditions alike, but only subdominant in hydric conditions. As is argued in the discussion, the abundance of hemlock in mesic and xeric conditions likely reflects the role of disturbance and land-use history. The abundance of hemlock in wetland soils, on the other hand, may be due to factors which could not be evaluated with the transplant experiment.

Several forest classes were observed to be abundant on wet and wetland soils, including hemlock, red maple, and white ash. These are the same species that were observed to have the lowest mortality during establishment in seasonally saturated soils, 73%, 56%, and 40% mortality, respectively. Each of these three species is predicted to dominate hydric soils in the absence of the other two. In competition, however, white ash is predicted to dominate in hydric conditions. This suggests that other processes govern the regeneration and persistence of hemlock and red maple in wetland soils. In particular, the regeneration of hemlock in wetland soils may be due to the establishment on nurse logs, a factor which was not considered in the transplant experiment.

Discussion

In models, simple explanations are sought for the patterns observed in nature. Gradient models have been widely used to evaluate whether landscape patterns result from simple physiological and morphological constraints on the acquisition and use of multiple limiting resources (Tilman, 1988; Smith and Huston, 1989). Our effort to calibrate and test such a model has yielded simple explanations for species distribution patterns while also revealing further complexity in the causes of landscape pattern. It has been found that interspecific variation in drought and shade tolerance can, in part, explain landscape patterns in the distribution and abundance of species. However, it is also found that the relationship between shade and drought tolerance is not governed solely by physiological and morphological constraints on carbon gain. Furthermore, it is found that some species' distribution patterns may reflect the legacy of disturbance and land-use history as much as the underlying heterogeneity in resource availability.

An emphasis on the physiology and allometry of carbon gain has eclipsed the role of survivorship in the development of vegetation models (Bazzaz, 1979; Kobe *et al.*, 1995). Previous landscape models have omitted a direct effect of water availability on mortality by assuming that the consequences of drought are expressed solely through chronic suppression of carbon gain (Smith and Huston, 1989; Huston, 1994). Thus, they assume that species differ in drought tolerance only insofar as they differ in growth. It has been found that differences in drought tolerance among the species studied are determined largely by interspecific variation in mortality rather than growth. These results have important consequences not only for the physiological underpinnings of drought tolerance but also for the relationship between drought tolerance and shade tolerance. The relationship between shade and drought tolerance may depend on the relative importance of growth and mortality in determining a species' ability to tolerate limiting resources.

Many studies have shown that the ability to grow and survive in the shade are associated with distinct resource utilization strategies. The ability to grow in the shade is associated with a strategy for maximizing carbon gain, including allocation patterns that maximize leaf area and leaf traits that maximize photosynthetic efficiency (Walters and Reich, 1996), which may predispose saplings to drought stress. However, the ability to survive in the shade is also associated with leaf traits and allocation patterns that minimize susceptibility to herbivores, pathogens, and various other sources of mortality (Kitajima, 1994; Kobe, 1997). These include thicker and tougher leaves and higher allocation to roots, storage, and defense; traits which do not necessarily preclude drought tolerance and may, in fact, promote drought tolerance. Thus, survival may be associated with a conservative resource utilization strategy that permits tolerance of multiple limiting resources, whereas traits that maximize carbon gain may predispose saplings to inevitable trade-offs. Consequently, the extent to which species conform to a trade-off between shade tolerance and drought tolerance may depend on the relative importance of growth and survival in determining the species ability to tolerate limiting resources.

Most of the species we studied conform to a trade-off between shade tolerance and drought tolerance, both in terms of performance measured in the field and in terms of their distribution across the landscape. Hemlock, however, is distinctly bimodal in its distribution and our field studies indicate that it is drought tolerant as well as shade tolerant. This exception may reflect differences in the relative importance of low-light carbon gain in the shade tolerance strategies of deciduous broadleaf and needleleaf evergreen species.

It is widely recognized that hemlock is most common in moist habitats. However,

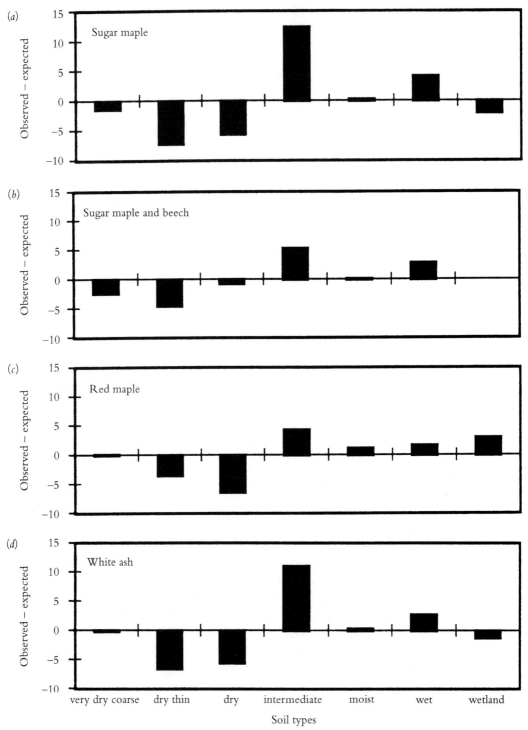

Fig. 2.9. Observed–expected frequency (shown standardized to percentage of pixels) of seven forest classes on seven soil types. (a) *sugar maple,* (b) *northern hardwoods (sugar maple and beech),* (c) *red maple,* (d) *white ash,* (e) *red oak,* (f) *hemlock, and* (g) *white pine.*

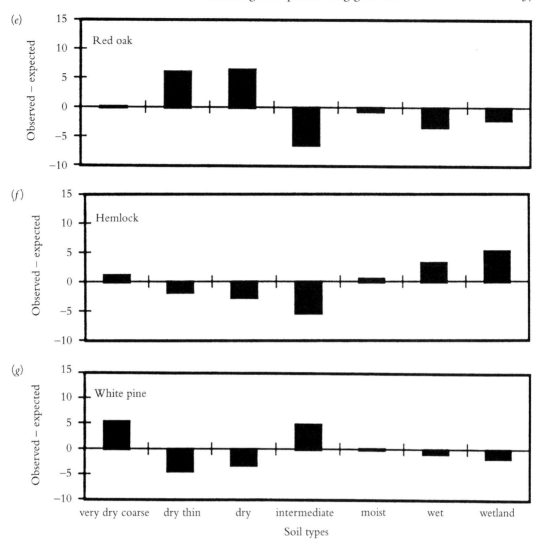

it is not widely recognized that hemlock is also locally abundant across a wide range of conditions. Indeed, there are two morphologically distinct ecotypes of hemlock associated with mesic and xeric habitats (Kessel, 1979). Furthermore, old-growth hemlock stands have been located in both intermediate and xeric habitats throughout southern New England, including xeric sites on sandy soils in GMF and shallow soils on steep slopes and ridgetops in western and central Massachusetts (Nichols, 1913; Dunwiddie *et al.*, 1996; David Orwig, personal communication). The wide amplitude and local abundance of hemlock suggest that competitive ability is but one determinant of a species' prevalence in a landscape, and that the current distribution and abundance of hemlock may reflect disturbance and land-use history.

Several studies suggest that the distribution and abundance of hemlock may be influenced by fire (Rogers, 1977; Kelty, 1986; Foster, 1995). Hemlock is particularly fire sensitive because it has shallow roots, thin bark and a thick, persistent leaf litter layer (Rogers, 1977). Moreover, hemlock does not exhibit rapid height growth and retains a deep canopy with foliage close to the ground (Pacala et al., 1995). Thus, the fact that hemlock is most abundant in moist habitats and only locally abundant in intermediate and xeric habitats may reflect the higher frequency of fire in intermediate and xeric habitats.

Finally, the bimodal distribution of hemlock may reflect the legacy of land-use history. Old-growth stand composition, historical records and palynological studies suggest that prior to European colonization hemlock was more abundant and more widely distributed in the upland portions of New England than it is today (Nichols, 1935; Foster, 1995). As a consequence of widespread agricultural clearing in the eighteenth and nineteenth centuries, trees were restricted to relict stands located on soils unsuitable for cultivation, particularly on excessively wet or dry soils. Following agricultural abandonment over the last 100 years, hemlock has not readily recolonized the portions of the landscape where it was formerly abundant (Kelty, 1986; Foster, 1995). This suggests that bimodal distribution of hemlock may in part reflect its limited dispersal ability (Ribbens et al., 1994).

Conclusions

There are two principal advantages to modeling landscape dynamics at the level of the individual. First, individual-based models provide the means to examine how large-scale pattern emerge from complex dynamics at smaller scales. Second, model parameters correspond directly to whole-plant processes that can be directly measured in the field. Thus, it has been shown how landscape patterns in the distribution and abundance of species reflect interspecific variation in drought and shade tolerance. However, there are several disadvantages to modeling landscape dynamics at the level of the individual. First, because the model is not analytically tractable it is not possible to fully explore parameter space and evaluate all the underlying assumptions of the model. Second, the sampling error grows with the complexity of the model and the error of the various submodels may interact non-additively. Finally, individual-based models are computationally intensive.

For these reasons, it has been sought to balance the need for biological realism with the need to minimize complexity. To further this goal, analysis and simplification of the model is continuing to evaluate the consequences and necessity of the model's complexity. To evaluate the robustness of the model predictions an error analysis is being conducted that translates the statistical uncertainty associated with each parameter into statistical uncertainty in the model's predictions. To further simplify the model a set of partial differential equations has also been derived, which approximates the first moments of the stochastic processes modeled by individual-based simulators. The initial results presented in this chapter demonstrate that the model strikes a reasonable balance between simplicity and biological realism, providing landscape-level predictions that can be traced to aspects of individual performance measured in the field.

Acknowledgments

We thank the Childs family for their hospitality and for use of the facilities at Great Mountain Forest. We thank Beth Hobby, Brandon Maio, Ed Roy, Kristin Gondar, and Ginger Pollack for help in the field and laboratory. We thank John Mickleson, Rich Kobe, Ben Bolker, George Hurtt and Miguel Zavala for their advice and assistance. Lastly, we gratefully acknowledge the support of the National Aeronautics and Space Administration (NAGW-2088, NAGW-3471, NA3703, NGT-30240). This chapter was written while the lead author was supported by a NASA earth system science fellowship.

References

Austin, M. P. (1985). Continuum concept, ordination methods, and niche theory. *Annual Review of Ecology and Systematics*, **16**, 39–61.

Baker, W. L. (1989). A review of models of landscape change. *Landscape Ecology*, **2**, 111–33.

Barling, R. D., Moore, I. D. and Grayson, R. B. (1994). A quasi-dynamic wetness index for characterizing the spatial distribution of zones of surface saturation and soil water content. *Water Resources Research*, **30**, 1029–44.

Bazzaz, F. A. (1979). The physiological ecology of plant succession. *Annual Review of Ecology and Systematics*, **10**, 351–71.

Botkin, D. B., Janak, J. F. and Wallis, J. R. (1972a). Rationale, limitations, and assumptions of a northeast forest simulator. *IBM Journal of Research and Development*, **16**, 106–16.

Botkin, D. B., Janak, J. F. and Wallis, J. R. (1972b). Some ecological consequences of a computer model of forest growth. *Journal of Ecology*, **60**, 849–72.

Botkin, D. D. (1993). *Forest dynamics: An ecological model.* Oxford University Press.

Canham, C. D., Finzi, A. C., Pacala, S. W. and Burbank, D. H. (1993). Causes and consequences of resource heterogeneity in forests: interspecific variation in light transmission by canopy trees. *Canadian Journal of Forest Research*, **24**, 337–49.

Curtis, J. T. and McIntosh, R. P. (1951). An upland forest continuum in the prairie–forest border region of Wisconsin. *Ecology*, **32**, 476–96.

Damman, A. W. H. and Kershner, B. (1977). Floristic composition and topographical distribution of the forest communities of the gneiss areas of western Connecticut. *Naturaliste Canadienne*, **104**, 23–45.

Dunwiddie, P., Foster, D., Leopold, D. and Leverett, R. (1996). Old-growth forests of southern New England, New York, and Pennsylvania. In *Eastern old-growth forests*, ed. M. B. Davis. Washington, DC: Island Press.

Ellenberg, H. (1954). Uber einige Fortschritte der Kausalen Vegetationskunde. *Botanischen Gesellschaft*, **65**, 351–62.

Foster, D. (1995). Land-use history and four hundred years of vegetation change in New England. In *Global land use change: a perspective from the Columbian encounter*, ed. B. L. I. Turner, A. G. Sal, F. G. Bernaldez and F. di Castri. Madrid, Consejo Superior de Investigaciones Cientificas.

Gauch, H. G. and Whittaker, R. H. (1972). Coenocline simulation. *Ecology*, **53**, 446–51.

Gonick, W. N. and Shearin, A. E. (1970). *Soil Survey of Litchfield County, Connecticut*. Washington, DC, US Government Printing Office.

Huston, M. A. (1994). *Biological diversity*. Cambridge: Cambridge University Press.

Kelty, M. J. (1986). Development patterns in two hemlock-hardwood stands in southern New England. *Canadian Journal of Forest Research*, **10**, 885–91.

Kessel, S. R. (1979). Adaptation and dimorphism in eastern hemlock, *Tsuga canadensis* (L.) Carr. *The American Naturalist*, **133**, 333–50.

Kitajima, K. (1994). Relative importance of photosynthetic traits and allocation patterns as correlates of seedling shade tolerance of 13 tropical trees. *Oecologia*, **98**, 419–28.

Kobe, R. K. (1997). Carbohydrate allocation to storage as a basis of interspecific variation in sapling survivorship and growth. *Oikos*, **80**, 226–33.

Kobe, R. K., Pacala, S. W., J. A. Silander, J. and Canham, C. D. (1995). Juvenile tree survivorship as a component of shade tolerance. *Ecological Applications*, **5**, 517–32.

Mickelson, J. G., Jr., Civco, D. L. and Silander, J. A. J. (1998). Delineating forest canopy species in the Northeastern United States using multi-temporal TM imagery and GPS referenced data. *Photogrammetric Engineering and Remote Sensing*, **64**, 891–907.

Nichols, G. E. (1913). The vegetation of Connecticut II. Virgin forest. *Torreya*, **13**, 1991–215.

Nichols, G. E. (1935). The hemlock–white pine–northern hardwood region of eastern North America. *Ecology*, **16**, 403–42.

Oliver, C. D. and Larson, B. C. (1996). *Forest stand dynamics*. New York: Wiley.

Pacala, S. W. and Deutschman, D. H. (1995). Details that matter: The spatial distribution of individual trees maintains forest ecosystem function. *Oikos*, **74**, 357–65.

Pacala, S. W., Canham, C. D., Saponara, J., Silander, J. A., Kobe, R. K. and Ribbens, E. (1996). Forest models defined by field measurements: estimation, error analysis and dynamics. *Ecological Monographs*, **66**, 1–43.

Pacala, S. W., Canham, C. D., Silander, J. A. and Kobe, R. (1995). Sapling growth as a function of resources in a north temperate forest. *Canadian Journal of Forestry*, **24**, 2174–83.

Parrish, J. A. D. and Bazzaz, F. A. (1976). Underground niche separation in successional plants. *Ecology*, **57**, 1281–8.

Pastor, J. and Post, W. M. (1986). Influence of climate, soil moisture, and succession on forest carbon and nitrogen cycles. *Biogeochemistry*, **2**, 3–27.

Phipps, R. L. (1967). Annual growth of suppressed chestnut oak and red maple, a basis for hydrological inference, US Geological Survey Professional Paper 485-C.

Pickett, S. T. A. and Bazzaz, F. A. (1978a). Germination of co-occuring annual species on a soil moisture gradient. *Ecology*, **57**, 169–76.

Pickett, S. T. A. and Bazzaz, F. A. (1978b). Organization of an assemblage of early successional species on a soil moisture gradient. *Ecology*, **59**, 1248–55.

Pickett, S. T. A. and White, P. S. (1985). *The ecology of natural disturbance and patch dynamics*. New York: Academic Press.

Reynolds, H. L., Hungate, B. A., Chapin, F. S. and D'Antonio, C. (1997). Soil heterogeneity and plant competition in an annual grassland. *Ecology*, **78**, 2076–90.

Ribbens, E., Silander, J. A. and Pacala, S. W. (1994). Seedling recruitment in forests: Calibrating models to predict patterns of tree seedling dispersion. *Ecology*, **75**, 1794–806.

Rogers, R. S. (1977). Forests dominated by hemlock (*Tsuga canadensis*): distribution as related to site and postsettlement history. *Canadian Journal of Botany*, **56**, 843–54.

Runkle, J. R., ed. (1985). Disturbance regimes in temperate forests. In *The ecology of natural disturbance and patch dynamics*. Orlando, Florida: Academic Press.

Shugart, H. H., Crow, T. R. and Mett, J. M. (1973). Forest succession models: a rationale and methodology for modeling forest succession over large regions. *Forest Science*, **19**, 203–12.

Smith, T. and Huston, M. (1989). A theory of the spatial and temporal dynamics of plant communities. *Vegetatio*, **83**, 49–69.

Sokal, R. R. and Rohlf, F. J. (1995). *Biometry*. New York: W.H. Freeman and Co.

Solomon, A. M. (1986). Transient response of forests to CO_2-induced climate change: Simulation modeling experiments in eastern North America. *Oecologia*, **68**, 567–79.

Tilman, D. (1982). *Resource competition and community structure*. Princeton, NJ: Princeton University Press.

Tilman, D. (1987). Secondary succession and the pattern of plant dominance along experimental nitrogen gradients. *Ecological Monographs*, **57**, 189–214.

Tilman, D. (1990). Constraints and tradeoffs: Toward a predictive theory of competition and succession. *Oikos*, **58**, 3–15.

Tilman, D. and Kareiva, P., eds. (1998). *Spatial ecology: the role of space in population dynamics and interspecific interactions*. Princeton, NJ: Princeton University Press.

Tilman, G. D. (1988). *Plant strategies and the dynamics and structure of plant communities*. Princeton, NJ, USA: Princeton University Press.

Walters, M. B. and Reich, P. B. (1996). Are shade tolerance, survival, and growth linked? Low light and nitrogen effects on hardwood seedlings. *Ecology*, **77**, 841–53.

Watt, A. (1925). On the ecology of British beech woods with special reference to their regeneration: II. The development and structure of beech communities on the Sussex Downs. *Journal of Ecology*, **13**, 27–73.

Watt, A. (1947). Pattern and process in the plant community. *Journal of Ecology*, **35**, 1–22.

Whitney, G. G. (1991). Relation of plant species to substrate, landscape position, and aspect in north central Massachusetts. *Canadian Journal of Forest Research*, **21**, 1245–52.

Whittaker, R. H. (1956). Vegetation of the Great Smoky Mountains. *Ecological Monographs*, **26**, 1–80.

Whittaker, R. H. (1967). Gradient analysis of vegetation. *Biological Reviews*, **42**, 207–64.

3

Spatial and temporal impacts of adjacent areas on the dynamics of species diversity in a primary forest

Jianguo Liu, Kalan Ickes, Peter S. Ashton, James V. LaFrankie and N. Manokaran

Introduction

During the last several decades increasingly extensive and intensive exploitation of primary tropical forests has resulted in historically unprecedented rates of deforestation. Clearcutting, conversion of forests to agricultural use or tree plantations, and construction of houses and roads have converted continuous forests into a mosaic of small remnant forest patches (Harris, 1984; Panayotou and Ashton, 1992), which are often not large enough to support local populations of many species (Powell and Powell, 1987; Malcolm, 1988; Harper, 1989; Bierragaard *et al.*, 1992; Viana and Tabanez, 1996). These forest fragments are dispersed across human-occupied landscapes and are often adjacent to managed tree plantations and non-forest areas (Schelhas and Greenberg, 1996). Significant species loss due to forest fragmentation and human activities is of major concern because species diversity is a foundation for sustainable development including production of both timber and non-timber goods (e.g., Lubchenco *et al.*, 1991; Panayotou and Ashton, 1992).

Most previous ecological research efforts have concentrated on ecological patterns and processes within a focal forest. During the past few years, however, many ecologists and government agencies have recognized the significance of studying the surrounding areas and the need for ecosystem management beyond ecological, political, and ownership boundaries (e.g., Christensen *et al.*, 1996). Ecosystem management requires a better understanding of how human disturbances such as timber logging influence ecosystem dynamics and how a focal ecosystem interacts with adjacent areas. Research on a focal ecosystem alone is usually not easy, but study of ecological impacts across boundaries is even more challenging because more variables must be considered. Interactive effects of numerous variables involved in studying landscape-scale phenomena, such as interactions between seed dispersal from outside and disturbances occurring inside a focal forest, are difficult to measure through conventional experiments or field observations. Spatially explicit models could provide a useful and complementary tool (Dunning *et al.*, 1995; Turner *et al.*, 1995*a*).

A large number of spatially explicit forest models have been developed during the last several decades (e.g., Fries, 1974; Ek *et al.*, 1988; Urban, 1990; Pacala *et al.* 1996). Distance-dependent models consider the effects of distances among trees. This modeling approach requires that each tree be placed at a point on a spatial coordinate plane. The basic assumption is that competition among individual trees is determined by factors such as inter-tree distance and tree size (e.g., biomass, height, or diameter at breast height). The first and probably best-known distance-dependent model was built by Newnham (1964). Following Newnham's lead, numerous distance-dependent models were developed for estimating forest growth and yield (e.g., Lee, 1967; Lin, 1969; Mitchell, 1969; Bella, 1970; Hatch, 1971; Fries, 1974; Hegyi, 1974; Larocque and Marshall, 1988; Wensel, 1990).

Traditional gap models (e.g., Botkin *et al.*, 1972; Shugart, 1984) assume that light utilization differs vertically among trees. Shorter trees receive less light because taller individuals intercept light from the canopy. Conventionally, gap models consider a plot to be homogeneous and do not explicitly take account of inter-tree distance effects. Recently, however, attention has been paid to horizontal differences and interactions (Smith and Urban, 1988; Busing, 1991; Pacala *et al.*, 1993). In the ZELIG model (Smith and Urban, 1988; Urban 1990), horizontal homogeneity at the plot scale (usually 10 m) is assumed. Many individual trees are assigned to a grid cell, but the location of an individual is not specified. One of the major differences between ZELIG and its predecessors is that in ZELIG adjacent cells (or plots) interact with each other by shading and seed dispersal, while plots in conventional models operate independently. The SPACE model developed by Busing (1991) uses a much finer spatial scale (0.5 m grid cell) than ZELIG. An individual tree occupies one or more grid cells depending on tree size. Pacala *et al.* (1993) adopted the approach of distance-dependent growth-yield models (e.g., Newnham, 1964) and placed all trees on a plane according to their x- and y-coordinates.

Most spatial models ignore the characteristics and contributions of the surrounding areas to the dynamics of a focal forest (Liu and Ashton, 1995). For example, widely used gap models introduce new individuals to the modeled gap area from an external, hypothetical, constant seed pool regardless of what type of adjacent areas actually exist (Botkin *et al.*, 1972; Shugart, 1984). This assumption needs to be re-examined, however, because adjacent non-forested areas, such as industrial or agricultural land, may not provide any seeds to a focal forest. At the other methodological extreme, a recent gap model by Pacala *et al.* (1993) assumed that all the seeds were produced inside the focal area. Many forest models avoid edge effects by wrapping the modeled area onto itself (e.g., Smith and Urban, 1988) or by treating forest edges as reflecting boundaries for seeds (Clark and Ji, 1995).

Whereas hundreds of models have been built in recent years specifically for forests in temperate (e.g., Shugart, 1984; Ek *et al.*, 1988) and boreal regions (e.g., Leemans and Prentice, 1987, Bonan *et al.*, 1990), relatively few have been developed for simulating the dynamics of tropical rainforests (e.g., Shugart and Noble, 1981; Van Daalen and Shugart, 1989; for a review see Liu and Ashton, 1995). Yet, tropical forests are the most biologically diverse on Earth and occupy 51.5% of the world's remaining forested area (Borota, 1991). The majority of the existing tropical forest models are stand models that are mainly used for predicting timber growth/yield, but which do not consider the dynamics of species richness (e.g., Vanclay, 1989).

An individual-based spatially explicit landscape model called FORMOSAIC was developed that treats the focal forest as part of the landscape mosaic, and which was applied to tropical forests (Liu and Ashton, 1998, 1999). FORMOSAIC accounts for not only the dynamics inside the focal area but also the ecological conditions of adjacent areas. This model tracks the characteristics of individual trees, including their position, size, growth, regeneration, and death. It considers both ecological and anthropogenic processes in a focal forest (e.g., recruitment, growth and death, windthrows, timber harvest), seed immigration from outside the focal forest, and other external factors such as damage done to the forest by animals like pigs. A major difference between FORMOSAIC and many other published forest models is that this model places a focal forest in a spatial location and explicitly identifies the surroundings.

In this chapter FORMOSAIC's structure, function, parameterization, validation, sensitivity and uncertainty analysis are first introduced. Then FORMOSAIC is applied to a case study from Pasoh Forest Reserve in Peninsular Malaysia involving current research on the effects of native, wild pigs on the understory plant community. Using FORMOSAIC the responses of tree species richness to interactions are evaluated among four factors: spatial scale, habitat surrounding a focal forest, harvest impact, and pig damage. In the case study, two major questions were asked: (i) how do parameters related to pig damage (e.g., percentage of area receiving pig damage, percentage of small trees killed within the damage area, and duration and timing of pig damage) change tree species richness in a focal forest? (ii) Do interactions among timber harvest, types of surroundings, and pig damage vary at different spatial scales or sizes of forest remnants? Finally, implications of the simulation results and of this modeling approach are discussed for managing forest remnants in landscape mosaics, and for understanding the mechanisms of remnant dynamics.

FORMOSAIC structure

FORMOSAIC is a spatially explicit, individual-based, stochastic model for simulating forest dynamics in landscape mosaics (Liu and Ashton, 1998). The model predicts population trajectories for individual species, species richness, stand density, and timber volume (basal area) in response to management practices, as well as biotic and abiotic factors, which influence tree recruitment, growth, and mortality.

FORMOSAIC is hierarchically structured at four levels: landscape, focal forest, grid cell, and tree location (Fig. 3.1). The landscape mosaics consist of a focal forest and surrounding areas, and the focal forest may be square or rectangular in shape. Possible surrounding areas include natural or plantation forests, clearcuts, agricultural fields, industrial land, roads, and residential buildings. In addition, the surrounding areas on the four compass sides east, west, north, and south of a focal forest may vary in structure and function. For example, Fig. 3.1 shows a case where there is an oil palm plantation on the west side of the focal forest, a species-rich forest on the east side, and non-forested areas on the north and south sides. Seeds can immigrate to the focal forest from the species-rich forest, but no seeds are available from the non-forested areas. Wild pig density is assumed to increase in response to year-round food availability in the oil palm plantation, or due to immigration from that area.

In recognition of the ecological heterogeneity within a forest, a focal forest is divided into a grid of 10 m × 10 m cells, each of which contains many individuals of different species. For example, in the 50-hectare permanent tree plot in Malaysia, there are usually 60–80 individual trees > 1.0 cm diameter at breast height (dbh), belonging to 30–50 tree species in a 10 × 10 m area (Manokaran *et al.*, 1990; Liu, unpublished data). In Fig. 3.1, the example focal forest is 0.25 ha in size and is divided into 25 grid cells. In FORMOSAIC, the size of a focal forest is only limited by computer capacity. Seeds and pigs can move among grid cells. A grid cell may receive seeds produced by trees inside the cell, from other grid cells, or from outside of the focal forest. At the tree level, the location of each tree is explicitly mapped using *x*- and *y*-coordinates. FORMOSAIC tracks recruitment, growth, and death for each individual tree.

FORMOSAIC was programmed in the object-oriented computer programming language C++ (e.g., Ellis and Stroustrup, 1990; see Liu 1993 for an ecological application using C++). It can be implemented in UNIX (e.g., SunSparc and Silicon Graphics workstations) and PC platforms (Windows). A graphical user-interface was developed using Tcl/Tk (Welch, 1995). The exact amount of simulation time required relies on many factors like computer speed, forest size, disturbance option, management scheme, as well as output type. For example, with an Ultra II Sun workstation (167 MHz), it took about 2 minutes to finish one annual simulation (i.e., time step = 1 year) on a 50-ha forest containing more than 300 000 individual trees, and about 2 seconds to complete an annual simulation on a 1-ha forest of some 8000 trees.

FORMOSAIC parameterization

The data for parameterizing FORMOSAIC were derived mainly from the 50-ha long-term study plot in the Pasoh Forest Reserve (2°59′ N, 102°18′ E), Peninsular Malaysia, established in 1985 (Manokaran *et al.*, 1990; LaFrankie, 1992*a*, *b*). The Reserve is a lowland dipterocarp forest (Symington, 1943). The plot was censused in 1987 and again in 1990. All trees were mapped and tagged and their *x*- and *y*-coordinates recorded. Demographic information included diameter growth,

Fig. 3.1. Hierarchical relationship of four spatial scales (landscape, focal forest, grid cell, and tree location (point)) considered in FORMOSAIC. At the landscape scale, this schematic diagram shows that a focal forest is surrounded by three types of adjacent areas (species-rich forest, oil palm plantation, and non-forest). Seeds can immigrate into the focal forest from the adjacent species-rich forest, but no seeds are available from the neighboring non-forest. Wild pigs can immigrate into the focal forest from the oil palm plantation. Considering computational convenience and ecological heterogeneity of a forest, the focal forest (0.25 ha) is represented by a grid of 25 cells. Each cell is 10 × 10 m in size and contains many individuals of different tree species. Tree location is mapped at the point level. The model tracks recruitment, growth and death of each individual tree.

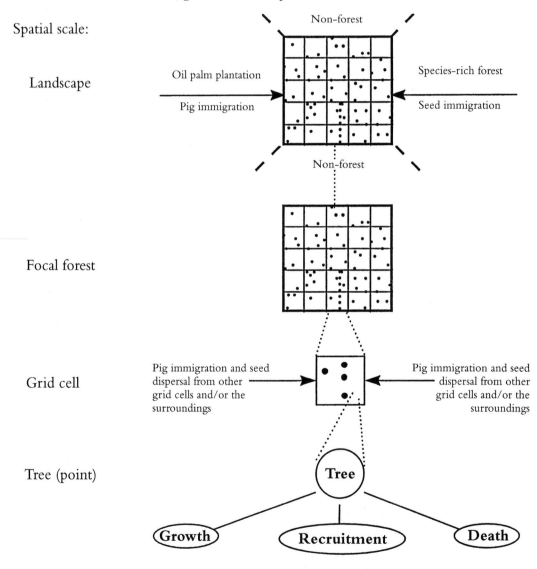

mortality, and recruitment. Height and crown diameter data were available from smaller subplots. Environmental data consisted of elevation, slope, and habitat. Elevation and slope were estimated on 20 × 20 m basis (i.e., all trees within a 20 × 20 m area were assumed to share the same elevation and slope). The habitat at each tree point was measured as the distance from the boundaries of swamps or streams. In the first census, over 800 tree species and more than 330 000 trees with dbh >1.0 cm were recorded in the plot.

Liu and Ashton (1998, 1999) detailed FORMOSAIC's recruitment, growth, and mortality functions and parameters, based on the empirical data from the 50-ha permanent plot. Because sample sizes in the mortality and recruitment analyses for a single species were usually small, species were grouped into four guilds: emergent, canopy, understory, and successional species. The classification was based on flora information (Whitmore, 1972a,b; Ng, 1978, 1989) and field knowledge of the genera including architecture, habitat, and life history (P. Ashton, personal observations; S. Thomas, personal communication). The same mortality and recruitment functions were used for all species of the same guild. In the growth analysis, rare species (<1 individual/ha) were also classified into the four guilds and a growth function for each guild was developed. All rare species in the same guild shared the same growth function. Separate growth functions were developed for each of 502 species which had relatively large population sizes (> 1 individual/ha). The growth rate of each individual tree was a function of tree size, neighborhood pressure (defined below), and specific local environmental conditions determined by elevation, slope, and distance from swamps and streams. Neighborhood pressure was measured as the total basal area of all trees in a grid cell, except the focal individual.

For mortality estimation, trees of each guild were classified into four dbh size categories: 1–5 cm, 5–10 cm, 10–30 cm, and > 30 cm. Empirical data showed that the three small size classes had significant positive relationships between mortality and tree density or basal area in a grid cell. Because the largest dbh size class (> 30 cm) did not demonstrate any relationship between mortality rate and tree density or basal area, the average mortality value was used for this class instead of a mortality function in FORMOSAIC.

Similarly, analysis from the 50-ha census data indicated that the number of new recruits in canopy, successional and understory species had negative relationships with tree density. Because the emergent species did not show any relationship with tree density or basal area, the average recruitment rate was used for this guild. FORMOSAIC assumes that recruits can be generated from seeds produced both inside a focal forest and in adjacent areas. A species in the surroundings with a higher population size is assumed to have a higher probability of providing more seeds to the focal forest (Liu and Ashton, 1998). The number of seeds input into a given grid cell is a function of seed dispersal characteristics, distance of the grid cell to the seed source area (other grid cells or adjacent area to the focal forest), availability of seeds in the seed source area, and the pre-programmed maximum number of recruits allowed in the grid cell because of competitive effects. For example, grid cells near boundaries of the focal forest will have more immigrants from outside of the focal forest, while more recruits near the center of the focal forest come from within the forest. The location of a seed dispersed into a grid cell is assumed to be random.

Because FORMOSAIC is individual based, it keeps information for each individual tree (e.g., species name, size, and location) until the individual dies, and use the guild information when necessary by matching the species' name with guild type. As a result, FORMOSAIC is able to examine species richness even though some information is guild-specific (Liu and Ashton, 1998).

Disturbance is very important in the dynamics of species richness (Connell, 1979). Thus, FORMOSAIC considers three major forms of disturbance: timber harvest, windthrow, and pig damage. The rationale to include them in FORMOSAIC and the modeling methods are discussed below.

Timber harvest

The cutting of wood for commercial use is pervasive in Southeast Asia, and few reserves are free from this pressure. Any model which cannot incorporate the effects of logging will have minimal applicability to real-world forest reserves. In Tropical Asian forests timber trees are classified as dipterocarps or non-dipterocarps, due to the unusual dominance of the family Dipterocarpaceae in species diversity and basal area in the canopy and emergent tree size classes. Non-dipterocarp timber species are those which can reach timber size but are in other families (Wyatt-Smith, 1952). The Dipterocarpaceae has about 10 genera and 170 species in Peninsular Malaysia and historically has contained the most important commercial timber species in Malaysia (Appanah and Weinland, 1990). As the demand for timber products has grown, however, the value of non-dipterocarp wood has increased and greater numbers of non-dipterocarp species are being harvested to meet this demand (Manokaran and Swaine, 1994). According to Appanah and Weinland (1993), 30 dipterocarp and 90 non-dipterocarp timber species were recorded in the 50-ha plot at Pasoh forest reserve. FORMOSAIC can change the minimum size for harvest, length of rotation cycle, species harvested, and harvest location so that consequences of various harvest strategies can be assessed.

Timber logging can have significant impacts on residual trees because of tree fall, road construction, and machinery movement (Whitmore, 1984; Schaetzl et al., 1989). For the sake of simplicity, it was assumed that a tree may fall randomly along one of the four compass directions: north, south, east, or west. It was further assumed that the size of the impact zone (IZ) was positively related to the size of a fallen tree:

$$IZ = LZ \times WZ$$

where L and W are two parameters (> 0) which determine the size of an impact zone. LZ is the length of the impact zone:

$$LZ = L \times H$$

where H is the height of a fallen tree, and WZ is the width of the impact zone:

$$WZ = W \times C$$

where C is the crown radius of a fallen individual.

Within the impact zone, a proportion (P) of the trees smaller than a fallen tree would be destroyed. FORMOSAIC allows the user to change the value of P, L and/or W to simulate different degrees of logging impact on species richness in a focal forest. The impacts due to road construction and machinery movement were not separated from tree falling. Rather, to simplify computation, these impacts were incorporated into the calculation of the impact zone and proportion of smaller trees destroyed.

Windthrow

Windthrow is an important force in shaping forest dynamics (Crow, 1982; Pacala et al., 1993), with the amount of tree damage changing in relation to windthrow frequency, intensity, and location as well as stand conditions such as tree height (Ruel, 1995; Mitchell, 1995). Although windthrows are thought to be fairly common at Pasoh (H. T. Chan, personal communication), there is little information available in regard to the frequency and intensity of windthrow damage in the study site. It was assumed that a windthrow would affect a certain area in which large trees would be blown over, and the falling of these large trees would in turn damage some proportion of smaller trees within the impact zone. The impact zone is calculated as the area determined by the height times the crown radius of a fallen tree. To allow for simulations of various windthrow impact scenarios, in FORMOSAIC it is feasible to change the value of windthrow frequency, location of windthrow impact within a focal forest, minimum dbh size of fallen large trees, proportion of large trees in the impacted area falling due to windthrow, and proportion of smaller trees killed by fallen large trees within the impact zone.

Oil palm

Oil palm was first planted in Malaysia in 1917 (Williams and Hsu, 1970), and the rate of planting greatly accelerated in the 1960s and 1970s. By 1980 there were a million hectares of oil palm plantations (Hartley, 1988) and the rate is steadily increasing as more and more rubber tree plantations are converted to oil palm. The trees fruit continuously. When harvested by workers using long poles, the fruit assemblage crashes to the ground and numerous small palm kernels roughly 1.5 cm in diameter are scattered by the impact. Not all fruits on the ground are taken by workers. Local hunters believe that pigs can survive and even thrive on a diet comprised almost entirely of the fallen, uncollected palm fruits. Various studies have showed that pig density is limited by food (Giffin, 1978; Singer, 1981;

Baber and Coblentz, 1986; Caley, 1993), and in Malaysia pig density was historically also probably regulated by food availability. The mast-fruiting phenomenon of many tree species in southeast Asia may have limited pig densities in Malaysia because of the lack of food availability in non-mast fruiting years. Now, however, the palm fruits of oil palm plantations have created an almost limitless food supply, and this is the probable cause of the unusually high pig density recently seen in Malaysia. Only hunting pressure outside of the reserves may be regulating the pig population densities. Oil palm was chosen as a surrounding matrix for focal forests in FORMOSAIC because it is already so prevalent in Peninsular Malaysia and quickly becoming more so in Sabah, Sarawak, and Indonesia, in addition to some South American countries. Also, Pasoh Forest Reserve has been virtually surrounded by mature oil palm plantations for almost 20 years.

Pig damage

An increasing number of studies have documented the powerful effects that large populations of mammalian herbivores can have on the plant community. Examples include white-tailed deer (Whitney, 1984; Frelich and Lorimer, 1985), elephants (Buechner and Dawkins, 1961; Barnes et al., 1994), beavers (Johnston and Naiman, 1990), and pikas (Huntly, 1987). In the Pasoh Forest Reserve, the damage to understory trees caused by wild pigs is extremely high, and has recently become the focus of several research projects (K. Ickes, unpublished data).

The wild pig (*Sus scrofa*) has a broad natural distribution, ranging from Northern Africa throughout Europe and Asia as far southeast as Peninsular Malaysia. As an invasive species they often reach high population densities and are being increasingly cited as a major threat to local species or ecosystems in many parts of the world (Bratton, 1974, 1975; Lacki and Lancia, 1986; Coblentz and Baber, 1987; Crome and Moore, 1990; Vtorov, 1993). Pigs are truly omnivorous and will eat acorns, worms, roots, tubers, seeds, fruits, leaves, twigs, grass, agricultural crops, and many other types of food (Giffin, 1978).

The reproductive biology of pigs is conducive to rapid population growth, which has also presumably contributed to their success as an invasive or pest species. In contrast to other ungulates, pigs have very high reproductive potential; puberty is reached at a very early age (as early as 7 months), females are polyestrous, litters are large (up to ten), and gestation and lactation periods are short, permitting up to two litters per year (Giffin, 1978; Singer, 1981; Graves, 1984). In most temperate habitats the breeding of wild pigs tends to be rather seasonal, but in wetter temperate habitats such as riparian forest and tropical areas they have been observed to breed twice a year or year-round (Giffon, 1978; Sweeney et al., 1978; Singer, 1981; Baber and Boblentz, 1986; Coblentz and Baber, 1987; Saunders, 1993; K. Ickes, personal observations).

Although there are no older estimates of pig densities in Peninsular Malaysia, communications with long-term field biologists and people living in rural settings regularly indicate that pig densities have increased dramatically over the past several decades (H. T. Chan, personal communication). Pig densities in Europe and Asia, including tropical southern Asia, are usually around 2.5 individuals/km^2 (Singer, 1981; Karanth and Sunquist, 1992, 1995). Preliminary results from the Pasoh Forest Reserve document densities of 20–80 pigs/km^2 (K. Ickes, unpublished data). This unusually high density may be due to three possible causes: (i) the population of tigers (the major predator of the pigs) has been declining since the 1960s; (ii) no hunting is permitted within Pasoh Forest Reserve; and (iii) the reserve is virtually surrounded by oil palm tree plantations, the fallen fruits of which provide a practically limitless food supply. Pigs can often be seen feeding in the palm plantation along the edge of the forest.

Pigs have various behavioral characteristics which affect the plant community in Malaysia, but only one has been used specifically in FORMOSAIC: nest building. At "normal" pig population levels this behavior probably does not contribute significantly to plant community dynamics. At pig densities of 10 to 40 times the historic levels, however, the impacts are quite pronounced, with possible long-term effects on plant species diversity and distribution.

In Malaysia, when close to delivering her litter, a female pig constructs a fairly large dome-shaped nest. Pigs seem to use a wide variety of vegetative matter for this construction, utilizing whatever materials are nearby; at Pasoh, they use almost entirely saplings (K. Ickes, unpublished data). To construct a nest, the female pig approaches a sapling, turns her head to the side to grasp it in her jaws, and then jerks her head powerfully upright. If the ground is water-logged, the entire sapling is usually uprooted, but if the soil is held tighter to the roots then the trunk is snapped off (K. Ickes, personal observation). In either case, the bole with all the foliage is dragged to the nest site and carefully added to the growing mound. When complete, the female crawls underneath and births her young (Medway, 1963; K. Ickes, personal observation).

Nest size is extremely variable (K. Ickes, unpublished data) and is presumably determined by the size of the pig. Not only does the number of trees used to construct each nest vary, but so does the size of the trees taken. Some nests have only 150 saplings (all less than 2 m in height), whereas others have over 500 saplings (some more than 4 m in height). Similarly, the area affected by a pig removing trees is quite variable. Based on a sample of ten nests in the 50-ha plot, an average of 310 trees were taken per nest from an area of roughly 180 m^2 (K. Ickes, unpublished data). This represented over 50% of the number of stems over 70 cm tall and less than 2.5 cm DBH in that same 180 m^2 area. Even if it is assumed that areas damaged by pigs when making nests do not overlap, then roughly 20% of the 50-ha plot would have lost 50% of its total understory cover. The number of pig nests is quite high in the 50-ha permanent plot. For example, over a one-year period 152 new nests were constructed in 25 hectares (K. Ickes, unpublished data).

In FORMOSAIC, it is possible to adjust the maximum size of trees which are damaged by the pigs, location of damage, duration and timing of damage, and proportion of trees damaged in the impacted area of a focal forest. The ability to change these parameters allows for the estimation of the impacts of pigs on the plant community at different pig densities and damage scenarios.

Validation, sensitivity and uncertainty analysis

Validation

Half of the census data (1987 and 1991) from the 50-ha plot was used for FORMOSAIC development. The other half of the data was reserved for model validation. Using the 1987 data as initial input and running the model for the same length of time period as the time difference between the two censuses, the simulation results from FORMOSAIC fit the 1991 census data well in terms of species richness, number of trees, and basal area at two spatial scales (0.25 ha and 2.5 ha) (Liu and Ashton, 1998). At the scale of 0.25 ha, p values (from paired t-tests) for mean species richness, stand density and basal area from simulations and observations were 0.85, 0.84, and 0.97, respectively. At the scale of 2.5 ha, the p values for the three indices were 0.16, 0.19, and 0.27, respectively.

Sensitivity analysis and uncertainty analysis

A sensitivity analysis is used to test how model output responds to small changes in parameters of interest (Jorgensen, 1986; Turner et al., 1994; Starfield and Bleloch, 1991). Uncertainty analysis is employed to identify how model results vary with large variances in parameters when the values of parameters have too much uncertainty. Both types of analysis are useful in modeling and simulation studies. Sensitivity analysis can be used to detect parameters that strongly influence the simulation results. In cases where empirical estimates of model parameters are uncertain or management parameters have a wide range, uncertainty analysis is helpful in detecting the degrees of impacts within a reasonable range of parameters (e.g., within two extremes). In addition, uncertainty analysis can detect non-linear relationships between an independent variable and a dependent variable over a wide range of values.

Sensitivity and uncertainty analyses were conducted on nine parameters related to windthrow, timber harvest impact, and immigration from the surroundings at the scale of 0.25 ha (Liu and Ashton, 1998). Sensitivity analysis and uncertainty analysis indicated that the minimum size of timber trees for harvest was the most sensitive parameter for species richness, basal area, and number of trees in a focal forest. Species richness was also very sensitive to duration of immigration from surrounding forests and minimum tree size felled by windthrows. The second and third most sensitive parameters for number of trees were proportion of smaller trees killed in the harvest impact zone and duration of immigration. For basal area, the two other most sensitive parameters were proportion of smaller trees killed in the harvest impact zone and minimum tree size felled by windthrows (Liu and Ashton, 1998). Uncertainty analysis indicated that in most cases the relationships between species richness (or basal area, or number of trees) and each of the nine parameters were non-linear.

Case study: species richness affected by type of surroundings, timber harvest, and pig damage

Simulation methods

In order to address our study questions, 39 different simulations were designed to be run by FORMOSAIC (Table 3.1). Any two simulations might differ in one or more factors including spatial scale of a focal forest, type of surroundings, harvest impact, and degree of pig damage.

Spatial scales

Because forest remnants differ in size, simulations were run at five spatial scales: 0.25 ha (50 m × 50 m), 1 ha (100 m × 100 m), 4 ha (200 m × 200 m), 9 ha (300 m × 300 m), and 25 ha (500 m × 500 m). To avoid shape effects, all simulated forests were squares. To best achieve a balance between computational time and the size of many forest remnants, most simulations were done at the scale of 4 ha.

Type of surroundings

To evaluate the impacts of different types of surrounding areas on species richness within a focal forest, we created two types of surroundings: species-rich forest and oil palm plantations on all four sides of a focal forest. A species-rich forest in this study refers to a forest with the same species richness and composition as the 50-ha permanent plot, which has more than 800 tree species (LaFrankie, 1992a). Although there are many species grown monoculturally in tree plantations in tropical countries, such as rubber trees, fruit trees and commercial timber trees (Lamprecht, 1989), in our simulations only single-species plantations of the oil palm were used, a very important economic species occupying vast amounts of land in Malaysia (e.g., Williams and Hsu, 1970; Hartley, 1988). Furthermore, it was assumed that a focal forest could receive seeds from the adjacent species-rich forests but that no seeds entered the focal forest from oil palm plantations.

Harvest impact

Dipterocarps and non-dipterocarps are usually harvested at a minimum size of 50 and 45 cm, respectively (Appanah and Weinland, 1990). Our simulations followed this criterion in determining timber trees eligible for harvest. The first harvest would take place at year 10, with a rotation length of 30 years after that. Our harvest strategy can be classified as selective cutting, as only timber trees of at least 45 cm were harvested. Under harvest, the size of the impact zone was $(2 \times H) \times (2 \times C) = 4 \times H \times C$ (both L and W were assumed to be 2, see the section on FORMOSAIC Parameterization). It was further assumed that 80% (P) of the smaller trees within the impact zone were killed. This heavy impact regime may

Table 3.1. *Simulation scenarios*

Scale (ha)	Type of surroundings	Harvest impact	Percentage of small trees damaged by wild pigs	Percentage of area receiving pig damage	Duration of pig damage (years)	Timing of pig damage (years)
0.25	Oil palm plantation	yes	60	100	100	1–100
0.25	Oil palm plantation	no	60	100	100	1–100
0.25	Species-rich forest	yes	0	0	0	1–100
0.25	Species-rich forest	no	0	0	0	0
1	Oil palm plantation	yes	60	100	100	1–100
1	Oil palm plantation	no	60	100	100	1–100
1	Species-rich forest	yes	0	0	0	0
1	Species-rich forest	no	0	0	0	0
4	Oil palm plantation	no	20	100	100	1–100
4	Oil palm plantation	no	20	100	50	1–50
4	Oil palm plantation	no	20	100	50	51–100
4	Oil palm plantation	no	20	25	100	1–100
4	Oil palm plantation	no	20	25	50	1–50
4	Oil palm plantation	no	20	25	50	51–100
4	Oil palm plantation	no	60	100	100	1–100
4	Oil palm plantation	no	60	100	50	1–50
4	Oil palm plantation	no	60	100	50	51–100
4	Oil palm plantation	no	60	25	100	1–100
4	Oil palm plantation	no	60	25	50	1–50
4	Oil palm plantation	no	60	25	50	51–100
4	Oil palm plantation	no	60	100	75	26–100
4	Oil palm plantation	no	60	100	25	76–100
4	Oil palm plantation	no	60	100	75	1–75
4	Oil palm plantation	no	60	100	25	1–25
4	Oil palm plantation	no	40	100	100	1–100
4	Oil palm plantation	no	100	100	100	1–100
4	Oil palm plantation	no	60	75	100	1–100
4	Oil palm plantation	no	60	50	100	1–100
4	Oil palm plantation	no	0	0	0	0
4	Oil palm plantation	yes	60	100	100	1–100
4	Oil palm plantation	no	60	100	100	1–100
4	Species-rich forest	yes	0	0	0	0
4	Species-rich forest	no	0	0	0	0
9	Oil palm plantation	yes	60	100	100	1–100
9	Oil palm plantation	no	60	100	100	1–100
9	Species-rich forest	yes	0	0	0	1–100
9	Species-rich forest	no	0	0	0	0
25	Oil palm plantation	yes	60	100	100	1–100
25	Oil palm plantation	no	60	100	100	1–100
25	Species-rich forest	yes	0	0	0	0
25	Species-rich forest	no	0	0	0	0

occur when powerful machinery such as crawler tractors are used (Panayotou and Ashton, 1992). The surrounding forests were intact. Only timber trees in a focal forest were eligible for harvest.

Pig damage

Four parameters were considered related to pig damage: percentage of trees < 3cm dbh damaged, percentage of area receiving pig damage, duration of pig damage, and timing of pig damage. In the simulations different percentages of small trees damaged by the wild pigs were set (0%, 20%, 40%, 60%, and 100% of the small trees within the pig damage area). Percentage of area receiving pig damage ranged from 0%, 25%, 50%, 75%, to 100% of the entire focal forest.

Duration of pig damage refers to the number of years that the wild pigs have been building nests in the focal forest. In our simulations, duration of pig damage was 0, 25, 50, 75 or 100 years. Twenty-five years was chosen as a duration interval because oil palm plantations' life cycle is about 25 years (Williams and Hsu, 1970). It was assumed that without abundant fruits from the surrounding palm plantations, pig densities would be low and therefore damage to small trees in nearby primary forests would be not significant. Damage duration of 0 years may be equivalent to a situation in which there are no oil palm plantations in the surroundings, and hence pig densities have not been artificially elevated.

Timing of pig damage defines when, or during which years, damage takes place. For example, pig damage could occur in different periods of time corresponding to the existence of oil palm plantations. If two 25-year rotations are thought to be the maximum yield possible for growing oil palm, after which time the surrounding land is left to regenerate to forest, pig damage with a duration of 50 years could be input for the first half of a 100-year simulation, with no damage occurring in the last 50 years.

Data for simulation initialization

To initialize a simulation, we used the environmental data of slope, elevation, and location of streams and swamps from the 50-ha plot at Pasoh. In addition, tree data including dbh, location, and species were obtained from the 1987 census. The 50-ha plot is rectangular, running 0–1000 m west to east, and 0–500 m south to north. Therefore, a simulated focal area of 50 m × 50 m would employ the known topographic relief and the 1987 tree survey information for the 0.25-ha area lying 0–50 m west to east and 0–50 m south to north within the 50-ha plot. Similarly, a simulated focal area of 500 m × 500 m would use information from the 25-ha area of 0–500 m west to east, and 0–500 m south to north within the 50-ha plot. A simulated focal area was assumed to be embedded in one of the two types of surroundings described previously.

Liu and Ashton (1998) reported that simulations of high frequency and intensity of windthrows for this forest usually lower species richness. In this study, it was assumed that windthrows had the same frequency and severity, starting at year 20 and occurring subsequently once every 40 years. When a windthrow took place,

20% of the trees with > 30 cm in dbh were assumed to be blown down and 20% of the smaller trees within the impact area (height × crown radius of a fallen tree) were presumably destroyed. Windthrows were included in all the simulations because they are thought to be an important force in shaping forest dynamics (Crow, 1982; Pacala et al., 1993) and have recently occurred in the Pasoh Forest Reserve (H. T. Chan, personal communication). Because the purpose of this paper is not to assess the impacts of windthrows but rather to evaluate the interactive effects of surroundings, timber harvest, and pig damage on species richness, the same windthrow frequency and severity were maintained for all simulations.

Calculation of average species richness and statistical tests
Each simulation step was 1 year and each run lasted 100 years. Those simulations at the scale of 9 ha or larger had five replicates. Smaller scales had ten replicates. Simulations at small spatial scales were replicated to a greater extent because FOR-MOSAIC is a stochastic model and small scales tended to have higher variations among replicates. The average species richness was calculated over the entire simulation period of 100 years and across replicates. To detect main effects of variables and their interactions on species richness and to test for significance of differences in average species richness among various simulation scenarios, ANOVA and Bonferroni tests were used for multiple comparisons in SYSTAT (SPSS Inc., 1996).

Simulation results

Average species richness decreased with increases in percentage of small trees damaged by wild pigs (Fig. 3.2). Species richness dropped rapidly between 0 and 20%, but as the percentage of small trees damaged increased from 20% to 100% the rate of decline in species richness was much lower. Species richness had a negative linear relationship with percentage of pig damage area (Fig. 3.3). With regard to the impact resulting from duration of pig damage, species richness was sharply reduced during the first 25 years, but showed little change in the latter 75 years (Fig. 3.4). There were very strong interactions among percentage of small trees damaged, percentage of area receiving pig damage, and duration of pig damage (Table 3.2).

Timing of pig damage was also important for species richness. Damage during the early years of simulations resulted in much lower average species richness after 100 years than damage during the late years of simulations (Fig. 3.5). It was clear that species richness was sharply lowered due to the first few years of pig damage (Fig. 3.6), while subsequent damage in later years did not influence species richness as much.

Species richness was scale-dependent and varied significantly due to the interactions among types of surroundings, timber harvest impact, and pig damage (Fig. 3.7). When a focal forest was surrounded by species-rich forests with no harvest impact and no pig damage, species richness was always the highest, regardless of how large a focal forest was. The combination of being surrounded by plantations

Fig. 3.2. Effects of percentage of small trees damaged by wild pigs on species richness in a 4-ha forest. In the simulations, timber trees were not harvested. Pig damage took place continuously in the entire forest. The focal forest was surrounded by oil palm plantations. Results are means (n = 10) for 100-yr simulations. Error bars indicate two standard errors.

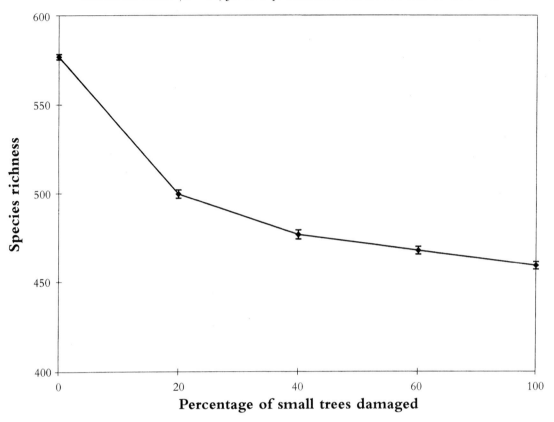

with harvest impact and pig damage led to the lowest species richness across spatial scales. Species-rich surroundings with harvest and no pig damage resulted in higher species richness than oil palm plantations with pig damage and no harvest at scales of less than 4 ha, but lower species richness at scales of 4 ha or larger.

Discussion

FORMOSAIC mimics forest dynamics in fragmented and heterogeneous landscape mosaics, which represent a common pattern of forest distribution (Shugart, 1984; Harris, 1984; Schelhas and Greenberg, 1996). Ecological processes and patterns may vary at different spatial scales (e.g., Levin, 1992; Turner et al., 1995b). FORMOSAIC is able to run simulations at multiple spatial scales and provides a tool to incorporate the effects of timber harvest and ecological processes inside a focal forest with those of seed and pig immigration from the surroundings of the

Fig. 3.3. Effects of percentage of area with pig damage on species richness in a 4-ha forest. In the simulations, timber trees were not harvested. Pig damage took place continuously, with 60% of small trees (<3 cm in dbh) damaged. The focal forest was surrounded by oil palm plantations. Results are means (n = 10) for 100-yr simulations. Error bars indicate two standard errors.

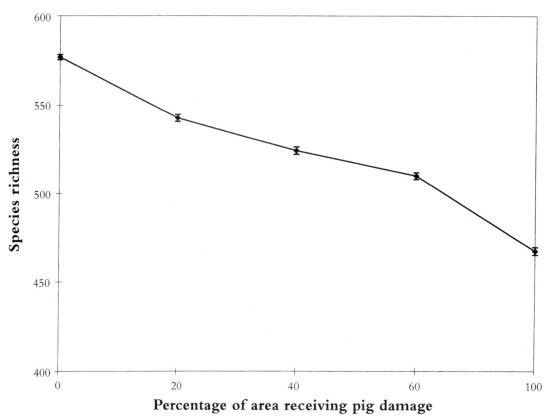

focal forest. Our simulations indicated that type of surroundings, harvest impact, and pig damage had major impacts on maintenance of species richness in a focal forest, and that their roles varied among spatial scales (Fig. 3.7).

FORMOSAIC is quite different from a typical gap model (Botkin et al., 1972; Shugart, 1984). First, it explicitly considers the impacts (e.g., seed and wildlife immigration) from the surroundings, whereas a typical gap model does not. Second, it simulates the dynamics of an entire forest (which may consist of both gaps and non-gaps at the same time), while a typical gap model simulates the dynamics of a gap only. Third, the functions for growth, mortality, and recruitment in FORMOSAIC are derived from demographic census data, while a typical gap model has light and moisture-driven functions that include detailed vertical layers through the canopy. Growth functions of many species in FORMOSAIC have slope, elevation and distance as independent variables in addition to tree size and

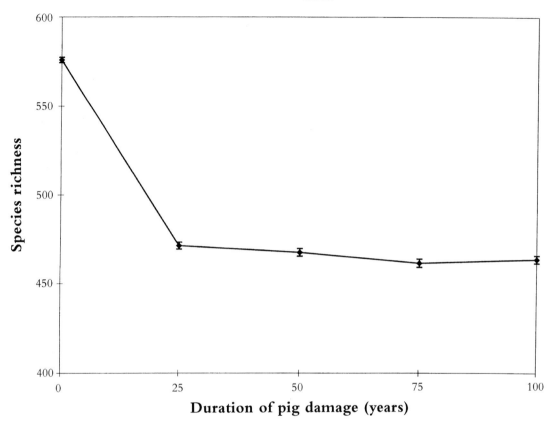

Fig. 3.4. Impacts of duration of pig damage on species richness in a 4-ha forest. In the simulations, timber trees were not harvested. Sixty percent (60%) of small trees (< 3 cm in dbh) were damaged in the entire forest. The focal forest was surrounded by oil palm plantations. Results are means (n = 10) for 100-yr simulations. Error bars indicate two standard errors.

neighborhood pressure. Fourth, FORMOSAIC considers the horizontal differences (e.g., location) among trees, even within the same grid cell, while a typical gap model does not.

Individual-based spatially explicit models like FORMOSAIC need large amounts of individual-based and spatial data collected in the field. The process of collecting fine-scale data in the field (in our case, the Pasoh forest) is quite time-consuming. However, this modeling approach is useful for in-depth study of long-term species diversity dynamics. Hopefully, the challenge of data collection can be met by future remote sensing techniques, although current remote sensing technologies are not able to classify tree species (especially understory species) as accurately as plant systematists, and the spatial resolutions regarding locations of individual trees in remote sensing imagery are not as accurate as field measurements.

Table 3.2. *Analysis of variance for the main and interactive effects (in a 2^3 factorial design) of percentage of small trees damaged by wild pigs (PSTD, 20% and 60%), percentage of area with pig damage (PAPD, 25% and 100%), and duration of pig damage (DPD, 50 and 100 years) on species richness in a 4-ha forest.*

Source	Sum-of-squares	DF	Mean-square	F-ratio	P
PSTD	1402.98	1	1402.98	313.88	< 0.001
DPD	27008.31	1	27008.31	6042.393	< 0.001
PAPD	21108.428	1	21108.428	4722.451	< 0.001
PSTD×DPD	1469.755	1	1469.755	328.819	< 0.001
PSTD×PAPD	1274.805	1	1274.805	285.204	< 0.001
DPD×PAPD	15242.653	1	15242.653	3410.139	< 0.001
PSTD×DPD×PAPD	966.398	1	966.398	216.206	< 0.001
Error	321.826	72	4.47		

When the duration of pig damage was 50 years, pig damage occurred from years 51–100. In the simulations, timber trees were not harvested. The 4-ha focal forest was surrounded by oil palm plantations.

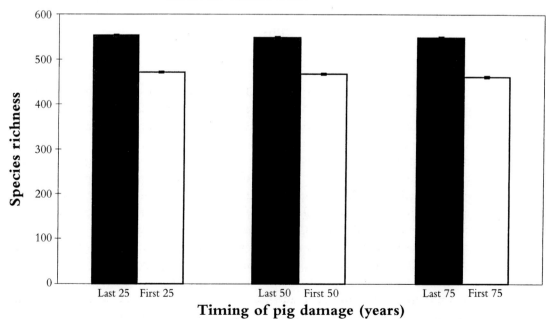

Fig. 3.5. *Relationship between timing of pig damage and species richness in a 4-ha forest. In the simulations, timber trees were not harvested. When pig damage took place, 60% of small trees (< 3 cm in dbh) were damaged in the entire forest. The focal forest was surrounded by oil palm plantations. Results are means (n = 10) for 100-yr simulations. Error bars indicate two standard errors.*

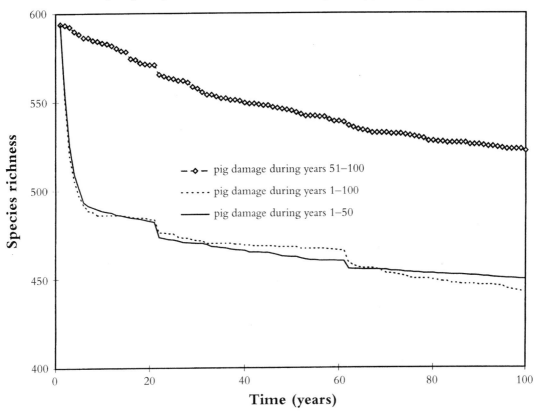

Fig. 3.6. Dynamics of species richness in a 4-ha forest with different timing of pig damage. In the simulations, timber trees were not harvested. When pig damage took place, 60% of small trees (< 3 cm in dbh) were damaged in the entire forest. The focal forest was surrounded by oil palm plantations. Results are means (n = 10) for 100-yr simulations.

Results obtained regarding impact of wild pigs on tree species diversity are not consistent with the widely supported "intermediate disturbance hypothesis," which states that species diversity peaks under intermediate disturbance (Connell, 1979). Simulation results in our case study, however, indicate a monotonic decrease of species diversity with increase in the degree of pig damage. We believe that the reason is that the wild pigs destroy only small trees. Damage to small trees creates small gaps, which do not provide sufficient niches for new species to colonize. However, in many simulations where large trees were harvested or large gaps were generated, FORMOSAIC did produce outputs supporting the "intermediate disturbance hypothesis." For example, an intermediate rotation length resulted in the highest species diversity (Liu and Ashton, 1998).

As is well known, realistic models have to depend on realistic data. However, it is not always easy to obtain adequate data for parameterizing and validating the models. In addition to logistic issues such as money and personnel, serious attention

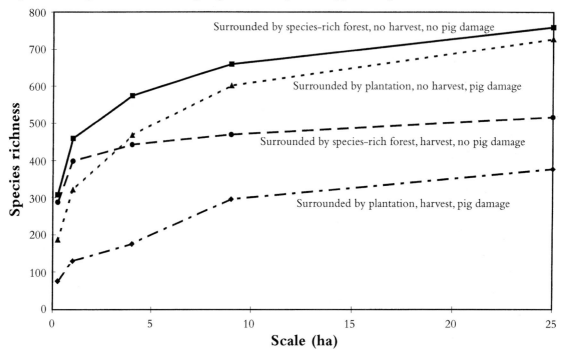

Fig. 3.7. Interactive effects of type of surroundings, harvest impact, and pig damage on average species richness along spatial scales. When pig damage took place, 60% of small trees (< 3 cm in dbh) were damaged in the entire forest. Results are means (n = 10 for spatial scales of < 9 ha and n = 5 for spatial scales of ≥ 9 ha) for 100-yr simulations.

should be paid to the methodology of data collection because studying ecological processes and patterns beyond boundaries are more complicated than studying a focal forest alone. In this study timing and duration of pig damage were used as surrogates to represent dynamic change of the surroundings. Such measures do not simulate the spatial dynamics of the whole complex of landscape mosaics. Better understanding and simulation of forest and wildlife dynamics in the adjacent areas could enhance model predictability. In the Pasoh Forest Reserve, field study on pig damage to small trees in the 50-ha permanent plot is under way (K. Ickes, unpublished data). Much more information on the relationships between pig damage (e.g., percentage of small trees damaged, area receiving pig damage, and duration and timing of pig damage) and pig population dynamics due to adjacent oil palm plantations is still needed.

Visitors to the Pasoh Forest Reserve are often shocked by the damage caused by the making of even one pig nest. Pigs are removing a substantial percentage of the understory in some areas (K. Ickes, personal observations), and this must have some impact on both plant and animal community ecology. For example, understory tree species often fruit year-round, with low numbers of fruit at any given time, and various bird species rely on the fruits of these trees for food.

Obviously, the quantity of available fruits for these birds is much lower in areas with multiple pig nests. Similarly, less foliage will mean fewer arthropods for insect eating birds and vertebrates. If pigs are preferentially selecting certain tree species for nests and avoiding others, then the relative abundance or distribution of those tree species could be substantially altered. The impacts of a dense pig population on the primary forests are thus potentially numerous and critical.

Our simulation results may have important implications for guiding future field study, for conserving tree species richness in primary forests, and for managing oil palm plantations. For example, because a low percentage of small trees damaged by wild pigs could reduce species richness dramatically (Fig. 3.2), it is necessary to accurately measure the damage caused by wild pigs. Oil palm plantations should not be near primary forests, or the oil palm fruits on the ground should be immediately collected so that they will not provide food for wild pigs, thus hopefully reducing pig population size.

In this case study an attempt was made to explicitly address the interactions among adjacent areas, timber harvest, and pig damage, as well as their impacts on species richness of a tropical forest. This approach could be also useful for understanding the dynamics of tree species richness in temperate and boreal forest zones, because there are many similarities (e.g., fragmentation, Harris, 1984) among various forest types. Simulations that study ecological processes and patterns within and across ownership and management boundaries have the potential to provide valuable information for resource management from a landscape perspective (Liu and Ashton, 1999).

Acknowledgments

We are very grateful to S. Appanah, Marco Boscolo, Y. Chen, Rick Condit, Saara DeWalt, Stephen Hubbell, Michael Huston, Bruce Larson, Chris Lepzcyk, Elizabeth Losos, Stephen Pacala, and Daniel Rutledge for their insight and help. We thank S. Appanah and S. Thomas for their unpublished data. Carrie Lenz, JoAnna Lessard, and Risa Oram provided assistance in preparing the manuscript. We also appreciate useful comments from David Mladenoff, William Baker, and two anonymous reviewers. The 50-ha forest plot at Pasoh Forest Reserve is an ongoing project of the Malaysian Government, initiated by the Forest Research Institute Malaysia (FRIM) through its Director-General, Dato' Dr Salleh Mohd. Nor, and under the leadership of N. Manokaran, Peter S. Ashton and Stephen P. Hubbell. Through the Pasoh project, FRIM is cooperating with the Center for Tropical Forest Science of the Smithsonian Tropical Research Institute in an international program of standardized long-term forest dynamics research in the tropics. This study is supported by the John D. and Catherine T. MacArthur Foundation, the Rockefeller Foundation, the Smithsonian Tropical Research Institute, and the National Science Foundation.

References

Appanah, S. and Weinland, G. (1990). Will the management systems for hill dipterocarp forest stand up? *Journal of Tropical Forest Science*, **3**,140–58.

Appanah, S. and Weinland, G. (1993). A preliminary analysis of the 50-hectare Pasoh demography Plot: I. Dipterocarpaceae. Research Pamphlet, No. 112. Forest Research Institute Malaysia, Kepong, Kuala Lumpur, Malaysia.

Baber, D. W. and Coblentz, B. E. (1986). Density, home range, habitat use, and reproduction in feral pigs on Santa Catalina Island. *Journal of Mammalogy*, **67**, 512–25.

Barnes, R. F. W., Barnes, K. L. and Kapela, B. (1994). The long-term impact of elephant browsing on baobab trees at Msembe, Ruaha national Park, Tanzania. *African Journal of Ecology*, **32**, 177–84.

Bella, I. E. (1970). Simulation of growth, yield, and management of aspen. PhD Dissertation, The University of British Columbia, Vancouver, Canada.

Bierragaard, R. O. Jr, Lovejoy, T. E., Kapos, V., dos Santos, A. A. and Hutchings, R. W. (1992). The biological dynamics of tropical rainforest fragments. *BioScience*, **42**, 859–66.

Bonan, G. B., Shugart, H. H. and Urban, D. L. (1990). The sensitivity of some high-latitude boreal forests to climatic parameters. *Climatic Change*, **16**, 9–29.

Borota, J. (1991). *Tropical forests: some African and Asian case studies of composition and structure*. Amsterdam: Elsevier.

Botkin, D. B., Janak, J. S. and Wallis, J. R. (1972). Some ecological consequences of computer model of forest growth. *Journal of Ecology*, **60**, 167–79.

Bratton, S. P. (1974). The effect of the European Wild Boar (*Sus scrofa*) on the high-elevation vernal flora in Great Smoky Mountains National Park. *Bulletin of the Torrey Botanical Club*, **101**, 198–206.

Bratton, S. P. (1975). The effect of the European Wild Boar, *Sus scrofa*, on Gray Beech forest in the Great Smoky Mountains. *Ecology*, **56**, 1356–66.

Buechner, J. K. and Dawkins, H. C. (1961). Vegetation change induced by elephants and fire in Murchison Falls National Park, Uganda. *Ecology*, **42**, 752–66.

Buschbacher, R. J. (1987). Deforestation for sovereignty over remote frontiers. In *Amazonian rain forests: ecosystem disturbance and recovery*, ed. C. F. Jordan, pp. 46–57. New York: Springer-Verlag.

Busing, R. T. (1991). A spatial model of forest dynamics. *Vegetatio*, **92**, 167–79.

Caley, P. (1993). Population dynamics of feral pigs (*Sus scrofa*) in a tropical riverine habitat complex. *Wildlife Research*, **20**, 625–36.

Christensen, N. L., Bartuska, A. M., Brown, J. H., Carpenter, S., D'Antonio, C., Francis, R., Franklin, J. F., MacMahon, J. A., Noss, R. F., Parsons, D. J., Peterson, C. H., Turner, M. G. and Woddmansee, R. G. (1996). The report of the Ecological Society of America committee on the scientific basis for ecosystem management. *Ecological Applications*, **6**, 665–91.

Clark, J. S. and Ji, Y. (1995). Fecundity and dispersal in plant populations: Implications for structure and diversity. *The American Naturalist*, **146**, 72–111.

Coblentz, B. E. and Baber, D. W. (1987). Biology and control of feral pigs on Isla Santiago, Galapagos, Ecuador. *Journal of Applied Ecology*, **24**, 403–18.

Connell, J. H. (1979). Tropical rain forests and coral reefs as open non-equilibrium systems. In *Population dynamics*, ed. R. M. Anderson, B. D. Turner and R. D. Taylor, pp. 141–63. Oxford: Blackwell Scientific Publications.

Crome, F. H. J. and Moore, L. A. (1990). Cassowaries in north-eastern Queensland: Report of a survey and a review and assessment of their status and conservation and management needs. *Australian Wildlife Research*, **17**, 369–85.

Crow, T. R. (1982). A rainforest chronicle: A 30-year record in structure and composition at El Verde, Puerto Rico. *Biotropica*, **12**, 42–55.

Dunning, J. B. Jr, Steward, D. J., Danielson, B. J., Noon, B. R., Root, T. L., Lamberson, R. H. and Stevens, E. E. (1995). Spatially explicit population models: Current forms and future uses. *Ecological Applications*, **5**, 3–11.

Ek, A. R., Shifley, S. R. and Burk, T. E. (1988). Forest growth modeling and prediction. *Proceedings of the IUFRO Conference*, SAF-87 12. Society of American Forestry.

Ellis, M. A. and Stroustrup, B. (1990). *The annotated C++ reference manual*. Reading, Massachusetts: Addison-Wesley Publishing Company.

Forman, R. T. T. and Moore, P. N. (1992). Theoretical foundations for understanding of boundaries in landscape mosaics. In *Landscape boundaries: consequences for biotic diversity and ecological flows*, ed. A. J. Hansen and F. di Castri, pp. 236–58. New York: Springer-Verlag.

Frelich, L. E. and Lorimer, C. G. (1985). Current and predicted long-term effects of deer browsing in hemlock forests in Michigan, U.S.A. *Biological Conservation*, **34**, 99–120.

Fries, J., ed. (1974). Growth models for tree and stand simulation. Research Notes 30, Department of Forest Yield Research, Royal College of Forestry, Stockholm.

Giffin, J. (1978). Ecology of the feral pig on the island of Hawaii. Final Report P-R Project W-15-3. Study No. II. Hawaii Department of Land and Natural Resources, Division of Fish and Game.

Graves, H. B. (1984). Behavior and ecology of wild and feral swine (*Sus scrofa*). *Journal of Animal Science*, **58**, 482–92.

Grubb, P. J. (1977). The maintenance of species-richness in plant communities: the importance of the regeneration niche. *Biological Review*, **52**, 107–45.

Harper, L. H. (1989). The persistence of ant-following birds in small Amazonian forest fragments. *ACTA Amazonica*, **19**, 249–63.

Harris, L. D. (1984). *The fragmented forest: island biogeography theory and the preservation of biotic diversity*. Chicago: IL: University of Chicago Press.

Hartley, C. W. S. (1988). *The oil palm*, 3rd edn. New York: Longman Scientific and Technical, co-published in the US with John Wiley & Sons, Inc.

Hatch, C. R. (1971). Simulation of an even-aged red pine stand in northern Minnesota. PhD Dissertation, The University of Minnesota.

Hegyi, F. (1974). A simulation model for managing Jack-pine stands. In *Growth models for tree and stand simulation*, ed. J. Fries, Research Notes 30, Department of Forest Yield Research, Royal College of Forestry, Stockholm, pp. 74–90.

Hone, J. (1990). Note on seasonal changes in population density of feral pigs in three tropical habitats. *Australian Wildlife Research*, **17**, 131–4.

Huntly, N. J. (1987). Influence of refuging consumers (Pikas: *Ochotona princeps*) on subalpine meadow vegetation. *Ecology*, **68**, 274–83.

Janzen, D. H. (1978). Seeding patterns of tropical trees. In *Tropical trees as living systems*, ed. P. B. Tomlinson and M. H. Zimmermann, pp. 83–128. Cambridge, UK: Cambridge University Press.

Johnston, C. A. and Naiman, R. J. (1990). Browse selection by beaver: Effects on riparian forest composition. *Canadian Journal of Forest Research*, **20**, 1036–43.

Jordan, C. F. (1987). *Amazonian rain forests*. New York: Springer-Verlag.

Jorgensen, S. E. (1986). *Fundamentals of ecological modelling*. Amsterdam: Elsevier.

Karanth, K. U. and Sunquist, M. E. (1992). Population structure, density and biomass of large herbivores in the tropical forest of Nagarahole, India. *Journal of Tropical Ecology*, **8**, 21–35.

Karanth, K. U. and Sunquist, M. E. (1995). Prey selection by tiger, leopard and dhole in tropical forests. *Journal of Animal Ecology*, **64**, 439–50.

Lacki, M. J. and Lancia, R. A. (1986). Effects of wild pigs on Beech growth in Great Smoky Mountains National Park. *Journal of Wildlife Management*, **50**, 655–9.

LaFrankie, J. V. (1992a). The 1992 data set for the Pasoh 50-ha forest dynamics plot. Miscellaneous Internal Report 26.9.92. Center for Tropical Forest Science, Smithsonian Institute for Tropical Research.

LaFrankie, J. V. (1992b). Estimating diameter growth from the 50-ha permanent plot data set at Pasoh Forest Reserve, Malaysia. Miscellaneous International Report 26.8.92. Center For Tropical Forest Science, Smithsonian Institute for Tropical Research.

Lamprecht, H. (1989). *Silviculture in the tropics*. Federal Republic of Germany: Deutsche Gesellschaft fur Technische Zusammenarbeit.

Larocque, G. and Marshall, P. L. (1988). Improving single-tree distance-dependent growth models. In *Growth modelling and prediction*, ed. A. R. Ek, S. R. Shifley and T. E. Burk, *Proceedings of the IUFRO Conference*, 23–27 August 1987, Minnesota, US Forest Service, NC-120, pp. 94–101.

Lee, Y. (1967). Stand models for lodgepole pine and limits to their application. PhD Dissertation, The University of British Columbia, Vancouver, Canada.

Leemans, R. and Prentice, I. C. (1987). Description and simulation of tree-layer composition and size distribution in a primeval *Picea–Pinus* forest. *Vegetatio*, **69**, 147–56.

Levin, S. A. (1992). The problem of pattern and scale in ecology. *Ecology*, **73**, 1943–67.

Lin, J. Y. (1969). Growing space index and stand simulation models for Douglas-fir and western hemlock in the Northwestern United States. In *Growth models for tree and stand simulation*, ed. J. Fries, Research Notes 30, Department of Forest Yield Research, Royal College of Forestry, Stockholm, pp. 102–8.

Liu, J. (1993). ECOLECON: An ECOLogical–ECONomic model for species conservation in complex forest landscapes. *Ecological Modelling*, **70**, 63–87.

Liu, J. and Ashton, P. S. (1995). Individual-based simulation models for forest succession and management. *Forest Ecology and Management*, **73**, 157–75.

Liu, J. and Ashton, P. S. (1998). FORMOSAIC: An individual-based spatially explicit model for simulating forest dynamics in landscape mosaics. *Ecological Modelling*, **106**, 177–200.

Liu, J. and Ashton, P. S. (1999). Simulating effects of landscape context and timber harvesting on tree species richness of a tropical forest. *Ecological Applications*, **9**, 186–201.

Lovejoy, T. E. and Oren, D. C. (1981). Minimum critical size of ecosystems. In *Forest island dynamics in man-dominated landscapes*, ed. R. L. Burgess and D. M. Sharp, pp. 7–12. New York: Springer-Verlag.

Lubchenco, J., Olson, A. M., Brubaker, L. B., Carpenter, S. R., Holland, M. M., Hubbell, S. P., Levin, S. A., MacMahon, J. A., Matson, P. A., Melillo, J. M., Mooney, H. A., Peterson, C. H., Pulliam, H. R., Real, L. A., Regal, P. J. and Risser, P. G. (1991). The sustainable biosphere initiative: An ecological research agenda. *Ecology*, **72**, 371–412.

McDonnell, M. J. and Pickett, S. T. A. (eds) (1993). *Humans as components of ecosystems: the ecology of subtle human effects and populated areas*. New York: Springer-Verlag.

Malcolm, J. R. (1988). Small mammal abundances in isolated and non-isolated primary reserves near Manaus, Brazil. *ACTA Amazonica*, **18**, 67–83.

Manokaran, N., LaFrankie, J. V., Kochummen, K. M., Quah, E. S., Klahn, J. E., Ashton, P. S. and Hubbell, S. P. (1990). Methodology for the fifty hectare research plot at Pasoh Forest Reserve. Research Pamphlet No. 104, Forest Research Institute Malaysia, Kuala Lumpur, Malaysia.

Manokaran, N. and Swaine, M. D. (1994). Population dynamics of trees in Dipterocarp Forests of Peninsular Malaysia. Forest Research Institute Malaysia, Kepong, Malaysia.

Medway, L. (1963). Pigs' nests. *Malay Naturalist Journal*, **17**, 41–5.

Mitchell, K. J. (1969). Simulation of the growth of even-aged stands of white spruce. *Bulletin 75*, Yale School of Forestry. New Haven, CT.

Mitchell, S. J. (1995). The windthrow triangle: A relative windthrow hazard assessment procedure for forest managers. *The Forestry Chronicle*, **71**, 446–50.

Murray, D. R., ed. (1986). *Seed dispersal*. Sydney: Academic Press.

Newnham, R M. (1964). The development of a stand model for Douglas fir. PhD Dissertation. The University of British Columbia, Vancouver, Canada.

Ng, F. S. P. (1978). *Tree flora of Malaya*, Vol. III. London: Longman Group Ltd.

Ng, F. S. P. (1989). *Tree flora of Malaya*, Vol. IV. London: Longman Group Ltd.

Nyland, R. D. (1996). *Silviculture: concepts and applications*. New York: McGraw-Hill.

Ott, L. (1988). *An introduction to statistical methods and data analysis*. 3rd edn. Boston: PWS-KENT Publishing Company.

Pacala, S. W., Canham, C. D. and Silander, J. A. Jr. (1993). Forest models defined by field measurements: I. The design of a northeastern forest simulator. *Canadian Journal of Forest Research*, **23**, 1980–8.

Pacala, S. W., Canham, C. D., Saponara, J., Silander, J. A. Jr, Kobe, R. K. and Ribbens, E. (1996). Forest models defined by field measurements: estimation, error analysis and dynamics. *Ecological Monographs*, **66**, 1–43.

Panayotou, T. and Ashton, P. S. (1992). *Not by timber alone: economics and ecology for sustaining tropical forests*. Washington, DC: Island Press.

Powell, A. H. and Powell, G. V. N. (1987). Population dynamics of euglossine bees in Amazonian forest fragments. *Biotropica*, **19**, 176–9.

Pulliam, H. R. (1988). Sources, sinks, and population regulation. *American Naturalist*, **132**, 652–61.

Ruel, J-C. (1995). Understanding windthrow: Silvicultural implications. *The Forestry Chronicles*, **71**, 434–45.

Saunders, G. (1993). The demography of feral pigs (*Sus scrofa*) in Kosciusko National Park, New South Wales. *Wildlife Research*, **20**, 559–69.

Schaetzl, R. J., Burns, S. F., Johnson, D. L. and Small, T. W. (1989). Tree uprooting: review of impacts on forest ecology. *Vegetatio*, **79**, 165–76.

Schelas, J. and Greenberg, R. (1996). *Forest patches in tropical landscapes*. Washington, DC: Island Press.

Schulte, A. and Schone, D., eds. (1996). *Dipterocarp forest ecosystem: towards sustainable management*. Singapore: World Scientific.

Shugart, H. H. (1984). *A theory of forest dynamics*. New York: Springer-Verlag.

Shugart, H. H. and Noble, I. R. (1981). A computer model of succession and fire response of the high altitude Eucalyptus forest of the Brindabella Range. Australian Capital Territory. *Australian Journal of Ecology*, **6**, 149–64.

Singer, F. J. (1981). Wild pig populations in the National Parks. *Environmental Management*, **5**, 263–70.

Smith, D. M. (1986). *The practice of silviculture*. 8th edn. New York: John Wiley.

Smith, T. M. and Urban, D. L. (1988). Scale and resolution of forest structural pattern. *Vegetatio*, **74**, 143–50.

SPSS, Inc. (1996). *SYSTAT 6.0 for Windows*. Chicago, IL.

Starfield, A.M. and Bleloch, A. L. (1991). *Building models for conservation and wildlife management*. 2nd edn. Edina, MN: Burgess International Group, Inc.

Sweeney, J. M., Sweeney, J. R. and Provost, E. E. (1978). Reproductive biology of a feral pig population. *Journal of Wildlife Management*, **43**, 555–9.

Symington, C. F. (1943). *Foresters manual of dipterocarps*. Malayan Forest Records No. 16. (New edn. 1974, Penerbit Universiti Malaya, Kuala Lumpur).

Turner, M. G., Wu, Y., Wallace, L. L., Romme, W. H. and Brenkert, A. (1994). Simulating winter interactions among ungulates, vegetation, and fire in northern Yellowstone Park. *Ecological Applications*, **4**, 472–96.

Turner, M. G., Arthaud, G. J., Engstrom, R. T., Hejl, S. J., Liu, J., Loeb, S. and McKelvey, K. (1995a). Usefulness of spatially explicit population models in land management. *Ecological Applications*, **5**, 12–16.

Turner, M. G., Gardner, R. H. and O'Neill, R. V. (1995b). Ecological dynamics at broad scales. *Science and Biodiversity Policy. BioScience Supplement*, S29–S35.

Urban, D. L. (1990). A versatile model to simulate forest pattern: a user's guide to ZELIG version 1.0. Department of Environmental Sciences, University of Virginia, Charlottesville.

Vanclay, J. K. (1989). A growth model for North Queensland Rainforests. *Forest Ecology Management*, **27**, 245–71.

Van Daalen, J. C. and Shugart, H. H. (1989). OUTENIQUA – a computer model to simulate succession in the mixed evergreen forests of the southern Cape, South Africa. *Landscape Ecology*, **2**, 255–67.

Viana, V. M. and Tabanez, A. A. J. (1996). Biology and conservation of forest fragments in the Brazilian Atlantic moist forest. In *Forest patches in tropical landscapes*, ed. J. Schelhas and R. Greenberg, Washington, DC: Island Press.

Vtorov, I. P. (1993). Feral pig removal: effects on soil microarthropods in a Hawaiian rain forest. *Journal of Wildlife Management*, **57**, 875–80.

Welch, B. B. (1995). *Practical programming in Tcl and Tk*. PTR, New Jersey: Prentice Hall.

Wensel, L. C. (1990). A bibliography of conference proceedings. In *Forest simulation systems*, ed. L. C. Wensel and G. S. Biging, *Proceedings of the IUFRO Conference*,

Bulletin 1927, Division of Agriculture and Natural Resources, University of California, pp. 405–10.

Whitmore, T. C. (1972*a*). *Tree flora of Malaya*, Vol. I. London: Longman Group Ltd.

Whitmore, T. C. (1972*b*). *Tree flora of Malaya*, Vol. II. London: Longman Group Ltd.

Whitmore, T. C. (1984). *Tropical rain forests of the Far East*, 2nd edn. Oxford: Clarendon Press.

Whitney, G. G. (1984). Fifty years of change in the arboreal vegetation of Hearts Content, an old-growth hemlock-white pine-northern hardwood stand. *Ecology*, **65**, 403–8.

Williams, C. N. and Hsu, Y. C. (1970). *Oil palm cultivation in Malaya: technical and economic aspects*. Kuala Lumpur, Malaysia: University of Malaya Press.

Wilson, E. O., ed. (1988). *Biodiversity*. Washington, DC: National Academy of Science Press.

Wyatt-Smith, J. (1952). *Pocket check list of timber trees* (Malayan Forest Records No. 17). Kuala Lumpur, Malaya: Caxton Press Ltd.

4

Scaling fine-scale processes to large-scale patterns using models derived from models: meta-models

Dean L. Urban, Miguel F. Acevedo and Steven L. Garman

Introduction

Ecologists and natural resource managers face a common scaling dilemma in many applications. Our conventional knowledge base is rather fine-scale, but many of the issues that now face us are of much larger extent, often played out at landscape to regional scales. As an extreme example, consider the potential effects of anthropogenic climatic change on forests. Our best mechanistic understanding of the effects of temperature, moisture, and CO_2 on tree growth is at the level of plant ecophysiology (i.e., leaves, seedlings, and perhaps single trees; see Strain and Cure, 1985; Bazzaz, 1990), while assessments of these effects are typically addressed at regional or even global scales (e.g., Smith and Tirpak, 1989; IPCC, 1996; Walker and Steffen, 1996). Other applications, while less extreme, do not escape this fundamental scaling mismatch. For example, forest managers now integrate their activities at the level of ecosystems and at scales of entire forests (i.e., landscapes), yet we still work most comfortably at the scales we know best; that is, stand-level prescriptions carried out on individual trees.

Ecologists are increasingly savvy about scale (Delcourt et al. 1983; Wiens, 1989; Levin, 1992). The basic scaling rule that trades off spatial resolution and detail for spatial extent is appreciated: detailed fine-scale studies are carried out on small study areas, while studies of much broader extent necessarily sacrifice details to emphasize coarser-resolution patterns. This trade-off comes at some expense; applications at disparate scales are divorced from one another empirically and sometimes conceptually. For example, many models that address forest dynamics at the scale of the stand (*ca.* 1–10 ha) simulate the behavior of individual trees (Botkin *et al.*, 1972a,b; Shugart, 1984; DeAngelis and Gross, 1992) or are based on field measurements of individual trees (e.g., FVS: Wykoff *et al.*, 1982; Dixon, 1994). At intermediate scales, point models are often implemented to represent "average stands" (e.g., PnET: Aber and Federer, 1992; Century: Parton *et al.*, 1987). These point models are in fact scale-indeterminate, but are typically interpreted as if they represent homogeneous stands (a few m² for grasslands, 10s to 100s m² for forests). At a still larger extent, global vegetation simulation models used in climate-change research simulate plant functional types, cover types, or other abstractions of forests, and include dynamics that are analogies for plant demography (e.g., VEMAP, 1995). These differences in state variables and dynamics are appropriate as a strict scaling rule, yet as a result these classes of models do not share a common empirical basis; nor, in some cases, do they even share a common conceptual model of how vegetation responds to environmental forcings.

Our goal is to devise a modeling approach that can bridge disparate scales while preserving a common empirical and conceptual basis across scales. Our approach begins with a fine-scale model (in this case, a gap model), and then uses it to derive new models as statistical abstractions of the fine-scale model. The derived models operate at coarser resolution and hence over larger spatial extent, but they retain those finer-scale details needed for larger-scale applications. Because the derived models are statistically derived from the gap model, they are in a sense models of the fine-scale model: meta-models.

In the following sections an overview is provided of the general approach to scaling a gap model up to landscapes, and then three models are presented as illustrations of the meta-modeling approach. Some of the methodological issues of parameterizing and testing such models are discussed, and the chapter closes with a prospectus of where this approach seems to be leading.

Scaling from trees to landscapes: three approaches

Several approaches can be used to extend a fine-scale model to applications at larger scales. Two of these approaches are rather intuitive: a sampling approach such as used in distributing a point model over a wide range of parametric conditions; or using a bigger computer to simulate larger areas. It is instructive to review these intuitive approaches to point out their strengths as well as some shortcomings that argue against their general use.

A sampling approach

A straightforward way to represent a heterogeneous landscape is simply to simulate each of the environmental conditions on a case-wise basis. This approach is well established with point models (e.g., Solomon, 1986; Burke *et al.*, 1990). In this, a set of parametric combinations is assembled as, perhaps, a factorial stratification over the parameters that drive the model. Each parameter set is then simulated separately, and the output of all cases is aggregated to provide a "landscape scale" integration of the model.

Importantly, this approach is entirely consistent with the way landscapes are sampled in field studies, for example, by stratifying sample quadrats across a study area to represent the possible range of combinations of topographic position, soil type, and so on (witness the huge literature on gradient analyses e.g., Whittaker, 1956, 1967).

The chief advantage of this approach is simplicity; it requires no special modifications to the basic model. The drawback is that the approach is not truly spatial. Stratified field samples cannot discover the effects of local spatial context except via sophisticated statistical techniques (e.g., Legendre and Fortin, 1989); models distributed as stratified points cannot address these issues at all. Thus, forest dynamics that reflect nearby conditions, such as upslope area contributing surface runoff to the water budget, cannot be simulated in this way. Similarly, the spatial effects of contagious processes such as seed dispersal or disturbance (e.g., fire, pests) are lost to the sampling approach. A model that merely samples a landscape removes forest dynamics from their spatial context.

Brute force

Another approach to modeling large areas is to get a bigger computer. Indeed, it is instructive to trace the recent history of gap models from this perspective. The original gap model, JABOWA (Botkin *et al.*, 1972b) was undertaken partly as an experiment in high-performance computing with the sponsorship of IBM. The model simulated a single plot, 10 m × 10 m in area, that could support as many as 100 individual trees. A forest stand was aggregated statistically by simulating several independent plots and averaging their output. By contrast, the gap model used today, Zelig version Facet (see below), simulates a forest stand as an interactive grid of as many as 2500 plots; the model runs on a UNIX workstation but is well within the computational reach of today's PCs. Using a distributed queuing system on a UNIX network (see below), hundreds of model grids are run routinely in the time it originally took JABOWA to run a single plot: computing muscle has increased by a factor of $\sim 10^5$.

An alternative to brute force is to use a bit more finesse in applying more powerful computers to the scaling issue. One especially promising approach is to reformulate the model to take advantage of parallel processing (Schwarz, 1993). Because many ecological applications can be framed as parallel problems, this approach is a compelling means of scaling a model to simulate large areas while retaining fine-scale details: the application of what we might term elegant force.

Meta-models

A third approach is to derive new models to operate at larger scales – but to do so in a manner that retains as much of the finer-scale information as required for the application of interest. Here this approach is illustrated by using a gap model to generate and parameterize new models that reproduce, as statistical constructs, selected behaviors of the gap model. The new models are themselves models of the gap model: meta-models.

There are two compelling features of this approach. First, because the meta-model is defined to reproduce the finer-scale model, the two models provide consistent results across scales. As will be illustrated, this is because they share the same empirical basis and conceptual foundations. Second, this approach provides a two-layered modeling package that allows the researcher to use the model that is best suited to the application. Although there will always be cases where the simplification implicit in the larger-scale model undermines the application, there is the finer-scale model to fall back on in these cases. Thus, the fine-scaled model is available when details are needed, and the coarser-resolution model is available when some simplification is desirable.

A general approach is described for using a fine-scaled model as a means to develop a larger-scale, coarser-resolution model (a meta-model). This approach is illustrated with the example of a cellular automaton, derived from the gap model Zelig and emphasizing contagious spatial processes as the interactions of interest at the landscape scale. The versatility of this general approach is then illustrated with two further examples: a semi-Markovian model that emphasizes transient successional dynamics, and a stage-structured model developed for applications involving timber management.

First, there will be an overview of the gap model, and a description of the approach developed to facilitate performing the large numbers of simulations required to generate a meta-model.

The gap model

As the base model for our meta-modeling efforts, we use the gap model Zelig version Facet (Urban and Shugart, 1992; Miller and Urban in press, Urban *et al.* in review). Facet is so named because climatic variables are adjusted for topographic position. The functional unit is the slope facet (Daly *et al.*, 1994), which is defined in the model as a grid of homogeneous slope, aspect, and elevation. As a research tool, the model is continuously evolving; the examples presented here are based on version 97.3 of the Facet model, or FM 97.3.

Model structure

All versions of Zelig (there are several) simulate a forest stand as a grid of tree-sized cells. Each cell corresponds to a conventional gap model plot (Botkin *et al.*, 1972a,b, Shugart and West, 1980; Shugart, 1984; Urban and Shugart, 1992). The grid is underlain by a raster soils map, with each cell assigned a soil type. Zelig models allow the grid cells to interact in that trees on a grid cell may shade or be shaded by trees on nearby cells. This zone of interaction depends on tree height and latitude (sun angle), but for temperate forest may range over 5–6 cells or more. Typical applications simulate grids that are 10×10 to 50×50 cells, corresponding to stands on the order of ~1–20 ha.

Forest ecosystems can be envisioned as coupled sub-systems, and gap models can be envisioned as coupled sub-models. There are five conceptual sub-models

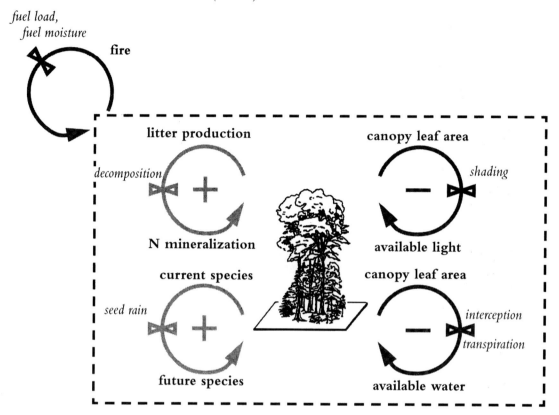

Fig. 4.1. *Schematic of system-level feedbacks as implemented in the full version of the gap model Facet. Each submodel tends to act as a positive (+) or negative (−) feedback on itself. The sub-models are coupled by physical relationships and by tree species life-history traits (see text).*

to FM (Fig. 4.1). These sub-models tend to behave either as positive or negative feedbacks in the model. For example, as the canopy develops, shading increases and this retards canopy growth (a negative feedback). In contrast, canopy development increases litterfall which increases N (nitrogen) mineralization, which further increases forest growth (a positive feedback). The sub-models are coupled by physical relationships. For example, litter moisture determined in the soil moisture sub-model affects decomposition rates and also affects flammability in the fire model. The sub-models are also coupled by life-history traits encoded as species parameters. For example, many trees that are shade tolerant are also drought intolerant and have high tissue N concentrations, effectively coupling the light, soil moisture, and decomposition submodels. In the examples used here, the primary emphases are the light regime, the soil water balance, and the fire model. A seed dispersal module under revision is not used in the examples presented here.

The physical template

FM 97.3 simulates the physical environment in terms of temperature, available light, soil moisture, and nutrient (nitrogen) status. As a Facet model, FM adjusts temperature and precipitation for elevation using locally regressed lapse rates (Running *et al.*, 1987; Daly *et al.*, 1994), and adjusts radiation for topographic position using standard geometric and micro-meteorological methods (Nikolov and Zeller, 1992). Here, an overview is provided of the light, soil moisture, and fire submodels, because these illustrate the methods of simplification in devising meta-models.

Light regime

FM summarizes vertical heterogeneity as height profiles in leaf area and available light, at a resolution of 1 meter. The crux of the light regime is a leaf-area profile defined for each grid cell, constructed by estimating total leaf area for each tree on the plot and distributing this leaf area along each tree's live crown (after Leemans and Prentice, 1987). The tree height and leaf-area allometries are based on regressions local to a study area (e.g., Gholz *et al.*, 1979; Garman *et al.*, 1995).

The leaf-area profile is used to estimate the light profile for each position (grid row, column, and height) within the modeled stand. FM partitions light into direct-beam and diffuse-sky components, and samples the forest canopy to estimate each component. This sampling is accomplished by constructing diagonal leaf-area profiles by "looking through" the vertical leaf-area array at a specified angle and direction. The "look angle" is specified as the height of a horizontal cross-section of a cell (i.e., a "slice" of the profile: the steeper the look angle, the thicker the slice). The "look direction" may be in any cardinal direction (N, E, S, W) or vertical. The direct-beam component is estimated by constructing a diagonal profile to the south, with a look angle derived from mean solar inclination angle as integrated hourly, daily, and monthly over the growing season (Bonan, 1989; Urban *et al.*, 1991). The diffuse-sky component is estimated with multiple samples of the sky, by constructing diagonal profiles at various look angles and directions. The geometry of these "looks" through the canopy is further adjusted by the canopy's being draped over terrain.

Light impinging through the diagonal leaf-area profile is attenuated exponentially. This approach results in a larger zone-of-influence (area shaded by a tree) as tree size increases or as solar angle decreases (Urban *et al.*, 1991; Urban and Shugart, 1992; Weishampel and Urban, 1996).

Soil water regime

The soil-water sub-model simulates multiple, multi-layer soils, with a soil type assigned to each cell. A soil type is defined by a number of mineral soil layers, each defined by its water-holding capacity at field capacity and wilting point. Litter and duff comprise an organic horizon, which varies in depth for each grid cell throughout the course of a simulation as duff accrues through annual litterfall and subsequently decomposes.

The model uses a Priestley–Taylor estimate of potential evapotranspiration (PET: Bonan, 1989). Because the PET estimate is driven by radiation and this can vary with topography, the routine adjusts the water balance for stands of different slope and aspect. PET is partitioned into two components. Transpiration depends on leaf area index. The fraction of PET not partitioned to transpiration may be drawn as evaporation from the litter and surface mineral soil. Actual evapotranspiration (AET) may be reduced from PET for three reasons: (i) precipitation and soil water storage may be insufficient to meet evaporative demand; (ii) leaf area may be low and hence limit drawdown to surface evaporation (i.e., deeper soil water is inaccessible); or (iii) cold temperatures may limit the canopy's capacity to transpire. The canopy does not "transpire" water directly, but does influence AET through interception (a linear function of woody surface area and LAI) and through its influence on surface evaporation.

The water balance operates on a monthly timestep, in that temperature, radiation, leaf area, and total precipitation are simulated per month. Temperature and precipitation vary stochastically during the simulation, based on interannual variances estimated from long-term data. Precipitation is partitioned stochastically into discrete events on a daily timestep; each event occurs either as rain or snow depending on a locally regressed function of temperature. The model simulates snow dynamics very simply. Snowpack accumulates as long as temperatures are below freezing; when temperatures warm, snow is melted at a constant rate.

The soil-water model accrues a tally of drought-days for each soil layer. A drought-day is a day for which soil water is at or below wilting point. The tally is integrated over the upper soil and used to constrain seedling establishment; a second index integrated over all soil layers is used to constrain the growth of established trees. In this latter integration, drought-days are weighted by the fraction of fine roots in each soil layer.

The current working version of the model does not simulate surface runoff nor lateral subsurface flow between plots. Thus, the model does not incorporate the hydrology of topographic convergence or divergence. An extension to the model toward this end is under development.

Tree demographics

Like most gap models, FM simulates the processes of seedling establishment, annual diameter growth, and mortality for each tree on each cell of the simulated grid. These processes are simulated with a common logic of specifying the maximum potential a tree might achieve and then reducing this potential to reflect suboptimal environmental conditions. Simple multiplier functions are used to describe these environmental responses.

Establishment

Seedling establishment is strongly keyed to light available at ground level and to the moisture status of the topsoil. Establishment probability may also be affected by litter depth (some species fare better on mineral soil). Each species has a maximum possible establishment rate, and in each simulation year species are filtered by light, temperature, and soil moisture multipliers that reduce the optimum rate. An environmentally filtered cohort of seedlings is then planted and tracked for a number of years until the seedlings become eligible to be established as trees.

Growth

Tree growth is modeled deterministically via a function that describes the maximum diameter increment that could be achieved by a tree of a given size under optimal environmental conditions. This function is itself driven by leaf area, with the result that a tree that prunes under shading naturally exhibits slower growth.

The optimal growth increment is further reduced to reflect the effects of shading, soil moisture, soil fertility (N), and temperature. Shading is modeled by integrating a shade response function over a tree's canopy, using the light regime computed as described above. Soil moisture affects tree growth by slowing growth as a species-specific maximum drought-day index is approached. Nutrient response is simulated in terms of the ratio of N supply to N demand for the plot. N supply is accounted as N mineralized as C is respired in the decomposition of litter and woody debris. If N supply is adequate, trees are unconstrained; otherwise, the growth of all trees is reduced accordingly.

The temperature response in this model assumes that there exists a cold temperature at which species response is low for physiological reasons, and that this occurs at reasonably cold temperatures (i.e., at high elevations in our systems). Conversely, at this latitude the effect of high temperature is largely expressed through its effect on the water balance – masking any direct effect on tree physiology. Thus, a one-sided temperature response curve is used that is disabled at warmer temperatures, where the soil moisture constraint becomes operative.

Environmental factors interact to constrain tree growth. An interaction is assumed between above- and below-ground constraints, but it is assumed that moisture and nutrients are so tightly interrelated that they are inseparable for our purposes. Thus, the overall constraint is the product of the temperature, light, and below-ground factors, where the below-ground factor is the minimum of moisture or nutrients.

Mortality

Trees may die for three reasons in the model. There is a low baseline rate of purely stochastic mortality that is estimated from expected species longevity; this annual probability is age and size independent. A second cause of mortality is lack of vigor, which is invoked when a tree fails to achieve a minimum growth threshold for more than two successive years. Trees of any size may be subjected to chronic drought stress on severe sites, or acute drought on any site in an extreme year.

Temperature effects are largely chronic rather than acute, and are a dominant constraint only on the "cold" edge of the range of a species. A third source of mortality is fire; crown scorch might kill trees directly through severe damage, or indirectly if partial crown loss results in loss of vigor.

The fire regime

The fire submodel in FM explicitly couples climate, forest condition, and fire behavior (Miller and Urban, in press). The sub-model simulates climate effects on forest productivity and species composition, generates fuel loads from forest condition, and simulates fuel moisture through the soil water balance, effectively coupling all other subsystems of the model (Fig. 4.2). Fuels are generated on a per-tree basis, using species-specific allometric relationships to predict leaf litter and branch biomass, portions of which are shed each year. Foliage lost through self-pruning as a stand develops also is added to the litter. Additionally, episodic inputs of debris are added to fuel classes when an individual tree dies: its bole, bark, branch, and foliage biomass are added to the appropriate fuel classes. Thus, unlike previous fire models, the fuels model in FM adjusts itself dynamically to changing forest composition, and also schedules litter inputs according to demographic processes.

The fuel moisture model in the fire regime is actually part of the soil moisture model: the litter/duff layer is the uppermost horizon of the soil. Thus, soil moisture and fuelbed condition are coupled explicitly in the model. Likewise, decomposition varies with moisture, and so the fuelbed's mass and moisture content also affect N mineralization directly.

Fire occurrence requires an ignition source and a burnable fuelbed. Ignition is a stochastic process; frequency of ignition (years between ignitions) is entered as a run-time parameter and used to generate an annual probability. Ignition points are selected randomly within the grid. From an ignition point, fires spread contagiously across the grid. For cells that are "ignited", fuel load, bulk density, and moisture are used to predict fireline intensity (Rothermel, 1972; Albini, 1976). Cells for which intensity falls below a threshold value "burn out". For burned cells, intensity is used to predict scorch height, and scorch height is used to predict tree mortality as a function of the proportion of each tree's crown that is burned. Fire also reduces fuels, and in the case of the duff layer, this may influence the ability of each tree species to successfully germinate. Species composition affects these submodels through species-specific litter production rates, species tolerance to fire (and regeneration after fire), and importantly, via the packing density of foliage litter which may vary substantially among species and dramatically affects fire behavior. In FM the fuel bulk density is tracked as a running average of the litter that accumulates on each plot, weighted according to species composition on the plot.

Fig. 4.2. Schematic of the fire model in FM, emphasizing couplings among forest condition, fuels, and fire behavior as governed by climate. Fuel loads reflect forest condition because litterfall is a function of tree demographics. Fuel moisture is computed from the soil water balance, thus coupling fire with climate. Fire frequency is internally generated by the model, with fire occurrence and fire intensity being functions of this fuel moisture and fuel load. Fire then affects forest condition via crown scorch (which might be lethal) and seedling establishment.

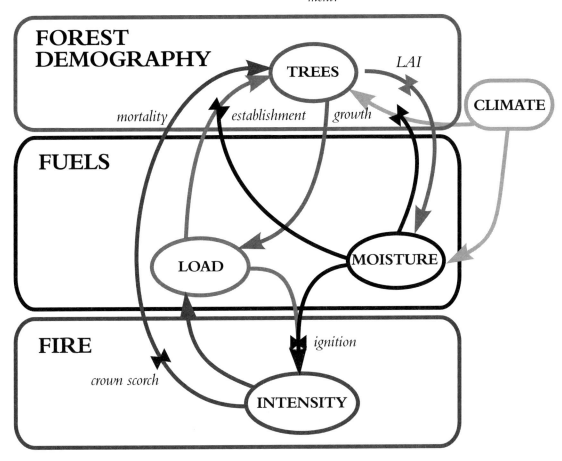

Extending the gap model to landscapes

There are two steps in our approach to extending a gap model to the landscape scale. The first step is implicit in the structure of the Facet model, which adjusts the grid for topographic positions defined by slope, aspect, and elevation (Fig. 4.3). The second step involves performing the large number of simulations needed to characterize the range of environmental heterogeneity represented within a landscape. For this, a distributed queuing system has been developed.

Fig. 4.3. Scaling the gap model ZELIG version Facet from the model plot (grid cell), to a forest stand (grid), to a slope facet on a landscape. Climatic drivers in the model are adjusted for slope, aspect, and elevation using locally regressed lapse rates for temperature and precipitation, and geometric models for radiation.

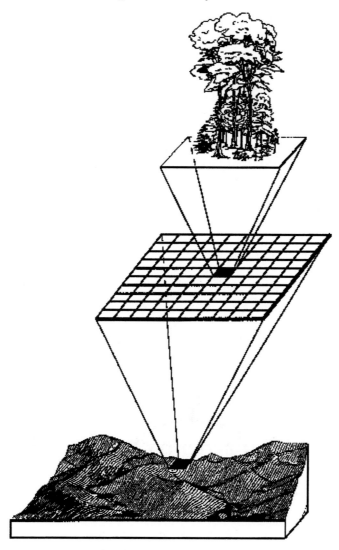

The distributed queuing system

A distributed queuing system (DQS) distributes simulations (jobs) to a number of client workstations in a network. Our system uses the Tcl/Tk tool command language toolkit (Ousterhout, 1994) to present the user with a series of three graphic user interfaces that pre-process the session, perform the actual simulations, and then post-process the collective output. In the first session, the user defines the combination of driving parameters for a set of simulations. A single simulation is defined by a combination of run-time parameters that include input Driver files (species parameters, climate and soils data), topographic position (slope, aspect, elevation, and a soils map for the grid), the number of years to simulate, and which output files to save. Run-time parameters may be held constant or varied according to a variety of sampling designs (uniform, stepwise incremental, or gaussian). The parameters are typically distributed, based on empirical distributions estimated from data in a geographic information system. This initial session results in the generation of a run list, which is a set of combinations of run-time parameters for each of a user-specified number of simulations (typically 100–500).

The second session with the DQS then distributes the runs across any number of client workstations in a network. The master server program on the host machine sends a job to a client, and when the job is finished the client copies the selected output file(s) back to a common output directory on the host machine. Client machines can be selected individually, or scheduled to avoid particular times of the day (run in background mode during office hours, resuming at night). Using a network of 15 Sun Sparc workstations we can perform 500 simulations in a few hours' clock time. This performance improves almost monthly as faster hardware becomes available.

The final session with the DQS consists of post-processing the collated output. UNIX awk scripts, generated internally by the post-processor, cull user-selected output variables to a new file which is formatted for use with graphics or statistics packages.

The DQS thus facilitates our conducting the many simulations needed to represent a landscape. More to the point here, this also allows us to perform the range of simulations needed to build and fully parameterize a meta-model.

Examples of meta-models

To illustrate our approach to building meta-models, the approach begins by "building" the cellular automaton MetaFor. A brief overview is then provided of the other two models, as a contrast to MetaFor and to provide some notion of the range of possibilities for meta-models.

MetaFor: a cellular automaton

A cellular automaton simulates dynamics in a raster grid by positing that the future condition of a given grid cell depends on its current state and the state of its neighbors (Hogeweg, 1988; Green, 1989). Neighbors are typically specified to include the four cells adjacent in the cardinal directions to the focal cell or eight neighbors (including also the diagonal neighbors); in fact, there is no reason why the neighborhood cannot be defined by whatever rules make sense for the application. Cellular automata are well suited for applications where the behaviors of interest are contagious processes or neighborhood interactions.

Table 4.1. *Structure of the cellular automaton MetaFor, as state variables, system dynamics, and parameterization scheme*

Component	Definition	Derivation
State variables	Cover type	Dominant tree species
	Age class	Simulation timestep (a counter)
Input map files	Elevation	Digital elevation model (DEM)
	Slope	From DEM
	Aspect	From DEM
Input data files	metaSpecies	Parameters derived from Facet
	metaSite	Climate coefficients (from Facet)
Processes	Establishment	Automaton; plus environmental constraint (from Facet)
	Mortality	Age-related or disturbance (from Facet)
	Fire	Automaton; conditioned on age and moisture (from Facet)

Structure and dynamics

MetaFor represents a landscape as a grid wherein each cell is assigned a cover type (one of several tree species) and age class (tallied in terms of model timesteps, in this case, decades). The cover and age maps are draped over a digital elevation model (DEM). The state dynamics consist of changes in cover type and age through time. The model simulates the demographic processes of establishment, aging (though not growth), and mortality, as well as a disturbance regime (fire). The contagious processes that are emphasized are seed dispersal (which affects species establishment) and fire, both of which are implemented as automata. Additionally, MetaFor conditions cell-by-cell dynamics according to species responses to environmental constraints of temperature and soil moisture, using the same response functions used in Facet. Environmental conditions are modeled as functions of topographic position, and are based on regressions of output from the Facet model's soil moisture module. Thus, MetaFor uses a physical template and demographics simplified from but consistent with those in Facet (Table 4.1).

Parameterization

Generation of the automaton model comprises several steps at which MetaFor algorithms or parameterizations are reconciled with the gap model. These steps include the calibration of the physical template (temperature and soil moisture), the definition of the neighborhood rules for establishment, and the fire spread model. Other components of MetaFor are simply reproduced from the gap model directly (e.g., the environmental response functions and mortality probabilities). Here, some of these procedures are outlined as illustrations of the general approach.

The physical template

The climate and soil moisture routines in Facet operate at monthly or submonthly timesteps, and require as input information on minimum and maximum temperature, precipitation, and radiation, as well as canopy leaf area, and soil texture for all layers in the soil profile. Long-term patterns in temperature and soil moisture are summarized by saving a temperature index (growing degree-days) and a moisture index (drought-days) from long-term (100-yr) simulations with a stand-alone version of the gap model's climate and soil moisture routines. These data are then used in a regression analysis, to build functions that predict the indices from topographic data stored in MetaFor. Growing degree-days are predicted from multiple regression on elevation, slope, and aspect. To incorporate stochastic interannual variability in temperature we also include the standard deviation of degree-days, which is predicted from the mean. Each simulation timestep, a random (gaussian) amount of variation in temperature is added to the mean degree-day index predicted for each grid cell. Thus, there is spatial as well as temporal variability in temperature in the landscape model.

Predicting soil moisture is confounded somewhat by the fact that temperature and moisture are themselves correlated: temperature decreases while precipitation increases with elevation in mountainous terrain; temperature is also a main component of evaporative demand. To attend this, degree-days are used for a given year (i.e., with the stochastic variation added) to partially predict drought-days. Another random deviate is then applied to add stochastic variation in drought-days that is unrelated to temperature (i.e., that due to year-to-year variation in precipitation). These relationships are derived by partial regression and used to generate temperature and soil moisture surfaces from the DEM files used as model input.

Demographic processes

Cell-based analogs of tree demography include the assignment of a species type to each unoccupied cell (establishment), the "aging" of these cells simply by accruing timesteps since establishment, and the clearing of cells after mortality (age-related or through disturbance). In MetaFor the processes of aging and mortality are quite simple; establishment is somewhat more complicated.

The colonization of an unoccupied cell by a species is conditioned on two factors: the physical environment and existing species composition within the neighborhood of the cell. The physical environment is specified as the cell's degree-day and drought-day indices, as computed each timestep with some stochastic variation. Species response functions on [0,1] are used to modify establishment probabilities; these functions are taken directly from Facet. Each species has its "environmental probability" of establishment calculated as the product of the temperature and moisture multipliers. A cell with an inhospitable environment (e.g., a cold alpine site or xeric outcrop) is likely to be unoccupied and will persist as a gap. The neighborhood effect on establishment, used to mimic seed dispersal, is estimated by tallying the proportion of the cells in the neighborhood that are occupied by each species (including "gap" cells that are unoccupied). The neighborhood is specified by the user and can consist of 4, 8, 12, or 24 cells. The inclusion of "gap" as a species in the neighborhood forces large gaps such as those created by fires to be colonized mostly by encroachment from the edges, rather than being recolonized immediately and entirely the year following a fire. Actual establishment probabilities are computed as the normalized product of the environmental constraints and neighborhood influences.

In effect, the dynamics of the landscape are rather straightforward after initial establishment: each cell either "ages" one timestep or it is cleared by mortality or disturbance.

Disturbance itself is a contagious process that is conditioned on site condition, specifically soil moisture (a proxy for fuel moisture) and time since establishment (a proxy for fuel load). At each timestep, fires may be ignited at stochastically selected points (cells). Fires spread from these ignition points probabilistically. Each neighboring cell has a probability of burning that is computed as the product of the moisture and stand age functions. A uniform-random number on [0,1] is drawn and the cell either burns or does not; if it burns, it is added to an array of cells comprising the current fire (a cluster of cells). This process is recursive and continues until a maximum fire size is reached (a computational check) or until no burnable cells can be found adjacent to the current fire.

Illustrations

The automaton generates realistic patterns in vegetation cover (Fig. 4.4, left: see color section). In this example, the landscape is a ~47 000-ha sample of Sequoia National Park in the Sierra Nevada of California, as a 1024 × 512 grid of 30-m cells. The model is capable of simulating much larger areas, but in this case we are limited by the availability of climate data (lapse rates are poorly defined east of the topographic divide).

The question naturally arises, "How well does this match the actual vegetation in the park?" This question is frustratingly difficult to answer in a straightforward way. It is known that the gap model matches field data, at least insofar as reproducing elevational and topographic trends in species basal area (Miller and Urban, in press; Urban *et al.*, in review). And, because the automaton is defined to reproduce the gap model, it is easy at one level to claim that the automaton thus also matches our data. But, in fact, only one image of the Park's vegetation is available, a map classified from a combination of satellite imagery, air photos, and ground data (Fig. 4.4, right). This map is flagrantly different from the predicted map, and yet any comparison of the two is misleading; neither map is a particularly valid picture of reality. The modeled map represents potential vegetation in the absence of any recent fires and with no other disturbances. The "real" map is a highly aggregated and interpolated composite of subjective cover types; the apparent homogeneity of huge expanses of the Park is clearly an artifact of the classification scheme used in generating the map. This broaches a quite general issue in landscape models, which is deferred to a later discussion.

Mosaic: a semi-Markov patch transition model

Another common way to model landscape dynamics is to simulate the transitions among discrete patch types, for example seral stages or land cover types (Johnson and Sharpe, 1976; Weinstein and Shugart, 1983; Baker, 1989). A Markov chain is a well-studied formalism for such models (Usher, 1992). In a first-order Markov model, the transitions depend only on the current state (i.e., history does not matter). A more complex and realistic model includes time lags making the transitions dependent on history; this extension renders the model semi-Markovian. The gap model Zelig has been used to generate and parameterize various versions of a semi-Markov model called Mosaic (Acevedo et al., 1995a,b, 1996).

The state space of a semi-Markov model is a set of n patch types; the state dynamics are transitions or conversions among these patch types. The heart of the model is the $n \times n$ transition matrix, the elements of which are the probabilities that a patch of a given type will undergo a transition to some other type. The transition will occur, however, after a time lag characteristic of each pair of states. In the aggregate, a Markov chain models the proportion of the study area that is in each of the states at a given time. For spatial applications, the model is implemented by applying the transition probabilities on a per-cell basis on a raster map. The result is a time series of new maps of the landscape, each map a stochastic realization of the model.

An issue in generating a semi-Markov model is how to estimate the transition probabilities and delays. If the number of patch types is large, or if some transitions are uncommon, it may be quite difficult to estimate these parameters. For complex transitions among landscape elements that undergo change over time scales of decades or longer, there may be no feasible way to measure these rates directly from readily available data. Our approach has been to use the gap model to estimate these parameters.

Structure and dynamics

The Mosaic models used are structured by defining a mosaic tile on the landscape as a homogeneous unit of arbitrary area (~1–10 ha). The state variable for the tile is not its dominant patch type, but rather, a frequency distribution of gap-sized elements within that tile that are in each patch type (Table 4.2). For example, a 1-ha tile would have 100 gap-model cells of 10 m $\times$ 10 m within it. The state dynamics of the model are the changes in the frequency distribution of within-tile types through time. Thus, some information is tracked on the within-tile heterogeneity of forest stands but without tracking the location of each gap-scale element within the larger tile (Fig. 4.5).

Table 4.2. *Structure of the semi-Markov model Mosaic, as state variables, system dynamics, and parameterization scheme*

Component	Definition	Derivation
State variables	Frequency distribution of cover types within tile	Classified by user (application specific), typically dominant species and/or age (size) class
Input map files	Elevation	Digital elevation model (DEM)
	Initial conditions	From GIS coverages, imagery
Input data files	Transition probabilities and transition delays	Fitted parameters (from Facet)
Processes	Establishment, succession	Semi-Markovian; biased by environment (from Facet)
	Mortality	Age-related or disturbance (transitions to 'gap') (from Facet)

Fig. 4.5. Schematic of the structure of the semi-Markov model Mosaic. Continuously varying species composition and age structure are classified to discrete cover types for each gap-scale element (cell) within a larger, 25-cell mosaic tile (in bold); the frequency distribution of these types within each mosaic tile in the landscape comprise the state of the system.

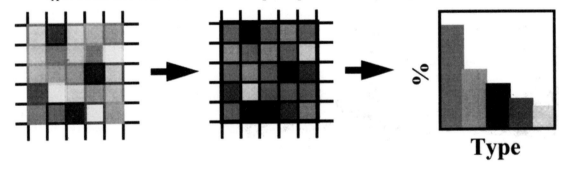

Parameterization

Building a semi-Markov model consists of defining the patch types, and then inserting a "patch type classifier" into the gap model. In the models we use, the types have been variously defined: by dominant species only (Acevedo *et al.*, 1996), or by combinations of dominant species in each of two height classes (Acevedo *et al.*, 1995a). The gap model is then run, and at every timestep each model plot is classified to patch type. As any model plot changes from one type to another, the transition is tallied. In the simple case of a first-order Markov chain and one model plot, these tallies can be used to derive the transition matrix directly. For semi-

Markov models, the calibration is more complicated. In the models developed, transitions may have fixed lags (latencies) concatenated with distributed delays, thus providing more realistic successional dynamics. This requires the estimation of several parameters for each transition (duration of the latency, if any, and parameters of the delay density function). These parameters are estimated by statistical analysis, including non-linear estimation procedures.

Illustrations

The first version of Mosaic was based on functional roles of trees, and was motivated in part by a desire to develop a tropical forest model that incorporated dynamics similar to that of a gap model while conceding that there was a lack of the life-history data to be able to parameterize a simulator as detailed as a gap model (Acevedo et al., 1996). This example is less applicable to landscape-scale applications, but it does highlight an important consideration for meta-models: if the model is sufficiently simple, there may be a tractable analytic solution. Indeed, Acevedo et al. (1996) illustrate a progression of models ranging from the detailed (and complicated) gap model, to a semi-Markov model that can simulate realistic successional dynamics but that also yields a concise analytic solution.

A second example of the semi-Markovian approach is a more complicated version of the model, with patch types defined as two-layered combinations of dominant species in the over- and understory (Acevedo et al., 1995a). This version of the model was implemented for the forests in the Pacific Northwestern United States, where old-growth issues are often framed in terms of the vertical structuring of forests. In this version, the transition probabilities are also conditioned, on elevation recognizing a major ecotone between lower elevation forest characterized by Douglas-fir and western hemlock, and high-elevation forests dominated by true firs (Fig. 4.6). The model reproduces patterns in species distribution across this ~6300-ha watershed in the central Oregon Cascades, in this example illustrating the dominance of the western hemlock plant association at lower and middle elevations (Fig. 4.7: see color section).

ZelStage: a stage-structured model

ZelStage was developed to investigate the effects of forest management and natural disturbances on stand dynamics and landscape patterns in the Oregon Cascade and Coastal mountain ranges. An important consideration in the development of this model was the ability to realistically simulate long-term stand dynamics for landscapes under alternative forest management practices. This argued for a modeling approach that tracked stem densities by species (the common currency of forest management), but further required an approach computationally modified for efficient, simultaneous simulation of multiple stands over a landscape.

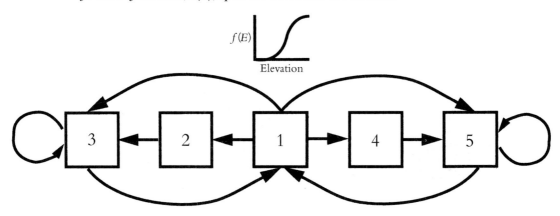

Fig. 4.6. Transition diagram for forests in a Pacific Northwestern version of the semi-Markov model Mosaic (after Acevedo et al., 1995a). Patch types are (1) gap, (2) young Douglas-fir/hemlock, (3) mature Douglas-fir/hemlock, (4) young true fir, and (5) mature true fir. The function of elevation, f(E), splits the model into two domains.

Structure and dynamics

ZelStage is a deterministic model made up of a stage-structured framework, statistical functions of growth and mortality, and algorithms taken directly from the gap model. The stage-structured component is the basis for tracking stems. Instead of dealing with individuals, ZelStage tracks the number of stems per hectare of each species in 3-cm diameter growth stages (size classes). This simplification reduces both storage requirements and amount of processing with only a nominal reduction in detail given the small size-class interval (Fig. 4.8).

Growth of stems among size classes is modeled using transition functions derived from simulation experiments with the gap model (Table 4.3, and see below). Transition functions formulated as linear and non-linear regression equations predict the proportion of stems advancing from one size class to another given the current size class, crown ratio, and cumulative leaf area index above the base of the crown. Mortality functions determine the proportion of stems that die from natural causes during a time step. Ingrowth is calculated using an approach similar to the gap model, but is deterministic.

ZelStage uses raster-based data layers to represent several levels of spatial organization of the forest as well as the environmental field for a landscape. An initial stand map indicates the stand code for each cell or group of cells of the landscape. This code is used as an index to the input stand table that designates the initial structure and composition of each stand. For forest management considerations, a harvest unit map is used to delineate aggregates of cells treated as unique management units. The spatial grain of input maps can be any size above 0.3 ha (this lower size is imposed for computational reasons explained below).

To implement forest management, ZelStage contains an event scheduler that allows the user to implement a range of stand-level treatments at any time during

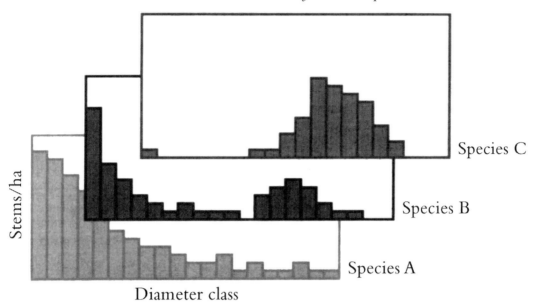

Fig. 4.8. *Schematic of the stage-structured model ZelStage, in which the state variables are 3-cm diameter classes for each tree species.*

Table 4.3. *Structure of the stage projection model ZelStage, as state variables, system dynamics, and parameterization scheme*

Component	Definition	Derivation
State variables	Diameter frequency distribution by species	Binning continuous distribution into discrete classes
Input map files	Stand map	GIS database
Input data files	Environmental parameters	GIS-based models and coverages
	Stand condition	GIS-linked data tables
Processes	Establishment	Constrained by stand condition and climate (from Zelig)
	Growth to next size class	Regression (from Zelig)
	Mortality	Regression (from Zelig)

the simulation. Commands are read from an ASCII script, and specify the year, type, and parameters of an event. Event parameters define the removal and retention of basal area, volume, and density, and planting of stems. Command scheduling arguments specify the harvest unit or individual stand to be effected, the species and size classes to consider, and the retention/removal strategy (e.g., from top down, bottom up, or proportional to the existing size-class distribution). Additionally, basic algorithms were incorporated into ZelStage so that harvest units can be selected to simulate dispersed and aggregated harvest patterns (from the Cascade model, Wallin *et al.*, 1994).

Parameterization

Data for the generation of transition functions are derived from controlled simulation experiments with a modified version of Zelig. An initial requirement was the calibration of the gap model for the environmental conditions of the area of interest. Field data sets for over 2000 stands in western Oregon were used for calibration. A data base of weather conditions (monthly means, variances of precipitation, temperature, solar radiation) has been developed for all of western Oregon and was used to determine environmental inputs for a simulation area.

For a given environmental field, the modified gap model simulates annual diameter growth of a stem for a 5-year period, given an initial diameter, crown ratio, and cumulative leaf area above the base of the crown. Mortality and weather conditions are simulated as stochastic processes, thus annual variability of stem growth and mortality are taken into account. Because the model is stochastic, replicates are used to derive samples of potential growth and mortality. The weather information specified in the simulation essentially defines the environmental domain of the resulting transition functions. For a user-selected species, the gap model automatically simulates growth and mortality for stems over a broad range of diameters, crown ratios, and leaf area indices. The initial diameter, simulated 5-yr diameter increment, occurrence of mortality, initial crown ratio, and leaf area index are output for analysis. A post-processing program combines all replicates for a species and determines the initial 3-cm size class of a stem, the proportion of stems that grew into a larger size class and the corresponding 3-cm size interval, and the proportion of stems that died.

For each species, transition and mortality functions are derived from step-wise linear and non-linear regression analysis of the proportion tallies. Variables considered for inclusion in a model include first- and second-order terms of mid-point diameter, leaf area index, and crown ratio, and all possible two-way interactions (Fig. 4.9). Separate transition functions are derived for predicting advancement to one, two, and three size-classes, if necessary.

Illustrations

An initial prototype of the ZelStage model was implemented for a mid-elevation, ~3000-ha watershed in the Oregon Coast Range. Transition functions for the four dominant tree species (three conifer, one hardwood species) were developed to handle growth and mortality of stems <120 cm in diameter. Average environmental conditions of the watershed were used to generate a single set of transition functions. Vegetative cover types of the watershed were derived initially from

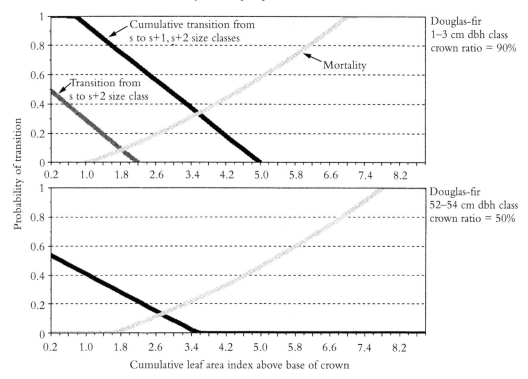

Fig. 4.9. *Examples of transition functions predicting the growth and mortality of stems in the stage-structured model ZelStage. Functions are species- and size-specific, and are modified further by leaf area and crown ratio.*

Landsat Thematic Mapper imagery. For simplicity, classes were aggregated into six cover types using a minimum mapping unit of 0.3-ha, for a final total of 1200 individual stands. An existing harvest unit map was used to designate 218 management units. Structure and composition of each stand type was initialized from representative field-plot data. Simulations were run over a 60-year period under a dispersed harvesting strategy (Wallin *et al.*, 1994), with 10%, 20% and 20% of the harvest units treated at years 1, 25, and 50, respectively. Stand treatments consisted of two levels of basal area removal (20%, 60%) with preference for larger stems. At year 60, stands were classified into hardwood, conifer, or mixed hardwood-conifer and saved as output cover-type maps.

The simulated landscapes illustrate the model's ability to simulate effects of overstory thinning (Fig. 4.10: see color section). Compared to the 20% removal, the 60% treatment resulted in an increase in mixed stands due to more favorable conditions for hardwood species (i.e., more light, less competition from conifer species). In general, these results corroborate current understanding of thinning effects in coastal watersheds, and lends credence to the ZelStage approach. In future applications, our intent is to develop and use transition probabilities for the suite of environmental conditions across watersheds in order to simulate gradient responses to forest management practices.

Discussion

Thus far the general approach of deriving landscape-scale models from a more detailed, fine-scale model has been illustrated. Three different models illustrate the approach, each emphasizing specific aspects of forest dynamics that might be important for particular applications. The models have complementary strengths and weaknesses. The automaton is especially appropriate for contagious processes or neighborhood interactions, and is extremely fast. This comes at the expense of rather crude representation of demographic processes, and consequently rather crude temporal dynamics. The semi-Markov model, in contrast, provides transient dynamics that rival those of the gap model, albeit for discrete patch types. An additional benefit of the Mosaic model is the inclusion of within-tile variability as represented by the frequency distribution of gap-scale elements; this approach is in marked contrast to most landscape-scale models, which assume homogeneity at the scale of the larger mosaic tile. Within this format, simpler versions lend the powerful summary and interpretative guide of analytic solutions, while more complex versions provide ever more realistic behaviors (at the cost of analytic simplicity). The stage-structured model retains nearly all of the information contained in the gap model, and thus is capable of reproducing gap-model behavior with striking fidelity yet in a computationally convenient form. The cost of this is the added burden of parameterizing the model, which must attend every size class for every species, each modified by various combinations of leaf area and crown ratio, and environmental conditions. For each model, there are challenging parameterzation problems that limit the practical applications of these models. Developing efficient algorithms for model parameterization will streamline this approach greatly.

Parameterization issues

The onus of parameterizing these models poses a compelling challenge to the meta-modeling approach. It should be emphasized that parameterization presents both statistical and computational challenges, the latter due to the sheer number of simulations involved. For this reason, we are actively pursuing methods to automate the parameterization of meta-models. There are, of course, issues that are specific to each of the model forms we have illustrated.

Automata are largely governed by the rules that bias cell fate according to neighboring cells, and the specification of the rule set still remains as much art as science. An issue particular to the Sierran study site is that a model with states defined as cover types (species) cannot easily simulate surface fires typical of this area (i.e., fires that burn through an area and leave the canopy intact). We are currently developing an automaton capable of simulating surface fires (Chang and Urban, 1997).

The parameterization of a semi-Markov model strictly requires a matrix of transition probabilities that are unbiased by the antecedent conditions of a patch type. In practice, the parameters we have derived may be influenced by preceding seral stages. A convenient parameterization entails a set of simulations in which each patch type is initialized by itself (i.e., a pure stand of a single type) and allowed to undergo transitions from that pure state (Ablan, 1997). This is a much more complicated parameterization scheme and we are currently improving the approach.

The stage-structured approach suffers numerical distractions stemming from the discretizing of continuous phenomena. These include cases where a tree might "grow through" more than one size class in a single timestep, an occurrence that must be attended in the regressions of transition functions. At the other extreme, stems can become "stuck" within very large size classes due to insufficient growth over the 5-year timestep. Employing variable size classes or time steps is a possible solution, but would complicate the model further. Similarly, ZelStage must be properly scaled to avoid transferring partial stems; the model works in integer trees. Thus, the minimum spatial resolution of the model is dictated by the stand area that will support sufficient tree densities to prevent partial stem transfers.

While more automated and robust approaches to many of these issues are pursued, there remains a set of issues that are general to all approaches to modeling landscapes. One especially compelling issue is that of model testing.

Testing landscape models

It comes as a tautology that models derived to simulate landscapes are difficult to test. After all, these models are developed, in part, because sufficient data to pursue empirical studies at these scales is lacked. Very little has been presented in the way of formal tests of the models illustrated here, which broaches a problematic issue in landscape modeling.

Landscape-scale data are typically available in the form of maps of vegetation or cover types, often classified from satellite imagery, air photos, and limited ground data. The vegetation map of Sequoia–Kings Canyon National Park (Fig. 4.4, right panel) is a typical example. Two points must be observed of this figure. First, there is only one image. This presents the logical difficulty of comparing the result of a stochastic simulation model to a real map which also can be considered to be a unique realization of a stochastic process (reality). Thus, even a "perfect" model should not be expected to reproduce the real map. The likelihood that this might happen by chance is vanishingly small, and decreases further if there are appreciable influences from initial conditions (which we probably do not know), or inertia or legacy effects due to events that happened long ago. Thus, a point-to-point comparison of model to data is largely futile. Likewise, in cases where there is another map of the real landscape, representing a later time period, one might be tempted to project the earlier map with the model and test it against the latter map. But, again, if there is considerable stochasticity to the real landscape, there is no particular reason to expect the prediction to match reality. One way to avoid this pitfall is to test the model at a higher level of abstraction than that of the maps themselves. For example, it is logical to compare the statistics of the model output with the statistics of the real map. Appropriate metrics might include any of the myriad metrics of landscape pattern (e.g., O'Neill et al., 1988; McGarigal and Marks, 1995; Riitters et al., 1995). Using this approach, one might assess the spatial processes in an automaton by comparing the spatial autocorrelation in model output to that observed in real maps. While these metrics are readily available and the tests are straightforward in concept, landscape modelers have been slow to develop appropriate tests for these models.

A second point to be made from Fig. 4.4 is that the vegetation cover is classified from a variety of data sources; that is, the "real" map is itself a model. Importantly, this model of vegetation typically includes some degree of interpolation and editing of boundaries of cover types or patches. Thus, there exists in the real map some (perhaps unknown) degree of smoothing. This means that the sizes, shapes, and internal homogeneity of patches are probably not realistic. Further, if the vegetation map has been classified with the aid of ancillary environmental data such as digital terrain data, tests of environmental patterns generated by the model may not be independent (or worse, might be circular). All of these issues may confound any formal comparison of the model output to the classified map.

Some of these issues may be resolved by remote sensing of data with higher information content. For example, a difficulty in testing many models stems from the fact that it has proven very difficult to classify individual species from remotely sensed imagery. Higher spectral resolution sensors may soon make it much more feasible to classify imagery to the species level (Martin et al., in press), at a spatial resolution compatible with models such as MetaFor. Similarly, new sampling designs are being actively devised to gather field data that are spatially compatible with landscape models.

This is not to excuse landscape modelers from the burden of formal model testing, but rather to underscore the importance of devising new and robust methods for making such assessments. Methods for testing spatial models have been suggested (e.g., Costanza, 1989; Turner et al., 1989), but landscape modelers have not yet taken full advantage of these. The use of neutral models might also be pursued (Gardner et al., 1987) in developing methods for evaluating model performance. Clearly, the credibility of landscape modeling demands further attention to model testing and evaluation.

Conclusions and prospectus

A typical evolution for models – and landscape models are no exception – is toward increasing complexity. This is a natural consequence of using a model: by applying it new things are discovered that might be refined or added to make it a more useful or realistic model. In our case, it is tempting to let the three meta-models evolve toward each other. For example, we might add spatial neighborhood interactions to Mosaic, or seed dispersal to ZelStage; or more realistic transients might be eked from MetaFor by incorporating elements of the other two models. Thus far this line of model evolution has been avoided. Instead, ways to build meta-models more efficiently and more robustly have been focused on. This entails, for example, devising automated and more streamlined schemes for model definition and parameterization. This, in turn, will allow more ready building of a greater number of variations on the models themselves; a larger family of meta-models geared to particular applications. Recognizing that each such model has its own strengths and weaknesses, variety allows us to choose the simplest model that will address the application at hand.

Acknowledgments

This work was supported in part by the National Science Foundation's program in Computational Biology, through a collaborative grant to Dean Urban (DIB-9630606 to Duke University), Miguel Acevedo (DIB-9615936 to University of North Texas), and Steve Garman (DIB-9615937 to Oregon State University).

References

Aber, J. D. and Federer, C. A. (1992). A generalized, lumped-parameter model of photosynthesis, evapotranspiration, and net primary production in temperate and boreal forest ecosystems. *Oecologia*, **92**, 463–74.

Ablan, M. (1997). Forest landscape dynamics: a semi-Markov modeling approach. PhD dissertation, Environmental Science, University of North Texas, Denton. 138 pp.

Acevedo, M., Urban, D. L. and Ablan, M. (1995a). Transition and gap models of forest dynamics. *Ecology Applications*, **5**, 1040–55.

Acevedo, M., Urban, D. L. and Ablan, M. (1995b). Landscape scale forest dynamics: GIS, gap, and transition models. In *GIS and environmental modeling: progress and research issues*, ed. M. F. Goodchild, L. T. Steyaert, B. O. Parks *et al.*, pp. 181–6. Fort Collins, CO: GIS World Books.

Acevedo, M. F., Urban, D. L. and Shugart, H. H. (1996). Models of forest dynamics based on roles of tree species. *Ecological Modelling*, **87**, 267–84.

Albini, F. A. (1976). Estimating wildfire behavior and effects. USDA Forest Service General Technical Report INT-30.

Baker, W. L. (1989). A review of models of landscape change. *Landscape Ecology*, **2**, 111–33.

Bazzaz, F. A. (1990). The response of natural ecosystems to the rising global CO_2 levels. *Annual Reviews of Ecology Systems*, **21**, 167–96.

Bonan, G. B. (1989). A computer model of the solar radiation, soil moisture, and soil thermal regimes in boreal forests. *Ecological Modelling*, **45**, 275–306.

Botkin, D. B., Janak, J. F. and Wallis, J. R. (1972a). Rationale, limitations, and assumptions of a northeastern forest growth simulator. IBM *Journal of Research and Development*, **16**, 101–16.

Botkin, D. B., Janak, J. F. and Wallis, J. R. (1972b). Some ecological consequences of a computer model of forest growth. *Journal of Ecology*, **60**, 849–73.

Burke, I. C., Shimel, D. S., Yonker, C. M., Parton, W. H., Joyce, L. A. and Lauenroth, W. K. (1990). Regional modeling of grassland biogeochemistry using GIS. *Landscape Ecology*, **4**, 45–54.

Chang, C. and Urban, D. (1997). A cellular automaton extending forest gap dynamics to landscape behavior. Paper presented at the 12th Annual Landscape Ecology Symposium, March 1997, Durham, North Carolina (manuscript in prep.).

Costanza, R. (1989). Model goodness of fit: a multiple resolution procedure. *Ecological Modelling*, **47**, 199–215.

Daly, C., Neilson, R. P. and Phillips, D. L. (1994). A digital topographic model for distributing precipitation over mountainous terrain. *Journal of Applied Meteorology*, **33**, 140–58.

DeAngelis, D. L. and Gross, L. J. (eds.) (1992). *Individual-based models and approaches in ecology*. New York: Chapman and Hall.

Delcourt, H. R., Delcourt, P. A. and Webb, T. (1983). Dynamic plant ecology: the spectrum of vegetation change in space and time. *Quaternary Science Review*, **1**, 153–75.

Dixon, G. (1994). Forest vegetation simulator, western Sierra Nevada variant (WESSIN). USDA Forest Service, Western Office Service Center, Fort Collins, CO.

Gardner, R. H., Milne, B. T., Turner, M. G. and O'Neill, R. V. (1987). Neutral models for the analysis of broad-scale landscape pattern. *Landscape Ecology*, **1**, 19–28.

Garman, S. L., Acker, S. A., Ohmann, J. L. and Spies, T. A. (1995). Asymptotic height–diameter equations for twenty-four tree species in western Oregon. Research Contribution 10. Forest Research Laboratory, College of Forestry, Oregon State University, Corvallis. 22 pp.

Gholz, H. L., Grier, C. C., Campbell, A. G. and Brown, A. T. (1979). Equations for estimating biomass and leaf area for plants in the Pacific Northwest. Research Paper 41, Forest Research Lab, Oregon State University, Corvallis.

Green, D. G. (1989). Simulated effects of fire, dispersal, and spatial pattern on competition within forest mosaics. *Vegetatio*, **82**, 139–53.

Hogeweg, P. (1988). Cellular automata as a paradigm for ecological modeling. *Applied Mathematics and Computations*, **27**, 81–100.

IPCC (Intergovernmental Panel on Climate Change). (1996). *Climate change 1995: the science of climate change*. Cambridge: Cambridge University Press.

Johnson, W. C. and Sharpe, D. M. (1976). An analysis of forest dynamics in the north Georgia piedmont. *Forensic Science*, **22**, 307–22.

Leemans, R. and Prentice, I. C. (1987). Description and simulation of tree-layer composition and size distribution in a primaeval *Picea-Pinus* forest. *Vegetatio*, **69**, 147–56.

Legendre, P. and Fortin, M. J. (1989). Spatial pattern and ecological analysis. *Vegetatio*, **80**, 107–38.

Levin, S. A. (1992). The problem of pattern and scale in ecology. *Ecology*, **73**, 1943–67.

Martin, M. E., Aber, J. D. and Congalton, R. (1999). Determining forest species composition using high spectral resolution remote sensing data. *International Journal of Remote Sensing* (in press).

McGarigal, K. and Marks, B. J. (1995). FRAGSTATS: Spatial pattern analysis program for quantifying landscape structure. Gen. Tech. Report PNW-GTR-351, USDA Forest Service, Pacific Northwest Research Station, Portland, OR.

Miller, C. and Urban, D. L. (1999). A model of surface fire, climate and forest pattern in Sierra Nevada, California. *Ecological Modelling, in press*.

Nikolov, N. T. and Zeller, K. F. (1992). A solar radiation algorithm for ecosystem dynamic models. *Ecological Modelling*, **61**, 149–68.

O'Neill, R. V., Krummel, J. R., Gardner, R. H., Sugihara, G., Jackson, B., DeAngelis, D. L., Milne, B. T., Turner, M. G., Zygmunt, B., Christensen, S. W., Dale, V. H. and Graham, R. L. (1988). Indices of landscape pattern. *Landscape Ecology*, **1**, 153–62.

Ousterhout, J. K. (1994). *Tcl and the Tk toolkit*. Reading, MA: Addison-Wesley Professional Computing.

Parton, W. J., Schimel, D. S., Cole, C. V. and Ojima, D. (1987). Analysis of factors controlling soil organic levels of grasslands in the Great Plains. *Soil Science Society American Journal*, **51**, 1173–9.

Riitters, K. H., O'Neill, R. V., Hunsaker, C. T., Wickham, J. D., Yankee, D. H., Timmins, S. P., Jones, K. B. and Jackson, B. L. (1995). A factor analysis of landscape pattern and structure metrics. *Landscape Ecology*, **10**, 23–40.

Rothermel, R. C. (1972). A mathematical model for predicting fire spread in wildland fuels. USDA Forest Service Research Paper INT-115. 40 pp.

Running, S. W., Nemani, R. R. and Hungerford, R. D. (1987). Extrapolation of synoptic meteorological data in mountainous terrain and its use for simulating forest evapotranspiration and photosynthesis. *Canadian Journal of Forestry Research*, **17**, 472–83.

Schwarz, P. A. (1993). *A suite of software tools for managing a large parallel programming project*. Ithaca, NY: Cornell Theory Center.

Shugart, H. H. (1984). *A theory of forest dynamics*. New York: Springer-Verlag.

Shugart, H. H. and West, D. C. (1980). Forest succession models. *BioScience*, **30**, 308–13.

Smith, J. B. and Tirpak, D. (1989). The potential effects of global climate change on the United States. Policy, Planning and Evaluation, PM-221. US EPA, Washington, DC.

Solomon, A. M. (1986). Transient response of forests to CO_2-induced climate change: simulation modeling experiments in eastern North America. *Oecologia*, **68**, 567–79.

Strain, B. R. and Cure, J. D., eds. (1985). Direct effects of increasing carbon dioxide on vegetation. DOE/ER-0238, Carbon Dioxide Research Division, US DOE, Washington, DC.

Turner, M. G., Costanza, R. and Sklar, F. H. (1989). Methods to evaluate the performance of spatial simulation models. *Ecological Modelling*, **4**, 1–18.

Urban, D. L., Bonan, G. B., Smith, T. M. and Shugart, H. H. (1991). Spatial applications of gap models. *Forestry Ecology Management*, **42**, 95–110.

Urban, D. L. and Shugart, H. H. (1992). Individual-based models of forest succession. In *Plant succession: theory and prediction*, ed. D. C. Glenn-Lewin, R. K. Peet and T. T. Veblen. London: Chapman and Hall.

Urban, D. L., Miller, C., Stephenson, N. L. and Graber, D. Forest pattern in Sierran landscapes: the physical template. (in review)

Usher, M. B. (1992). Statistical models of succession. In *Plant succession: theory and prediction*, ed. D. C. Glenn-Lewin, R. K. Peet and T. T. Veblen, pp. 215–48. London: Chapman and Hall.

VEMAP participants (J.M. Melillo and 26 others). (1995). Vegetation/ecosystem modeling and analysis project (VEMAP): comparing biogeography and biogeochemistry models in a continental-scale study of terrestrial ecosystem responses to climate change and CO_2 doubling. *Global Biogeochemical Cycles*, **9**, 407–37.

Walker, B. and Steffen, W., eds. (1996). *Global change and terrestrial ecosystems*. IGBP Book Series, No. 2. Cambridge: Cambridge University Press.

Wallin, D. O., Swanson, F. J. and Marks, B. J. (1994). Landscape pattern response to changes in pattern generation rules: land-use legacies in forestry. *Ecological Applications*, **4**, 569–80.

Weinstein, D. A. and Shugart, H. H. (1983). Ecological modeling of landscape dynamics. In *Disturbance and ecosystems*, ed. H. A. Mooney and M. Godron, pp. 29–45. New York: Springer-Verlag.

Weishampel, J. F. and Urban, D. L. (1996). Coupling a spatially explicit forest gap model with a 3-D solar routine to simulate latitudinal effects. *Ecological Modelling*, **86**, 101–11.

Whittaker, R. H. (1956). Vegetation of the Great Smoky Mountains. *Ecology Monographs*, **26**, 1–80.

Whittaker, R. H. (1967). Gradient analysis of vegetation. *Biology Review*, **49**, 207–64.

Wiens, J. A. (1989). Spatial scaling in ecology. *Functional Ecology*, **3**, 385–97.

Wykoff, W. R., Crookston, N. L. and Stage, A. R. (1982). User's guide to the stand prognosis model. USDA Forest Service Intermountain Research Station, General Technical Report INT-133. 112pp.

5

Simulating landscape vegetation dynamics of Bryce Canyon National Park with the vital attributes/fuzzy systems model VAFS/LANDSIM

David W. Roberts and David W. Betz

Introduction

In recent years modeling landscape or regional forest vegetation has been attempted with a variety of approaches (e.g., semi-Markov models: Acevedo *et al.*, 1995, 1996; vital attributes-based: De Vasconcelos and Zeigler, 1993; Mladenoff *et al.*, 1996; Roberts, 1996*a,b*; gap model-based, e.g., Zelig: Urban *et al.*, 1991; Burton and Cumming, 1995; Hansen *et al.*, 1995; or SORTIE: Pacala *et al.*, 1996; and grid-based: Baker, 1989, 1992, 1995). Individual models vary with respect to the level of biological detail, the nature of the environmental attributes included in the model, and in the temporal and spatial scale of the models. The Vital Attributes/ Fuzzy Systems (VAFS) modeling approach (Roberts, 1996*a,b*) is a mechanistic, spatially explicit yet parsimonious design that captures the most important elements of vegetation dynamics and distribution while remaining simple to parameterize. VAFS/LANDSIM models individual species age-class distributions at specific sites and calculates a range of community and landscape statistics from the basic site-specific species age-class dynamics. The landscape version of the model (VAFS/LANDSIM) is described and demonstrated below in an application to Bryce Canyon National Park. Roberts (1996*b*) describes an application of the landscape model to artificial landscapes.

In VAFS/LANDSIM there are three primary components of interest: (i) the biological characteristics of species – the vital attributes, (ii) the physical environment, and (iii) the environment-specific disturbance regime. Variability is generally represented by a simple classification rather than by a detailed mathematical parameterization. In many cases of interest, detailed information is not available for species and environments, and relatively simplistic representations are sufficient to realistically portray environmental variability, vegetation composition and structure, as well as vegetation response to disturbance. The model was developed under principles of parsimony and homogeneity of complexity, so that all mechanisms should be implemented as simply as possible and that complexity of subcomponents of the model should not vary tremendously.

The VAFS models employ fuzzy set mathematics, a generalization of set theory first applied to community ecology by Roberts (1986), and fuzzy systems theory, a fuzzy set generalization of dynamical systems theory first applied to ecology by Roberts (1989). A detailed understanding of these topics is not required to understand the VAFS models, and will not be presented here. In general, fuzzy systems theory is employed in VAFS/LANDSIM to synthesize higher-level community and landscape dynamics from the lower-level mechanistic site-specific species dynamics, and the necessary information to understand and interpret the model will be presented briefly.

Model description

Vital attributes – the biological characteristics of species

Species are defined in the VAFS models by a set of "vital attributes". Vital attributes were defined by Noble and Slatyer (1977, 1980) as the set of autecological characteristics necessary to predict a species' behavior in environments with recurrent disturbance. These vital attributes pertain generally to the means of dispersal and establishment of species, their disturbance tolerance or avoidance mechanism, and the age at which specific life history stages are achieved. Roberts (1996a) modified the set of vital attributes proposed by Noble and Slatyer (1980) to incorporate a slightly more detailed competitive mechanism and to give a broader range of disturbance intensities and responses. The vital attributes are designed to be parameters that are simple to estimate yet which capture significant details of the life history of different species. Specifically, the vital attributes employed in the VAFS models are:

(i)   reproductive method: dispersed seed, stored seed, or vegetative;
(ii) age of first reproduction: age of propagule development or time required to accumulate sufficient energy to reproduce vegetatively;
(iii) longevity: age of senescence for individuals not subjected to disturbance;
(iv) shade tolerance: a ranking (1–5) of competitive ability on an ordinal scale described below;
(v)   disturbance tolerance: a ranking (0–5) of relative tolerance to disturbance measured on an ordinal scale described below.

Reproduction

Reproduction by species is controlled by a combination of available propagules, suitable environmental conditions, and favorable competitive conditions. For species with dispersed seeds, seeds are produced in each time step for each species with sexually mature age classes present. Dispersed seeds are assumed to be viable only in the time step they are produced, and can germinate on the site where they are produced, or disperse to adjacent sites as described in the landscape section below. Species with stored seeds produce seeds similarly to dispersed seed species, but the seeds can remain viable for specific periods of time, even when the individuals that produced the seeds are no longer alive. In addition, stored seeds do not disperse off site.

Vegetative reproducers have the most complex reproductive vital attribute. Vegetative reproduction depends on stored carbohydrate, which in turn depends on the recent carbon balance of the plant. Living, above-ground non-reproducing individuals older than the age of maturity accumulate carbohydrate stores for every time step, up to a species-specific maximum. Under favorable shade conditions (generally requiring severe disturbance to reduce the shade) they reproduce vegetatively, which draws down their carbohydrate store. If the vegetatively reproduced individuals (sprouts or suckers) reach maturity, they are considered to have replenished the carbohydrate expended in sprouting and begin to accumulate carbohydrate stores again, up to the species-specific maximum. Repeated, short-term disturbance can prevent vegetative reproducers from replenishing their carbohydrate stores, leading to local extinction. Alternatively, in the absence of disturbance, when the youngest age-class of vegetative reproducers reaches the age of longevity, the species exists only below-ground as a root system. The carbohydrate storage is decremented each time step the species exists only below-ground, and the species can persist below-ground for periods depending on its stored carbohydrate reserve. If a species fails to reproduce before its carbohydrate reserves are exhausted, it goes locally extinct.

All species are assigned an age of first reproduction. For dispersed seed or persistent seed species, trees younger than the specified age are incapable of producing propagules. For vegetative reproducers, the age of reproduction indicates the number of years required to accumulate sufficient carbohydrate to replenish the carbohydrate expended in reproducing.

Shade tolerance

Given propagules, species reproduce only under favorable light conditions in suitable environments, described in the environment section below. Each species has a shade tolerance ranking from 1 through 5 (with ties if appropriate). Trees are assumed to cast shade proportional to their shade tolerance, based on the assumption that trees retain leaf area to the point where their lower crowns reach their photosynthetic compensation point. Stand shade is calculated as the maximum shade rank for a species whose age classes sum to at least 70. Species with shade tolerance from 1 through 4 can establish on suitable environments if they have propagules available and the stand shade is less than or equal to their shade tolerance. Species with shade tolerance 5 require some shade to establish. The net effect of the shade tolerance mechanism is to produce a competitive hierarchy (Horn, 1976) where any species with shade tolerance from 1 to 4 can establish on open sites, but where more shade-tolerant species eventually out-compete the less shade-tolerant species. The requirement for shade by species of shade tolerance 5 simulates a sensitivity to exposure (high soil surface temperatures and frost heaving) typical of many very shade-tolerant species.

Longevity

Species are assumed to have a maximum longevity determined by their morphology, physiology and life history characteristics. In the model, longevity is specified as the expected maximum age of individuals across a range of sites, and is generally less than the absolute maximum age recorded for rare individuals.

Disturbance tolerance

Species growing in environments subject to recurrent disturbance generally exhibit a range of morphological and life history characteristics that enable them to persist or re-establish in the event of a disturbance (Gill, 1975; Pickett et al., 1987). The disturbance tolerance attribute is intended to represent the ability to persist through disturbance, as re-establishment is handled by the reproductive characteristics described above. In the current VAFS/LANDSIM implementation the disturbance modeled is wildland fire, although the concept is more general (see Mladenoff and He, this volume for an example using wind disturbance). Following the nomenclature of Pickett et al. (1987), VAFS/LANDSIM distinguishes between disturbance intensity (measured in physical terms such as heat) and severity (measured in biological terms such as survival). In the model, disturbance intensity is a function of environment and site history, while disturbance severity is an interaction of disturbance intensity, species-specific fire tolerance, and age-class distributions by species. Species achieve fire tolerance generally by the development of thick bark that protects the cambium, and by a habit of self-pruning the crown to raise the lower portions of the crown above flame height in the event of a fire. Both of these characteristics are size dependent; thick bark is a relative characteristic with individuals of larger diameter having thicker bark, and crown height is dependent on the height of the individuals. Accordingly, while fire tolerance is generally a species-specific trait, it is also strongly size-dependent for individuals.

In VAFS/LANDSIM, fires vary in intensity as a function of environment and time since the last disturbance. Fire intensity is classified into five fire classes from 1 (low intensity) to 5 (high intensity). In the event of a fire of intensity lower than or equal to the fire tolerance of a species, individuals older than X survive the fire, where X is given by:

$$X = (6 - (\text{fire tolerance} - \text{fire intensity})) \times 10$$

If fires of intensity greater than the fire tolerance of a species occur, all individuals of the species are killed. Vegetative reproducing species survive the fire as belowground systems regardless of intensity. The net effect is to remove younger age classes of susceptible species in the event of fire, with the minimum age to survive a fire of given intensity a species-specific trait.

The vital attributes of ten species typical of southern Rocky Mountain landscapes are given in Table 5.1.

Environmental variability

In the VAFS model the environment is assumed to vary in any of a number possible ways: precipitation and temperature regimes, elevation, aspect, slope,

Table 5.1. *Species vital attributes*

Species name	Longevity	Age of reproduction	Shade tolerance	Fire tolerance	Reproductive method
Populus tremuloides	140	10	2	0	V
Quercus gambelii	300	50	2	0	V
Juniperus osteosperma	400	60	2	1	D
Pinus edulis	400	60	2	1	D
Juniperus scopulorum	400	50	2	1	D
Pinus ponderosa	400	40	2	4	D
Pinus flexilis	400	40	3	2	D
Pseudotsuga menziesii	300	40	3	3	D
Picea pungens	400	40	4	0	D
Abies concolor	300	40	5	0	D

Reproductive methods: V = vegetative reproduction, D = dispersed seed reproduction.

topographic position, soil characteristics, parent material, geology, etc. However, rather than include such explicit environmental variability, the environment is classified into a number of environment types or classes, called "habitat types" (Daubenmire and Daubenmire, 1968; Pfister *et al.*, 1977) that reflect the range of environmental variability in a simplified, but consistent manner. Due to the generally high correlation among environmental factors in specific regions, the number of types required to cover a broad range of environments is not terribly large. Habitat types are based on potential natural vegetation and are assumed to represent a narrow range of environmental variability, such that the range of possible vegetation composition and typical disturbance regime within each habitat type is known. Where habitat type classifications do not already exist, the environment can be classified into environment types for the purposes of the model fairly easily.

For each habitat type the model must know:

  (i)   the reproductive potential of each species of interest;
  (ii) the mean fire return interval;
  (iii) the rate of fuel accumulation.

The reproductive potential for species is expressed as the probability of successful establishment during one decade given an available seed source and suitable competitive conditions, as described above. In this sense, the habitat types operate as an "environmental sieve" (Harper, 1977) and only allow species with suitable physiology to establish. Within a given habitat type some species may be excluded due to temperature while others are excluded due to moisture or nutrient status, without requiring explicit consideration of the exact mechanism for each case.

The mean fire return interval of habitat types is assumed to correlate with a gradient from xeric to mesic or hydric conditions. Generally, more xeric habitat types have shorter fire return intervals (higher fire frequencies) than more mesic types. However, some very xeric environments have such low productivity that fires are relatively infrequent due to lack of fuel. Fire return intervals for western coniferous habitat types can be estimated by analysis of fire scars on standing trees or analysis of age-class distributions of fire-dependent species.

The mean fire return interval of individual sites depends on on-site ignitions as well as fires igniting in other portions of the landscape and spreading into the site as described in the section on disturbance regime below. Accordingly, the mean fire return interval of individual sites depends on the habitat types of the site and its neighbors. In the current version of VAFS/LANDSIM, probability of fire is an exponentially increasing function of time since the last fire, so that sites which recently burned have reduced probability of fire compared to sites of the same habitat type that have not burned recently.

Given the occurrence of fire, the intensity of the fire is determined by the existing fuel load. The fuel accumulation rate is habitat type-specific, and is modeled as a simple increasing step function with five steps or classes. In general, fuel is assumed to accumulate at a rate determined by the net primary productivity and decomposition rates of specific habitat types. Accordingly, the length of time spent in each fuel load class is habitat type-specific, and not all habitat types produce enough fuel to achieve the highest fuel load classes. Each fuel load class produces a fire of a given intensity when a fire occurs, and the intensity classes are keyed according to the species-specific response as outlined above. The net result is that fuel accumulation curves are set by estimating the amount of time required to accumulate sufficient fuel to kill specific age classes of specific species. Typically these severity-based estimates can be obtained from historic and anecdotal evidence for specific regions and habitat types.

The environmental attributes of a series of environments representative of the southern Rocky Mountains are given in Table 5.2.

Landscapes

Within the VAFS/LANDSIM model, landscapes are modeled as a spatially explicit collection of individual sites, where each site represents a unique location in the landscape. For each site the model maintains:

(i) the habitat type (as described above);
(ii) the number of years since the last fire;
(iii) the presence or absence of ten-year age-classes by species;
(iv) the area (m^2) of the site;
(v) a list of neighboring sites, with the length of adjoining edge (m) for each neighbor.

The design of VAFS/LANDSIM employs a topological concept of space, where space is expressed by neighbor relations; the actual coordinates of specific locations do not need to be known. In general, the assumption is that the relevant aspects of landscapes are the spatial autocorrelation of reproduction and disturbance, and

Table 5.2. *Characteristics of representative habitat types*

Attribute	Habitat types				
	PIED/JUOS	PIPO/ARPA	PSME/ARPA	PSME/BERE	ABCO/BERE
	Fire return interval				
	250	30	50	75	200
Intensity	Fuel accumulation curve				
Class 1	20	20	20	20	20
Class 2	100	50	50	40	40
Class 3	200	100	100	75	60
Class 4	400	∞	500	100	80
Class 5	∞	∞	∞	200	100
Species	Probability of reproduction				
POTR	0.00	0.00	0.00	1.00	1.00
QUGA	0.00	0.00	0.00	0.00	0.00
JUOS	0.50	0.00	0.00	0.00	0.00
PIED	0.50	0.00	0.00	0.00	0.00
JUSC	0.00	0.50	0.50	0.10	0.00
PIPO	0.00	1.00	1.00	1.00	1.00
PIFL	0.00	0.00	0.20	0.50	0.50
PSME	0.00	0.00	0.75	1.00	0.85
PIPU	0.00	0.00	0.00	0.00	0.20
ABCO	0.00	0.00	0.00	0.00	0.75

Habitat types listed in order from most xeric to most mesic (see Youngblood and Mauk, 1985 for abbreviations). Fuel curve values are years to reach fire intensity classes 1–5. The probability of reproduction values are for establishment within a decade. Species codes are: POTR = *Populus tremuloides*, QUGA = *Quercus gambelii*, JUOS = *Juniperus osteosperma*, PIED = *Pinus edulis*, JUSC = *Juniperus scopulorum*, PIPO = *Pinus ponderosa*, PIFL = *Pinus flexilis*, PSME = *Pseudotsuga menziesii*, PIPU = *Picea pungens*, ABCO = *Abies concolor*, ARPA = *Arctostaphylos patula*, and BERE = *Berberis repens*.

that topological space captures these aspects more succinctly than Cartesian systems.

The list of neighboring sites connects each site functionally to the landscape. Seed sources for reproduction of species must occur on site or in neighboring sites. Accordingly, species can be excluded from areas of suitable environment by lack of seed. However, species can disperse across the landscape into suitable sites, grow to maturity, and disperse to adjacent sites repeatedly. The rate of movement across the landscape is thus controlled largely by the age of maturity of the species and the size of individual sites in the landscape. In the present implementation all species disperse only to adjacent sites, but see Mladenoff and He (this volume) for a more detailed dispersal routine.

Fires igniting at a given site can spread into neighbors, and from those neighbors into their neighbors recursively. The maximum spread of fires is specified by the user at run time, up to a maximum of fifth-order neighbors. When fires occur, fire intensity is determined by the fuel load present, which is in turn determined by the fuel accumulation rate of the site and the time since the last fire at that site. The net result of the fire occurrence routine is fires of variable size and shape with variable intensity within a given fire. Thus the location of a specific site and the characteristics of its neighbors play a significant role in the site's disturbance regime and successional development. Mesic sites surrounded by more xeric sites burn more often than mesic sites surrounded by other mesic sites because they are subjected to fires spreading from their neighbors more often.

VAFS/LANDSIM performs landscape analysis in a manner similar to that first developed by Romme (1982), calculating patchiness, patch diversity, and landscape gamma diversity from a synthesis of the individual site characteristics. At each time step for each site in the landscape, the distribution of age classes by species is calculated. The age-class distribution is used to calculate alpha diversity for each site, and then used to classify the site to fuzzy community types as follows. First, the least shade-tolerant species present on the site is identified as the indicator species. Next, the relative abundance of each species is calculated. The fuzzy community types are binomials where the first element of the binomial is the indicator species, and the second element is the dominant species, where the relative dominance of that species (from 0.0 to 1.0) is the membership value of that plot in that fuzzy community type (Roberts, 1989). Fuzzy community types sum to 1.0 for each stand. The fuzzy community types for each site are multiplied by the area of that site and summed up over the entire area to determine the landscape abundance of each community type at each point in time.

To calculate patchiness, the dominant community type for each site is first calculated from the fuzzy community types by majority rule. Subsequently, each site is compared to each of its neighbors in turn. If neighboring sites are the same community type at that time step, they are merged into a patch, and all contiguous sites of the same community type are added to that patch. Patchiness is calculated as the number of patches found compared to the theoretical maximum number of patches that would occur if all sites were individual patches. The list of patch sizes is then used to calculate patch diversity using a Shannon–Weiner diversity index.

Gamma diversity is calculated from the fuzzy community type distributions of each site. The fuzzy set membership of each site in each community type is multiplied times the area of the site divided by the area of the entire landscape and summed over the entire landscape. Because this value is calculated over all replicates, it represents within-site alpha diversity, among-site beta-diversity, and among-replicate variability all in a single index.

The model runs on a ten-year time step, based on the assumption that changes in community composition for single years are likely to be small when considered from a landscape to regional scale. For each time step, the model calculates growth, reproduction, and mortality for each species in each site. Growth is calculated by incrementing the age-class distribution for each age-class present for that site.

Non-disturbance mortality is calculated by removing age classes greater than the longevity of individual species.

Model outputs

The model produces a broad range of outputs from site-specific to landscape level. Specifically, for each site the model produces: (i) an age class distribution by species, (ii) presence of propagules and remaining years of vitality for stored seed or root systems for vegetative reproducers, (iii) current fuel load, (iv) the community type distribution classified as fuzzy sets, and (v) the alpha-diversity calculated from the species abundances. At the landscape level the model calculates (i) patch number, (ii) patch diversity, (iii) gamma diversity, (iv) species abundance, and (v) community type abundance.

Model implementation

VAFS/LANDSIM is written in standard FORTRAN 77 and is suitable for simulation on most computing platforms. The model is generally run in an interactive mode from a command line, although batch mode processing is possible on UNIX or VMS systems. The interface is conversational and performs routine error checking on all inputs. The code was designed to be efficient and relatively fast. A flow diagram of program logic is given in Fig. 5.1. The simulations presented below (14 250 ha, 10 species, 25 environment types, 826 polygons, 500 years, 10 replicates) take approximately 5 minutes on a moderately fast UNIX workstation (Sun Sparc Ultra 1).

The model requires a small set of input files: a species vital attribute file – SPECIES.DAT, an environment data file – HT.DAT, and an initial conditions file, which can be named anything. All files are simple ASCII data files that are compatible across all systems. The vital attributes file defines any number of species up to the maximum for which the model is dimensioned (specified by the MAXSPC parameter). Each species is given an abbreviated name (up to six letters), a longevity (nearest 10 years), and age of first reproduction (nearest 10 years), a shade tolerance (1–5), a fire tolerance (0–5), and a reproductive habit (V,D, or S). This file completely defines the ecology of each species, except for its habitat type-specific reproductive probability, given in the environment file.

The environment file (HT.DAT) specifies all aspects of each classified environment type up to the maximum number of habitat types as specified by the MAXHT parameter. This files specifies the habitat type-specific: (i) mean fire return interval (ii) fuel accumulation step function, and (iii) probability of reproduction within a decade for each species.

The initial conditions file specifies numerous details about each site in the model. Specifically, for each site the file must provide: (i) the site number, (ii) its area, (iii) its habitat type (corresponding to entries in HT.DAT), (iv) the number

Fig. 5.1. *Flow chart of the logic and program units for model VAFS/LANDSIM.*

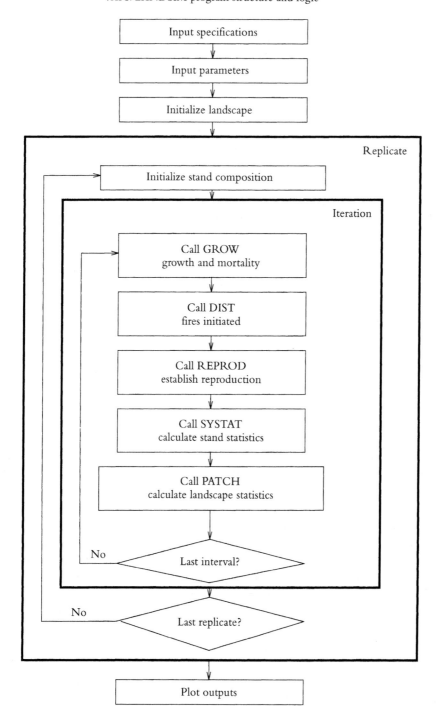

of years since the last fire, (v) the number of first-order neighbors, (vi) the list of neighbors and the length of edge between the site and each respective neighbor, and (vii) the age class distribution of each species in the stand at time 0.

Determining the identity of each first-order neighbor, and the length of edge between the site and its neighbors, is critical to the functioning of the model. Because the model was designed to integrate with the ARC/INFO GIS system, importing the landscape characteristics from ARC is relatively simple. Specifically, an ARC/INFO line coverage of the landscape to be modeled must exist, and the polygon to the left and right of each arc is written to a file. This list of polygons is reassembled by a FORTRAN program into a list of neighbors and edges for VAFS/LANDSIM. Alternatively, a set of FORTRAN utilities is provided to calculate this information from simple ASCII files where the north and east coordinates of each vertex of each polygon are given in a list.

Determining the initial age class distribution by species for each site is subject to difficulties, and a simple approach is to assume a typical age class distribution for each combination of successional community type and habitat type. Alternatives include running the model from a simple initial condition for a specified period of time and saving the output as an input file for subsequent simulation. In some cases, inventory data may be available to establish initial conditions, but this is rarely true on a landscape level.

VAFS/LANDSIM allows the user to specify characteristics of the disturbance regime at run time. Users can vary the mean fire return interval of all habitat types proportionately, specify details of how the probability of fire is calculated at each time step, and specify the relative size of fires. Specifying small or large fires results in VAFS/LANDSIM adjusting the site-specific probability of ignition to maintain the same fraction of the total landscape burned whether fires are numerous and small or few and large.

VAFS/LANDSIM produces detailed numeric output in a simple ASCII file for importation into numerous plotting packages. In addition, VAFS/LANDSIM also provides a number of utilities to import the results of a simulation back into ARC/INFO for display of results or further landscape analysis. While the import step is not as seamless as would be ideal, the power of the ARC/INFO environment, and the natural relation of the topological space in VAFS/LANDSIM to the arc-node format of ARC/INFO make the connection desirable. In addition, users have the options of specifying the amount of detail provided in the output, including patch size distributions and patch maps. Finally, VAFS/LANDSIM will save the final state of the model as an input file for further simulation so that changes in fire regime and the effects of prescribed burning can easily be achieved.

Model behavior

The model has been tested at a variety of scales from individual stand-level simulations in homogeneous environments (Roberts, 1996a, Will-Wolf and Roberts, 1993), to synthetic landscapes of variable heterogeneity (Roberts, 1996b), to the

Stand-level

Roberts (1996a) extensively tested the vital attributes approach to stand-level simulation modeling by simulating the vegetation dynamics of a typical southwestern US mixed conifer forest consisting of *Populus tremuloides*, *Pinus ponderosa*, *Pseudotsuga menziesii*, *Picea pungens* and *Abies concolor*. Simulations were conducted with 100 replicates for each of six different mean fire return intervals from 25 years to no fires. Evaluation consisted of tracking the changes in relative species abundance as a function of time, beginning always with an early seral community of all *Populus tremuloides*. Propagules were assumed available for all dispersed seed species at all points in time. In general, the results show that the average of the replicates eventually reaches an equilibrium community composition, and that the equilibrium composition is unique for each fire return interval. Specifically, at short fire return intervals (less than 100 years) stands are dominated by the fire-tolerant species *Pinus ponderosa*, with increasing dominance of this species as the fire return interval is shortened. At longer fire return intervals, stands ranged from mixed communities (at 250 years) to stands dominated by the shade tolerant species *Abies concolor* at 500 years or in the absence of fire. Roberts (1996a) concluded that the competitive advantage switched from fire tolerance at intervals between 100 and 250 years to shade tolerance at longer intervals. Interestingly, while all fire regimes exhibited a unique equilibrium composition, the time required to reach equilibrium exhibited a modal response and varied from approximately 200 years at return intervals of 250 or 500 years to greater than 400 years at shorter or longer return intervals.

In additional unpublished evaluations, we determined that the equilibrium community composition was independent of the initial conditions, but that the time required to reach equilibrium was not. In general, communities often took longer to reach equilibrium starting from "climax" conditions of pure *Abies concolor* than from the early seral community conditions evaluated in Roberts (1996a).

Roberts (1996a) conducted extensive sensitivity analysis of the vital attributes of species by systematically altering specific attributes of a single species and simulating vegetation dynamics at the six fire return intervals with the altered and other original species. The results indicated that the vital attributes generally exhibit what appears to be appropriate and sensitive ecological behavior. Specifically, (i) the model is sensitive to fire tolerance rankings in environments with short fire return intervals, (ii) it is moderately sensitive to shade tolerance rankings at all fire return intervals, and (iii) it exhibits bimodal sensitivity to longevity with greater sensitivity at short and long fire return intervals than at intermediate intervals. This is primarily a reflection of sensitivity to longevity of fire tolerant species at short intervals and shade tolerant species at long intervals. Surprisingly, the model was relatively insensitive to age of first reproduction, but this attribute varied only over a small range.

Will-Wolf and Roberts (1993) simulated Oak-Maple upland forests in Wisconsin consisting of seven species: *Populus grandidentata, Quercus alba, Quercus borealis, Fraxinus americana, Acer rubrum, Tilia americana,* and *Acer saccharum.* The results indicated that effects of the initial conditions of the stands persisted for periods from 150 to 250 years under most fire regimes, and that stands reached a dynamic equilibrium composition determined by the fire regime between 250 and 400 years. In a series of simulations intended to analyze the potential for maintaining significant amounts of *Quercus alba* in the landscape, it was determined that fire return intervals from 20 to 50 years with moderately intense fires were most successful; most other regimes led to dominance of *Acer saccharum.* The results suggested that the widespread dominance of oak forests may be an artifact of historical burning practices rather than a result of natural fire regimes in this part of the country.

Landscape level simulations

Roberts (1996*b*) examined the behavior of the model at the landscape scale on synthetic landscapes representative of southwestern US xeric to mesic forest conditions. Two landscapes were constructed, termed the "gradient" landscape and the "random" landscape. Each landscape consisted of 400 sites located on a hex grid wrapped on a torus. Accordingly, each site had exactly six neighbors, and the landscape had no edge. Each landscape consisted of 100 sites of each of four habitat types ranging from a xeric *Pinus ponderosa* savanna type to a mesic mixed-conifer forest supporting *Populus tremuloides, Pinus ponderosa, Pseudotsuga menziesii, Picea pungens,* and *Abies concolor.* In the gradient landscape the four habitat types were arranged in concentric circles from mesic to xeric (from the center out), and on the random landscape the four habitat types were placed at random. The gradient landscape was intended to represent a relatively homogeneous landscape with little local variability, while the random landscape represented a highly heterogeneous landscape. The two landscapes were employed in an analysis of the interaction of disturbance regime and environmental heterogeneity by systematically increasing or decreasing the fire return interval of the four habitat types proportionately.

The results exhibited a strong interaction between disturbance regime and environmental heterogeneity. On the random landscape, maximum landscape patchiness was obtained with intermediate disturbance intervals, but the response of landscape patchiness to disturbance interval was sinusoidal with the extreme values for return interval giving intermediate results. On the gradient landscape the results were more subtle, with maximum patchiness achieved at intermediate or shorter intervals, although extremely short intervals appear to homogenize the landscape and lead to low patchiness. Patch diversity exhibited similar response to variability in disturbance regime, and was generally more variable.

Application to Bryce Canyon National Park

Bryce Canyon National Park occupies an area of approximately 14 250 ha on the east face of the Paunsaugunt Plateau in south central Utah. The Park represents the divide between the Colorado River basin to the east and the Great Basin to the west, but is generally considered more similar to the Colorado Plateau than the Great Basin. The Park is composed of four geologic strata, consisting primarily of limestone, sandstone, and shales. The topography varies from broad areas of little relief to stunningly steep cliff faces with erosional remnants ("hoodoos") for which the Park is famous. The climate has components of the classical Colorado Plateau regime (summer precipitation maximum with frequent thunderstorm activity), as well some affiliation with the Great Basin (increased winter precipitation, mostly as snow). Maximum precipitation occurs in August (57 mm), with a May–June drought, and total annual precipitation is approximately 434 mm. The temperature regime exhibits the strong seasonal variability typical of the southwest, with monthly mean temperatures ranging from −5.9 °C in January to 17 °C in July. Daily temperature variability is very high, however, with daily maximum temperatures above 32°C possible from June through September, and frosts at least one day a month for every month except August.

The vegetation of Bryce Canyon follows an elevation sequence from sagebrush steppe (*Artemisia tridentata* and *A. nova*) through pinyon-juniper woodland (*Pinus edulis* and *Juniperus osteosperma*) through ponderosa pine (*Pinus ponderosa*) savanna to mixed-conifer forests with *Pinus ponderosa, Quercus gambelii, Juniperus scopulorum, Pseudotsuga menziesii, Populus tremuloides, Picea pungens* and *Abies concolor*. Especially harsh sites with poor soil development support a community of *Pinus flexilis* and *Pinus longaeva* (Roberts et al., 1988). The flora and vegetation of Bryce Canyon National Park are described in detail by Buchanan (1960), Graybosch and Buchanan (1983), Hallsten and Roberts (1988), and Roberts et al. (1988). In all, ecologists recognize approximately 20 potential natural vegetation types throughout the Park. In the following simulations, only the 13 types that typically support woodland or forest vegetation are explicitly simulated. Shrub and grassland types are included, and play a critical role in the Park disturbance regime, but age-class distributions are only maintained for tree species. Vital attributes were estimated for all tree species except *Pinus longaeva*, which plays a minor role in the Park. The ten tree species simulated have the potential to form 55 successional community types (see Roberts, 1989 or 1996b for the successional pyramid procedure), but only 25 successional community types are commonly observed; for any particular habitat type the range of possibilities is generally much smaller.

The composition and structure of southwestern forest vegetation is generally very sensitive to the prevailing disturbance regime (Cooper, 1960; Weaver, 1967, 1974; Kilgore, 1981; Covington and Moore, 1992, 1994a, b), and modified fire regimes have been proposed as responsible for changes in forest age-class distributions and composition in Bryce Canyon National Park (Stein, 1988a, b). In the following simulation attempts were made to determine the consequences of changes in the historical fire regime of Bryce Canyon and the possible future composition and structure under varying fire regimes.

Reconstructing the historic vegetation of Bryce Canyon

Bryce Canyon National Park is relatively young. Before 1928 Bryce Canyon was part of the Sevier and then Powell National Forests, and briefly managed by the Forest Service as Bryce Canyon National Monument (Scrattish, 1985). Before National Park status was achieved, the area was actively grazed by livestock and subjected to limited timber harvesting by area settlers. Accordingly, fire suppression within the boundaries of the existing Park was probably not effective until the 1930s due to administrative and access problems. However, the active grazing before that period probably served to reduce fine fuels and limit the number of fires. Pre-settlement use of the area by Native Americans is relatively poorly known, but generally consisted of occasional use for hunting (Scrattish, 1985).

To recreate the historic vegetation of Bryce Canyon National Park a three-part strategy was employed: (i) historical photograph relocation, (ii) analysis of age-class distributions of existing vegetation, and (iii) inference from similar regions. Site-specific information is generally of limited availability, and the length of period of analysis is relatively short, but qualitative assessments of historic vegetation are possible.

Bryce Canyon archives were searched for photographs with sufficient detail to allow relocation, and scenic photographs ($n=4$) as well as photographs from Buchanan (1960) ($n=15$) were selected for relocation; methods followed Rogers *et al.* (1984). Although photographs date back only to the 1930s or 1950s, some indication of change on the landscape is evident (Figs. 5.2 and 5.3). At lower elevation, more xeric sites (e.g. Fig. 5.2), species composition has not changed, but stand density has increased as younger age classes have become more prevalent. At higher elevation, more mesic sites (e.g. Fig. 5.3), density has increased, and stand composition has shifted toward more shade tolerant species such as *Abies concolor*.

On each of 160 approx. 400 m^2 (tenth-acre) plots Roberts *et al.* (1988) cored the largest (presumed oldest) tree of each species for assessment of stand age class distributions. At lower elevations, where *Pinus ponderosa* is the climax dominant, large trees were all *Pinus ponderosa* and varied in age from approximately 100 to 485 years. Evidently, stand-replacing fires were not uncommon in this zone, even given the relatively high fire resistance of *Pinus ponderosa*. At higher elevations, where *Pseudotsuga menziesii* and *Abies concolor* are the climax dominants, *Pinus ponderosa* had fewer young individuals, and *Pseudotsuga menziesii* and *Abies concolor* exhibited ages from less than 100 to approximately 250 years. Somewhat surprisingly, the fire-sensitive *Abies concolor* was older than 200 years on several sites, and several large individuals with fire scars were observed.

In general, results at Bryce Canyon are consistent with the rest of the Rocky Mountains. Fire return intervals in lower elevation ponderosa pine forests are often

Fig. 5.2. Change in low elevation Pinus ponderosa *forests. Photo* (a) *was taken in 1953. Photo* (b) *was taken in 1988.*

on the order of a few years to a few decades, with higher elevation forests ranging up to a few hundred years (Cooper, 1960; Weaver, 1967; Habeck and Mutch, 1973; Mueggler, 1976; Arno, 1980; Kilgore, 1981; Gruell, 1983, as well as numerous local studies).

All of Bryce Canyon National Park was mapped to current, existing successional community type as well as potential natural vegetation (habitat type) by a combination of field-based mapping, aerial photograph interpretation, and statistical predictive modeling (Roberts *et al.*, 1988; Roberts and Cooper, 1989). The successional community type map was overlaid with the potential vegetation map to produce a set of unique stands, and the resulting map was digitized into ARC/INFO. Stands larger than 40 ha were subdivided at logical fire breaks (i.e., streams, roads, and changes in aspect) to obtain 826 polygons with relatively consistent sizes of approximately 15–20 ha. Based on the historic reconstruction and general guidelines from the literature each polygon was assigned an initial age-class distribution by species representative of the vegetation at approximately 1880.

Simulating historic vegetation

The reconstructed historic vegetation was entered into the VAFS/LANDSIM model as the initial conditions at time 0, and simulated for 500 years under five different fire regimes, with mean return intervals of 0.25×, 0.5×, 1.0×, 2×, and 4× of the estimated natural return intervals. Because the effects of land management have been to lengthen the fire return interval, results will be presented below only for the longer return intervals. In the following analyses, an envelope is drawn that encloses the central 80% of the simulated values for each combination of initial condition and fire return interval at each point in time. In addition, the current conditions are plotted as a dotted line throughout the simulation period for comparison.

Patchiness

Landscape patchiness for three return intervals (1×, 2×, 4×) is compared to the simulated trajectory for the existing vegetation (Fig. 5.4). Landscape patchiness was estimated to be significantly lower historically, as frequent fires homogenized the lower elevation vegetation to the most fire-resistant species. During the course of the simulations, the historic vegetation would achieve the existing vegetation patchiness in approximately 250 years under the 1× and 2× fire regimes, but would be unable to achieve the existing patchiness under the 4× regime. The 1× and 2× regimes are relatively similar, but with the 1× scenario achieving higher ultimate patch diversity.

Patch diversity

Landscape patch diversity is presented in Fig. 5.5. In general, the trends for patch diversity are very similar to patch number, indicating relatively consistent patch sizes in the landscape. Bryce Canyon is notable for the "hoodoos" or breaks along

Fig. 5.3. Change in mid-elevation mixed conifer forest. Photo (a) (below) was taken in 1953. Photo (b) (opposite) was taken in 1988.

the length of the Park, that serve as relatively efficient fire breaks and possibly prevent fires from getting extremely large.

Gamma diversity

Surprisingly, landscape gamma diversity appears relatively insensitive to fire return interval, with low among-replicate variability and relatively flat trends with time (Fig. 5.6). None of the simulations beginning with the historic conditions achieves gamma diversity similar to the present conditions of the Park, indicating that the reconstructed historic landscape is probably overly-simplified compared to the actual historic conditions.

Alpha diversity

In contrast to the landscape indices given above, alpha diversity is calculated for each polygon at each point in time, and is most easily understood as a map at a point in time. Fig. 5.7 (see color section) shows the simulated alpha diversity after

100 years from initial historic conditions for three fire return intervals, compared to the current conditions. In contrast to patch diversity or patchiness, the patterns of landscape alpha diversity appear to recover much faster, achieving conditions somewhat similar to current conditions within about 100 years. The best match of simulated alpha diversity to actual (not shown) depends on the fire return interval, and occurs from about 120 to 150 years. However, none of the simulations achieves alpha diversities as high as observed at present.

Interestingly, different areas within the Park seem to achieve conditions similar to the present under different regimes. The northern end of the Park recovers to approximately current conditions under the 4× regime (Fig. 5.7(c)), while the southern end of the Park recovers more realistic conditions under shorter intervals (1× or 2×, Fig. 5.7(a) and 5.7(b), respectively). This appears to indicate that the estimated mean fire return intervals are less accurate in some vegetation types than others, or that the Park has responded non-linearly to the imposed fire suppression

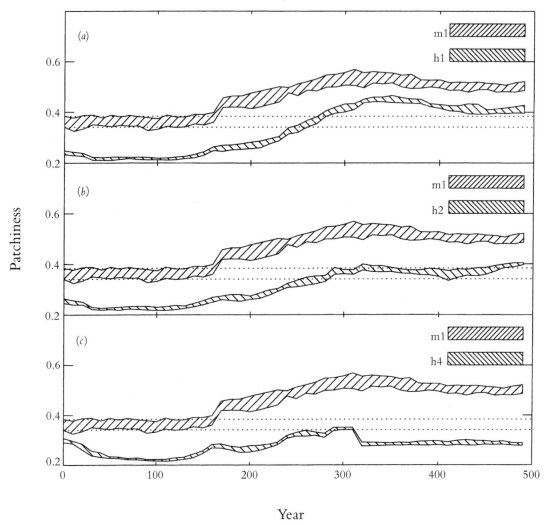

Fig. 5.4. *Landscape patchiness for 500 years comparing three fire return intervals from historic and modern initial conditions. (a) Fire return interval 1×, m1 = modern, h1 = historic (b) Fire return interval 2×, m1 = modern, h2 = historic (c) Fire return interval 4×, m1 = modern, h4 = historic. The envelopes bound 80% of the simulation results. The dotted line bounds the existing conditions.*

of the last century. None of the simulated patterns achieves a realistic alpha diversity for the north–central portion of the Park due to limitations in our database for the historic distribution of *Juniperus scopulorum* in *Pinus ponderosa* woodlands.

Species abundances

Because fire suppression has significantly lengthened the mean fire return inter-

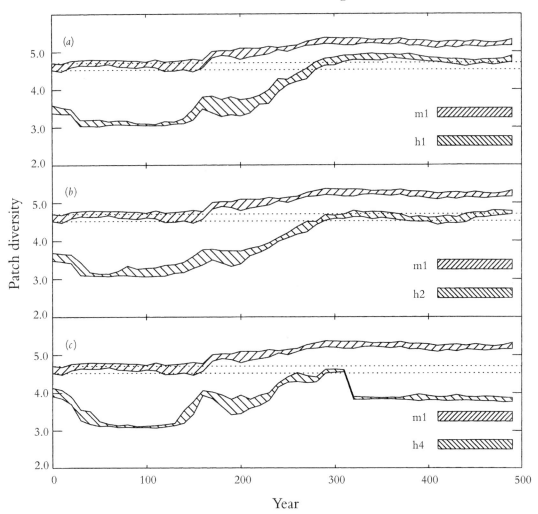

Fig. 5.5 Landscape patch diversity for 500 years comparing three fire return intervals from historic and modern initial conditions. (a) Fire return interval 1×, m1 = modern, h1 = historic (b) Fire return interval 2×, m1 = modern, h2 = historic (c) Fire return interval 4×, m1 = modern, h4 = historic. The envelopes bound 80% of the simulation results. The dotted line bounds the existing conditions.

val for the Park, results are shown for the 4× fire regime, which may most clearly match the effects of suppression. Similar to the landscape indices, species dominance took much longer to approach current conditions than the anticipated 100 years. Fig. 5.8 (see color section) shows the dominant species by polygon for *Pinus ponderosa*, *Pseudotsuga menziesii* and *Abies concolor* in red, green, and blue respectively. The clearest approximation to the existing pattern occurred between 200 and 300 years after initial conditions, rather than 100. In addition, none of the simulations predicted *Pseudotsuga menziesii* in the north

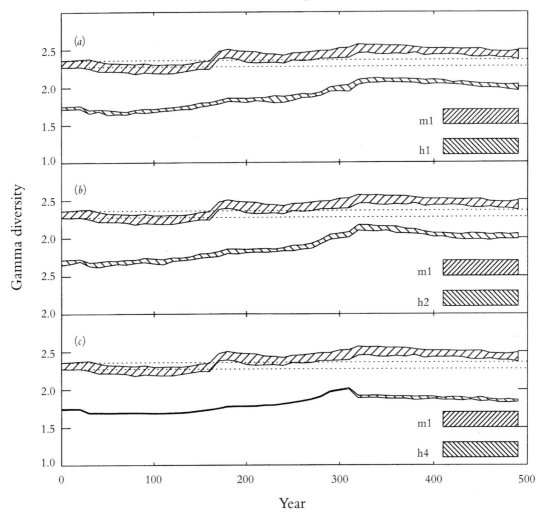

Fig. 5.6. *Landscape gamma diversity for 500 years comparing three fire return intervals from historic and modern initial conditions. (a) Fire return interval 1×, m1 = modern, h1 = historic (b) (FRI) Fire return interval 2×, m1 = modern, h2 = historic (c) Fire return interval 4×, m1 = modern, h4 = historic. The envelopes bound 80% of the simulation results. The dotted line bounds existing conditions.*

end of the Park at any time, due to a lack of seed in the initial conditions file.

Discussion

Landscape simulation modeling with the VAFS/LANDSIM model

VAFS/LANDSIM was designed to simulate real landscapes efficiently for significant periods of time, and to produce output that is interpretable at the stand and landscape levels. Clearly, the model is successful by these criteria. The simulations presented here required relatively modest computer requirements and produced significant interpretable output on multiple spatial and temporal scales. In addition, the results of the simulations provided a wealth of information about the vegetation distribution and dynamics in Bryce Canyon. Species vital attributes are relatively simple to parameterize, and the environmental partitioning represented by habitat types proved reasonably successful in representing the spatial pattern of environmental variability. Accordingly, the model should be relatively easy to parameterize and run on other landscapes in other regions.

The interface to the ARC/INFO GIS programs allows ecologists and resource managers to incorporate many existing landscapes into the simulation model, and to capitalize on the significant effort of developing GIS databases. Where such systems do not exist, the interface to the simpler MOSS GIS standards ensures that VAFS/LANDSIM is still relatively useful. On the other hand, some significant limitations in landscape modeling became apparent during the simulations presented here, and are discussed below.

Validating landscape simulation models

Simulating landscape vegetation dynamics for a real landscape with significant environmental heterogeneity is a daunting task. Simulating the same landscape from a reconstructed historic condition is significantly more complicated. Ideally, the results of the simulation for year 100 would have matched the existing conditions, but clearly in many cases they do not. There are several reasons for the discrepancy.

The historic reconstruction employed here developed only qualitative guidelines to stand composition and structure, in contrast to the relatively detailed stand-level initial conditions employed in the model. Consequently, the initial conditions employed by the historic simulations undoubtedly contained gross simplifications, and unfortunately, probably significant bias. This is undoubtedly true with regard to the distribution of *Juniperus scopulorum*, which is highly uncertain. Juniper invasion has been a controversial topic in the western US for some time now, and the extent of post-fire suppression invasion of juniper is unknown for Bryce Canyon. In addition, the existing conditions that were used for comparison to the historic simulations are only known qualitatively, and appreciable error in the stand composition and structure of specific locations undoubtedly exists. Accordingly, comparing the historic simulation results to the current condition confounds the error of simulation with the errors of initial and existing conditions. Given the relatively simple design and configuration of the VAFS/LANDSIM model, establishing initial conditions for more detailed models will likely prove even more difficult. It is suggested that, in addition to traditional sensitivity analysis of parameter values, all models be tested for sensitivity to initial conditions. If highly accurate spatially explicit data are required for initialization, then simulations of real landscapes are likely to prove extremely difficult and expensive to perform. In contrast, if general trends of qualitatively realistic landscapes are sufficient for the purpose, then existing landscape simulation models will prove adequate.

Acknowledgments

This research was supported by a grant from the University of Wyoming – National Park Service Research Center. Don Bragg helped prepare the figures. We would like to thank Bill Baker, Niklaus Zimmermann, and Mike Wendel for review comments. The VAFS/LANDSIM model is available at http://www.nr.usu.edu/~dvrbts/ and by anonymous ftp at ftp.nr.usu.edu/public/landsim.

References

Acevedo, M. F., Urban, D. L. and Ablan, M. (1995). Transition and GAP models of forest dynamics. *Ecological Applications*, **5**(4), 1040–55.

Acevedo, M. F., Urban, D. L. and Shugart, H. H. (1996). Models of forest dynamics based on roles of tree species. *Ecological Modelling*, **87**(1–3), 267–84.

Arno, S. F. (1980). Forest fire history in the northern Rockies. *Journal of Forestry*, **78**, 460–5.

Baker, W. L. (1989). Effect of scale and spatial heterogeneity on fire-interval distributions. *Canadian Journal of Forest Research*, **19**(6), 700–6.

Baker, W. L. (1992). Effects of settlement and fire suppression on landscape structure. *Ecology*, **73**(5), 1879–87.

Baker, W. L. (1995). Longterm response of disturbance landscapes to human intervention and global change. *Landscape Ecology*, **10**(3), 143–59.

Buchanan, H. (1960). The plant ecology of Bryce Canyon National Park. PhD Dissertation. University of Utah.

Burton, P. J. and Cumming, S. G. (1995). Potential effects of climatic change of some western Canadian forests, based on phenological enhancements to a patch model of forest succession. *Water Air and Soil Pollution*, **82**(1–2), 401–14.

Cooper, C. F. (1960). Changes in vegetation, structure, and growth of southwestern pine forests since white settlement. *Ecological Monographs*, **30**, 129–64.

Covington, W. W. and Moore, M. M. (1992). Post settlement changes in natural fire regimes: implications for restoration of old-growth ponderosa pine forests. In *Old-growth forest in the Southwest and Rocky Mountain Regions*, ed. Kaufmann, M. R. and W. H. Moir, pp. 81–99. USDA Forest Service Rocky Mountain Forest and Range Experimental Station General Technical Report RM-213.

Covington, W. W. and Moore, M. M. (1994a). Post settlement changes in natural fire regimes and forest structure: Ecological restoration of old-growth ponderosa pine forests. In *Proceedings of the American Forests scientific workshop*, ed. Sampson, D. L. and D. Adams, pp. 153–81. November 14–20, 1993, Sun Valley, ID. New York: The Haworth Press Inc.

Covington, W. W. and Moore, M. M. (1994b). Southwestern ponderosa forest structure and resource conditions: Changes since Euro-American settlement. *Journal of Forestry*, **92**, 39–47.

Daubenmire, R. F. and Daubenmire, J. (1968). Forest vegetation of eastern Idaho and western Washington. Washington Agricultural Experimental Station Technical Bulletin, 60.

De Vasconcelos, M. J. P. and Zeigler, B. P. (1993). Discrete-event simulation of forest landscape response to fire disturbances. *Ecological Modelling*, **65**(3–4), 177–98.

Gill, A. M. (1975). Fire and the Australian flora: a review. *Australian Forestry*, **38**, 4–25.

Graybosch, R. A. and Buchanan, H. (1983). Vegetative types and endemic plants of the Bryce Canyon Breaks. *Great Basin Naturalist*, **43**, 701–12.

Gruell, G. E. (1983). Fire and vegetation trends in the northern Rockies: interpretations from 1871–1982 photographs. USDA Forest Service Intermediate Forest and Range Experimental Station General Technical Report INT-158.

Habeck, J. R. and Mutch, R. W. (1973). Fire-dependent forests in the northern Rocky mountains. *Quaternary Research*, **3**, 408–24.

Hallsten, G. P. and Roberts, D. W. (1988). Additions to the vascular flora of Bryce Canyon National Park, Utah. *Great Basin Naturalist*, **48**, 352.

Hansen, A. J., Garman, S. L., Weigand, J. F., Urban, D. L., McComb, W. C. and Raphael, M. G. (1995). Alternative silvicultural regimes in the Pacific Northwest: Simulations of ecological and economic effects. *Ecological Applications*, **5**(3), 535–54.

Harper, J. L. (1977). *The population biology of plants*. Academic Press.

Horn, H. S. (1976). Succession. In *Theoretical ecology: principles and practice*, ed. R. M. May, pp. 187–204. Oxford: Blackwell Scientific Publ.

Kilgore, B. M. (1981). Fire in ecosystem distribution and structure. Western forests and scrublands. In *Proceedings of the Conference Fire Regimes and Ecosystem Properties*. USDA Forest Service Washington Office General Technical Report WO-26.

Mladenoff, D. J., Host, G. E., Boeder, J. and Crow T. R. (1996). LANDIS: a spatial model of forest landscape disturbance, succession, and management. In *GIS and environmental modeling: Progress and research issues*, ed. M. F. Goodchild, L. T. Steyaert and B. O. Parks. Fort Collins, CO, USA: GIS World Books.

Mueggler, W. F. (1976). Ecological role of fire in western woodland and range ecosystems. In *Use of prescribed burning in western woodland and range ecosystems. A symposium*, pp. 1–10. Utah State University Press.

Noble, I. R. and Slatyer, R. O. (1977). Post fire succession of plants in Mediterranean ecosystems. In *Proceedings of the symposium on environmental consequences of fire and fuel management in Mediterranean ecosystems*, ed. H. A. Mooney and C. E. Conrad, pp. 27–39. USDA Forest Service General Technical Report WO-3.

Noble, I. R. and Slatyer, R. O. (1980). The use of vital attributes to predict successional changes in plant communities subject to recurrent disturbance. *Vegetatio*, **43**, 5–21.

Pacala, S. W., Canham, C. D., Saponara, J., Silander, J. A., Kobe, R. K. and Ribbens, E. (1996). Forest models defined by field measurements: estimation, error analysis and dynamics. *Ecological Monographs*, **66**(1), 1–43.

Pfister, R. D., Kovalchik, B. L., Arno, S. F. and Presby, R. C. (1977). Forest habitat types of Montana. USDA Forest Service Intermountain Forest and Range Experimental Station General Technical Report INT-34.

Pickett, S. T. A., Collins, S. L. and Armesto, J. J. (1987). Models, mechanisms, and pathways of succession. *Botanical Review*, **53**, 335–71.

Roberts, D. W. (1986). Ordination on the basis of fuzzy set theory. *Vegetatio*, **66**, 123–31.

Roberts, D. W. (1989). Fuzzy systems vegetation theory. *Vegetatio*, **83**, 71–80.

Roberts, D. W. (1996a). Modelling forest dynamics with vital attributes and fuzzy systems theory. *Ecological Modelling*, **90**, 161–73.

Roberts, D. W. (1996b). Landscape vegetation modelling with vital attributes and fuzzy systems theory. *Ecological Modelling*, **90**, 175–84.

Roberts, D. W. and Cooper, S. V. (1989). Vegetation mapping and inventory: Concepts and techniques. In *Land classifications based on vegetation: applications for resource managers*, ed. D. Ferguson, P. Morgan and F. D. Johnson, pp. 90–6. USDA Forest Service Intermountain Research Station General Technical Report INT-257, Ogden, Utah.

Roberts, D. W., Wight, D. and Hallsten, G. (1988). Plant community distribution and dynamics in Bryce Canyon National Park. Final report PX 1200-7-0966. On file at Utah State University.

Rogers, G. F. (1982). *Then and now. A photographic history of vegetation change in the central Great Basin*. Salt Lake City, UT: University of Utah Press.

Romme, W. H. (1982). Fire and landscape diversity in subalpine forests of Yellowstone National Park. *Ecological Monographs*, **52**, 199–221.

Stein, S. J. (1988a). Explanations of the imbalanced age structure and scattered distribution of ponderosa pine within a high-elevation mixed-conifer forest. *Forest Ecology and Management*, **25**, 139–53.

Stein, S. J. (1988b). Fire history of the Paunsaugunt Plateau in southern Utah. *Great Basin Naturalist*, **48**, 58–63.

Urban, D. L., Bonan, G. B., Smith, T. M., Shugart, H. H., Mohren, G. M. J. and Kienast, F. (1991). Spatial applications of gap models. Modeling forest succession in Europe. Proceedings of a Workshop, "Modeling forest dynamics in Europe", held October 1988 in Wageningen, Netherlands. *Forest Ecology and Management*, **42**, 95–110.

Weaver, H. (1967). *Fire and its relationship to ponderosa pine*. Proceedings of the Tall Timbers Fire Ecology Conference, **7**, 127–49.

Weaver, H. (1974). Effects of fire on temperate forests: Western United States. In *Fire and ecosystems*, ed. Kozlowski, T. T. and Ahlgren, C. E., pp. 279–319. New York: Academic Press.

Will-Wolf, S. and Roberts, D. W. (1993). Fire and succession in oak-maple upland forests: A vital attributes modeling approach. In *John Curtis and Wisconsin plant ecology*, ed. Fralish, J. S., R. P. McIntosh and O. L Loucks, pp. 217–36. Wisconsin Academy of Sciences, Arts and Letters Special Issue.

6

Design, behavior and application of LANDIS, an object-oriented model of forest landscape disturbance and succession

David J. Mladenoff and Hong S. He

Introduction

Modeling forest landscape change is challenging because it involves the interaction of a variety of factors and processes, such as climate, succession, disturbance, and management. These processes occur at various spatial and temporal scales, and the interactions can be very complex on heterogeneous landscapes. However, simulation models make it possible to examine assumptions about landscape change explicitly by defining complex processes and their interactions logically and mathematically. More importantly, modeling allows us to deduce results that otherwise cannot be investigated due to their complexity, such as landscape change over long time periods and the ecological ramifications of large disturbances, or diverse management regimes.

The variety of approaches taken to model forest landscapes reflect the diverse backgrounds and objectives of individual researchers (Mladenoff and Baker, Chapter 1). LANDIS has been refined (Mladenoff *et al.*, 1996) as a forest landscape model that integrates forest succession, windthrow, fire, and forest management. LANDIS is a tool to study species-level responses and changes in forest landscape pattern with varied natural and anthropogenic disturbances. LANDIS addresses several needs, including to (i) simulate large (10^4–10^6 ha) landscapes that are heterogeneous in terms of site conditions or environment (landtypes), and initial vegetation conditions at the tree species level, (ii) simulate interaction of dominant forest disturbance regimes, such as fire, windthrow, and harvesting, with species-level forest succession, (iii) adapt to a range of possible scales and map input-data of varied resolutions, and (iv) include spatially explicit ecological interactions, and mechanistic realism, while having modest input parameter needs. These requirements are similar to a degree for most forest landscape models, and cannot all be optimized. The particular needs being addressed by the model drive how these requirements are balanced. These needs are framed by temporal and spatial scale (landscape extent and resolution), data availability, and parameter information for large areas.

The LANDIS model

General characteristics

LANDIS is a spatially explicit and stochastic model that simulates forest landscape change over long time domains and large, heterogeneous landscapes. LANDIS has several key characteristics (Mladenoff *et al.*, 1996; He and Mladenoff, 1999). LANDIS uses a cell-based, or raster data format, a widely used data structure for spatial analysis and modeling (e.g., Green, 1989; Baker *et al.*, 1991; Turner *et al.*, 1994; Keane *et al.*, 1996; Gardner *et al.*, 1996, Chapter 7; Urban *et al.*, Chapter 4). In general, the raster data format is more efficient computationally than the vector, or polygon format (Gao *et al.*, 1996). This makes it possible to incorporate greater mechanistic complexity (Mladenoff *et al.*, 1996). Raster data allow direct input of large-scale, satellite-based forest classification maps (e.g., Wolter *et al.*, 1995), a major source of species input data for large-scale simulations (He *et al.*, 1998). With the raster data format, cell size can be controlled and varied to reflect different spatial resolutions. This is of particular interest, since very often either the question investigated or input data availability imply a certain appropriate cell size. Also, the corresponding operations on vegetation pattern and environmental data layers allow multi-scaled issues to be examined, since aggregating or disaggregating cells are among the standard operations of raster GIS data (e.g., Arc/Info Grid (ESRI, 1996; ERDAS, 1994).

Spatial interactions, such as seed dispersal based on potential distances rather than polygon neighborhoods, can be more accurately simulated with raster data format than vector data (Mladenoff *et al.*, 1996). The LANDIS model is conceptually related to two existing approaches, the plot-level JABOWA-FORET "gap" models (Botkin *et al.*, 1972; Botkin, 1993; Shugart, 1984), and the landscape-scale LANDSIM model (Roberts, 1996; Roberts and Betz, Chapter 5). Both of these previous approaches simulate species succession, although their scale and mechanistic detail differ considerably (Mladenoff and Baker, Chapter 1).

Within each cell, LANDIS tracks the presence/absence of species age cohorts at 10-year time steps rather than the actual number of individual trees. This differs from most gap models, except FORCLIM (Bugmann, 1996), that track individual trees. Use of FORCLIM suggests that realism is not significantly reduced by tracking age cohorts rather than individuals for large-scale applications. Additionally, computational loads are greatly reduced, because actual species abundance, biomass, or density are not being simulated. If such detailed information is desired, it can be added to model output from the growth and yield relationships available in the literature, through a lookup-table relationship, or by linking with an ecosystem process model. In this context, species presence/absence information is relatively more scale-independent than quantitative data. Varying cell sizes has less effect on the way that species information is recorded than does tracking individuals, up to certain model design limits. This provides the basis for the model to be useful at different scales by varying cell size and appropriately scaling spatial interactions such as seed dispersal. A major purpose of LANDIS is to simulate large landscapes,

Table 6.1. *Species life history parameters that drive the model*

Species	sLong	sMat	sC	fireT	effD	maxD	vegP	spAge
Abies balsamea	150	25	5	1	30	160	0	0
Acer rubrum	150	10	3	1	100	200	0.5	150
Acer saccharum	300	40	5	1	100	200	0.1	240
Betula alleghaniensis	300	40	4	2	100	400	0.1	180
Betula papyrifera	120	30	2	2	200	5000	0.5	70
Carya cordiformis	300	30	3	2	30	1000	0.5	220
Fraxinus americana	200	30	4	1	70	140	0.1	70
Picea glauca	200	25	3	2	30	200	0	0
Pinus banksiana	70	15	1	2	20	40	0	0
Pinus resinosa	250	35	2	4	12	275	0	0
Pinus strobus	400	15	3	3	100	250	0	0
Populus grandidentata	90	20	1	2	−1	−1	1.0	90
Populus tremuloides	90	15	1	2	−1	−1	1.0	120
Prunus pensylvanica	30	10	1	1	30	3000	0	0
Prunus serotina	200	20	2	1	30	3000	0.5	140
Quercus alba	400	40	3	4	30	3000	0.5	300
Quercus ellipsoidalis	200	35	2	4	30	3000	1.0	300
Quercus macrocarpa	300	30	2	5	30	3000	1.0	220
Quercus ruba	250	25	3	3	30	3000	0.5	250
Quercus velutina	300	30	2	3	30	3000	1.0	220
Thuja occidentalis	350	30	4	1	45	60	0.5	400
Tilia americana	250	15	4	2	30	120	0.5	250
Tsuga canadensis	450	30	5	3	30	100	0	0

sLong – longevity (years), sMat–age of sexual maturity (years), sC – shade tolerance class (1–5), fireT – fire tolerance class (1–5), effD – effective seeding distance (m), maxD – maximum seeding distance (m), vegP – vegetative reproduction probability, spAge – maximum age of vegetative reproduction (years).

where available input data may be coarse or parameters poorly estimated. A species presence/absence approach avoids any false precision of predicting species abundance measures that may occur with inadequate input data or parameter information.

Our model is similar to LANDSIM (Roberts, 1996) in that successional dynamics are based on species vital attributes or life history characteristics, along with other ecological parameters relating to disturbance and site characteristics (Table 6.1). Similarly, the model is currently based on a 10-year time step. The LANDIS model differs from LANDSIM in several ways that opt for greater mechanistic detail in spatial interactions, with some corresponding increase in computational load. LANDSIM operates on fixed polygon maps, and spatial interactions such as seed dispersal operate on fixed polygon neighborhoods rather than actual distances (Roberts, 1996). This approach is well suited to areas such as mountainous regions where large and steep environmental gradients are amenable to mapping of relatively discrete vegetation patches and habitats, and where such polygons may constitute management units. The LANDIS model was designed to also operate in an environment where vegetation patterns and environmental gradients are less discrete, and to allow flexibility in spatial representation. Because LANDIS operates in a cell-based mode, vegetation patches are not fixed polygons, and can aggregate and disaggregate in response to spatial patterns of stochastic disturbance and succession.

LANDIS simulates disturbances in addition to simulating succession (He and Mladenoff, 1999). Any simulated disturbance is a result of spatially explicit interactions of environment variables, vegetation information, and the nature of the disturbance itself. Simulating windthrow disturbance, an important factor in many forest systems, in combination with fire, appears not to have been previously modeled (Mladenoff *et al.*, 1996). The interaction of these two disturbance types can provide feedbacks in the model and influence resulting successional pathways.

C++, an object-oriented programming language was used in developing LANDIS (Mladenoff *et al.*, 1996; He *et al.*, 1996, He *et al.*, 1999*b*). Programming of the model using hierarchical classes provides flexibility in model design and computational efficiency. The modeling system also includes a graphical user interface (GUI), as well as a freestanding spatial analysis package (APACK) that can be used with map output from LANDIS or other sources (Mladenoff and Dezonia, 1997).

Ecological dynamics

Site (cell) level representation

In LANDIS, a landscape is conceptualized as a lattice or grid of cells of equal size. These cells can be conceived of as an array of sites making up the larger landscape. Any given cell is unique in terms of the environment or landtype (defined below) on which it resides, the species present, the age cohorts of the species, the disturbance history, and fuel regime. Either cell or site is used in our discussion to refer to these units, depending on whether an ecological or map format emphasis is intended. It is also important to keep in mind that the cell size is user defined and can vary. Various ecological processes may simultaneously occur in each cell through time, including species establishment through competitive succession and seeding, wind and fire disturbances, fuel accumulation and decomposition. These processes constantly alter the species information and drive the vegetation dynamics in the cell (Fig. 6.1).

In succession, several parameters are treated as categorical inputs, rather than modeled explicitly, similar to LANDSIM (Roberts and Betz, Chapter 5; Table 6.1). In addition, each species has an *establishment coefficient*, that expresses species relative ability to grow on different site categories or *landtypes*, and are differentiated based on species relative responses to soil moisture, climate, and nutrients. The species establishment coefficients are not themselves modeled within

Fig. 6.1. Succession and disturbance dynamics of LANDIS model at a given site.

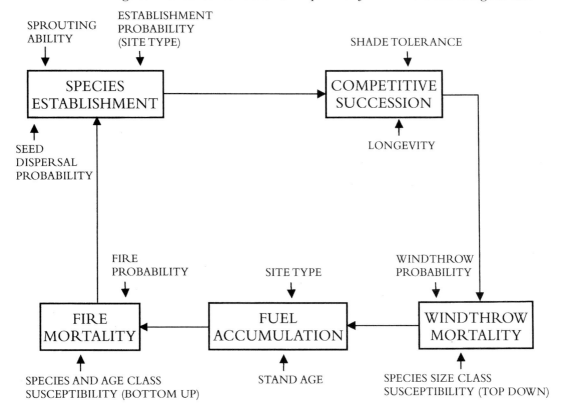

LANDIS. They can be estimated empirically or derived from a gap model with ecosystem-process drivers (He *et al.* 1999a). Similarly, fire characteristics such as severity vary with landtypes, and are based on interactions of productivity, decomposition, and time since last fire. These are also represented as categorical variables. The use of categories at the landscape-scale to represent more computationally complex phenomena at the site level allows the effects of ecological processes to be scaled-up (Roberts and Betz, Chapter 5; He and Mladenoff 1999).

Landscape-scale representations and dynamics

Ecological dynamics such as seeding, and fire and wind disturbances, occur at spatial extents that exceed a single cell, and cannot be defined at the site level. At landscape scales, heterogeneous landscapes are often stratified as ecoregions or landtypes (Fig. 6.2). Landtypes are processed from other GIS layers, and are scaleable corresponding to the question investigated and the data availability. At coarser regional scales (e.g., 10^4–10^6 ha), climatic zones, soil series associations, and general physiography can be used to delineate landtypes, while at finer scales (< 10^4 ha), high resolution data such as digital elevation models (DEMs) and finer soil maps can be used. In LANDIS, landtypes are used to stratify the environment

Fig. 6.2. Ecological processes, data input and output in LANDIS.

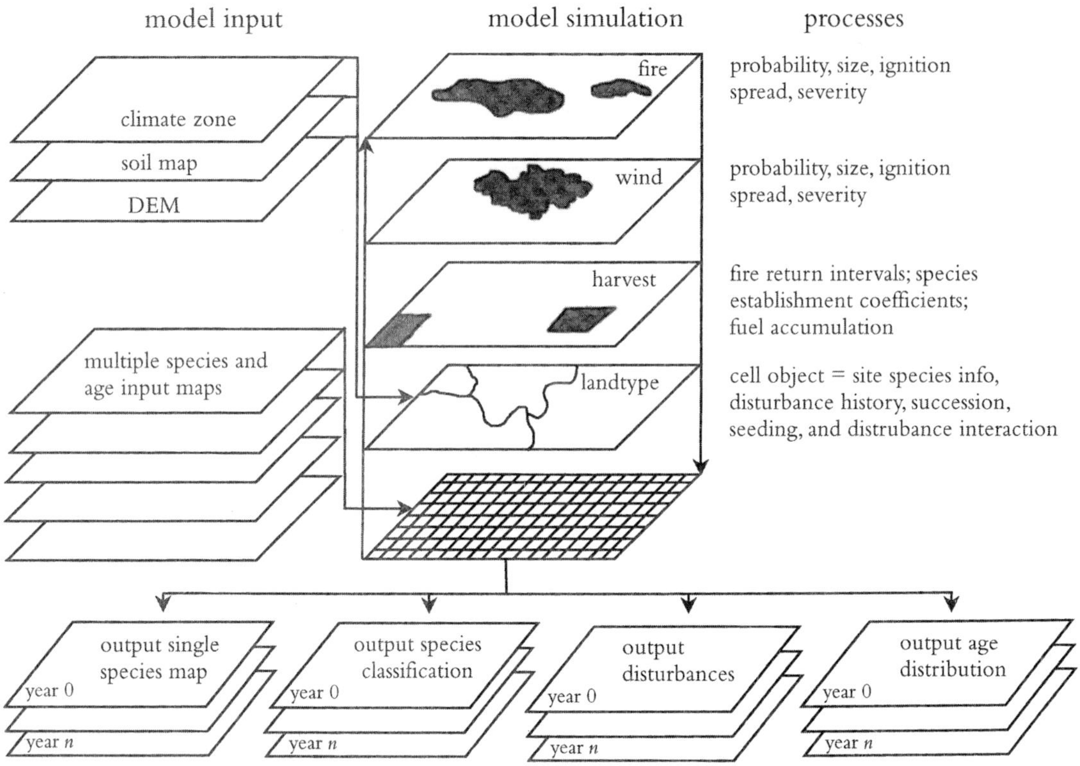

into areas with relatively consistent species establishment, fire characteristics, and fuel accumulation regimes.

Unlike most gap models that assume constant or random seed rain (Shugart, 1984), seed dispersal is explicitly simulated across the landscape in LANDIS. The available seed source is calculated from vegetation patterns by the model, and species seeding distances are explicitly defined. Seed dispersal curves (negative exponential) depict the probability of seed to travel certain distances based on existing literature. These derived curves are implemented in the model as simplified categorical parameters, again reflecting the presence/absence representation of species in cells. The dispersal categories represent effective and maximum seeding distances for each tree species. Within effective seeding distance, seed has high probability of dispersal. On the other hand, the chance for seed to travel farther than its maximum seeding distance is very low (He et al., 1996).

In LANDIS, fire and wind disturbances are simulated based on the historical distribution of disturbance sizes and mean return intervals. This information can be derived from the literature on the region or can be explicitly studied (e.g., Canham and Loucks, 1984; Frelich and Lorimer, 1991; Heinselman, 1973, 1981). Fire and windthrow disturbances may occur at various locations on the landscape, with each event varying in time and form (e.g., extent, shape, severity). Neither the time when a disturbance occurs nor the pattern and shape of the disturbance is deterministic. However, in real landscapes fire and wind disturbances are not purely stochastic events since some landtypes may be more susceptible to disturbance than others. At the landscape scale, spatial interactions between disturbance processes, environment and vegetation dynamics can be more precisely defined, and patterns of re-occurrence based on assigned distributions can be more realistically portrayed over time (He and Mladenoff, 1999).

Fire is simulated as a bottom-up disturbance; fires of increasing severity affect smaller tree age-classes first. Fires of increasing severity affect increasingly larger age-classes. Fire severity is determined by fuel availability, which is based on time since the last fire and landtype characteristics that influence production and decomposition.

Windthrow is predominately a top-down disturbance; with probability of tree cohorts being affected by a wind event increasing with tree age and size. Therefore generally the oldest, tallest canopy tree are affected first by windthrow, and increasingly more severe wind events can remove more of the canopy (younger cohorts lower in the canopy). The time since a windthrow event can also influence the potential fire severity class, depending on decomposition dynamics of the particular landtype. Interactions between these two disturbances can be interesting and complex. Generally, windthrow becomes more important on landtypes with longer-lived species, and where fire frequency is low.

Model structure and object-oriented design

The challenge in designing a model like LANDIS is balancing the representative ecological processes affecting landscape change at appropriate spatial resolutions, with computational efficiency in order to simulate large, heterogeneous landscapes. One relatively new and useful approach involves object-oriented modeling and design (Rumbaugh *et al.*, 1991; Coad and Yourdon, 1991; Varhol, 1992; Paepcke, 1993; Sigfried, 1996; Yourdon and Argilar, 1996). Object-oriented modeling and design techniques allow conceptualizing problems by using objects organized around real-world concepts. It facilitates the modeling process through several important features, including *modularity*, *abstraction*, and *encapsulation*. This approach may also result in a program that is easier to maintain or modify than computer code using more conventional, iterative programming and modeling approaches.

Modularity

The best way to simplify the problem-solving process is to divide a large problem into small, manageable parts or modules. One way to achieve a modular solution is by identifying within a problem various smaller components, called objects, that combine both data and the operations on the data. For example, the problem of forest landscape change can be broken down into species, succession, disturbance, management, environment, etc. An object-oriented approach produces a modular solution, that is, a collection of objects that interact, rather than a sequence of actions (Carrano, 1995).

Abstraction

Abstraction separates the purpose of a module from its implementation. This feature makes it easy to focus on the essential, inherent aspects of an entity and ignores its incidental properties. In model design, this means focusing on what an object is and does, before deciding how it should be implemented. Subsequently, the use of abstraction preserves the freedom to make decisions as long as possible by avoiding premature commitments to details (Rumbaugh *et al.*, 1991). When designing a modular solution to a problem, each module begins as a box that states what it does, but not how it does it. No one box may "know" how any other box performs its task; each box may know only the task of other boxes (Fig. 6.1). Therefore, modularity and abstraction complement each other. Modularity breaks a solution into modules; abstraction specifies each module clearly before implementing it in a programming language.

Encapsulation

Encapsulation consists of separating the external interface from its internal state. The external interface allows objects to communicate with other objects. The internal state of an object is used by that object to perform its duties (Fig. 6.1). Encapsulation prevents a program from becoming so interdependent that a small change has massive ripple-effects (Rumbaugh *et al.*, 1991). The implementation of an object can be changed without affecting the applications that use it. In object-oriented design, encapsulation is facilitated through the design of the *abstracted data type* (ADT). The ADT allows us to define a unique data structure for a specific object, and implement operations on the data. This is important since different objects require different data structures and operations. For example, a fire object may use fire probability, mean fire return interval, disturbance size, mean, maximum and minimum disturbance size as its data structure, and ignition and spread as its operations (He and Mladenoff, 1999). Encapsulation is not unique to object-oriented languages, but the ability to combine data structure and behavior in a single ADT makes encapsulation cleaner and more powerful than in designs using conventional languages that typically separate data structure and behavior.

Model structure

Two issues are often involved in landscape model design: spatial scale and representing the objects to be modeled. No model can comprehensively address all scales. Usually the scale range is decided at the model design stage, and this will affect the way we represent the real world objects in the model (He *et al.* 1999b). Assumptions often have to be made at various levels such as among objects and within objects. For example, between species and fire objects, there are multiple causes that result in mortality. However, in modeling assumptions have to be defined

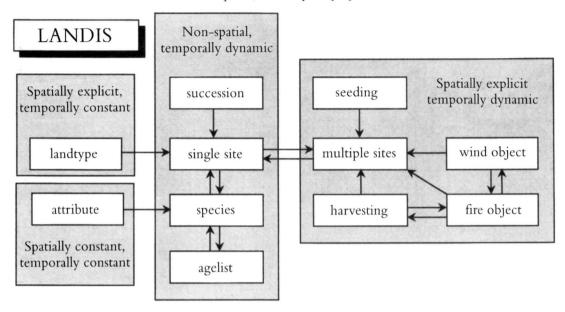

Fig. 6.3. LANDIS overall modular design. LANDIS modules within white boxes are grouped as shaded blocks representing model components that can be spatially explicit or non-spatial, and temporally dynamic or constant.

which simplify the removal of a certain species from the landscape. Within an object, the data type of a given parameter is another type of assumption, since ecological parameters are in the form of nominal (e.g., presence/absence), ranked, interval, or continuous numeric forms. Choosing a logical or mathematical representation for these parameters also entails assumptions and must be based on available information.

Due to the need for simplification and various stochastic components, landscape models are often not designed to predict particular events in a local and deterministic sense; rather they are designed to examine general patterns produced due to different sets of assumptions and interactions. Thus for a given landscape model, having the ability to update assumptions with new knowledge and re-calibrate each component of the model is essential. The objected-oriented design approach facilitates calibration, modification, and testing of the model through modular manipulation (He et al., 1999b).

The overall LANDIS design includes objects and components that may be of several types: they may be spatially explicit or non-spatial, and temporally constant or dynamic (Fig. 6.3). For example, fire, wind, harvest, and seeding are processes operating at spatial extents larger than a single site, especially when operating at smaller cell sizes. These processes are both spatially explicit and temporally dynamic. Succession occurring on every site where forest species exist is a temporally dynamic process. Through succession, the species list and agelist change with time. Landtype is spatially explicit and temporally constant. Usually landscape attributes encapsulated by landtype do not change within the simulation time span. The species attributes are non-spatial and temporally constant (Fig. 6.3). Since the actual interaction among these objects can be fairly complex, it is difficult to present all the possible interactions among all processes. Rather, an endeavor is made to conceptualize the integration of ecological processes at the site or cell level and to scale up to landscape levels.

LANDIS basic components and objects

Succession

Succession dynamics in each cell is a spatially constant, temporally explicit component of LANDIS, but interacts with several spatial components. For each cell or site, *succession* interacts with *species*, species *attributes*, and species age cohorts (*agelist*) objects for every simulation step (Fig. 6.3). *Succession* is a competitive process driven by species life history parameters (Table 6.1). It comprises a set of logical rules modified from the LANDSIM model (Roberts, 1996; Mladenoff *et al.*, 1996). Species competitive ability is mainly the combination of shade tolerance, seeding ability, longevity, vegetative reproduction capability, and the suitability of the landtype to a given species. Within *succession*, species birth, growth, and death are performed under the rules. For example, shade-intolerant species cannot establish on a site where more shade-tolerant species are present. On the other hand, the most shade-tolerant species are delayed in being able to occupy an open site. Without disturbance, shade tolerant species will dominate the landscape given that other attributes are not highly limiting and the environmental condition is generally suitable.

Seeding

Seeding is a spatially and temporally explicit component of LANDIS involving multiple sites or cells. *Seeding* interacts with *species*, *attributes*, *agelist*, and *landtype* objects for every model iteration (Fig. 6.3). Seeding activity can be conceptualized as the following expression:

$$Seeding = f(sMat, effD, maxD, rD, sC, eC, rP)$$

where $sMat$ is the species sexual maturity age, $effD$ is effective seeding distance, $maxD$ is maximum seeding distance, rD is the actual distance of a site from the seed source, sC is the species shade tolerance class, eC is the species establishment coefficient (Table 6.1) (Mladenoff *et al.*, 1996), rP is a random probability 0–1.

Seed can disperse from any cell or site on the landscape where a seed source is available, and a cell can receive seed from other cells. For any given cell, the *seeding* routine checks if any species' *agelist* contains a cohort older than its $sMat$. If such a species exists, *seeding* then looks up the $effD$ and $maxD$ stored in species *attributes*. In general, seed has a high probability of reaching sites within $effD$, one of the

Fig. 6.4. LANDIS site object, showing abstracted data type (ADT) operations.

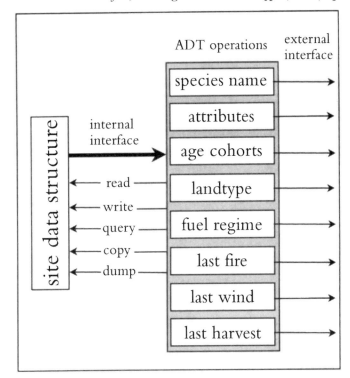

attributes measuring species seeding ability. Beyond this range, seed dispersal probability decreases exponentially with distance, but is expressed in the model as discrete distance classes rather than continuously. Once seed successfully arrives at a given cell at distance rD, rP is drawn from a uniform random number pool, and comparison between eC and rP is then made. The seed can establish in that cell only if $eC > rP$. Several seeding routines are available in LANDIS emphasizing various seeding parameters, and they can be chosen for the appropriate cell sizes (He et al., 1996).

Site object

The *site* object is a conceptual representation of the basic landscape unit (cell) being modeled (Fig. 6.3). The landscape is composed of multiple *sites* or cells with each containing unique information. *Site* interacts with all other objects and components querying site specific information. Responses to the queries are implemented as the operations on the object or external interface of *site* (Fig. 6.4).

Succession, seeding, wind and *fire* disturbance, and *harvesting* all interact with *site*. The *succession* routine queries the *site* for the information related to species list, age cohort, species *attributes* including mature age, longevity, shade tolerance class, and vegetative reproduction probability. The *fire* object queries for information such

Fig. 6.5. LANDIS species object, showing abstracted data type (ADT) operations.

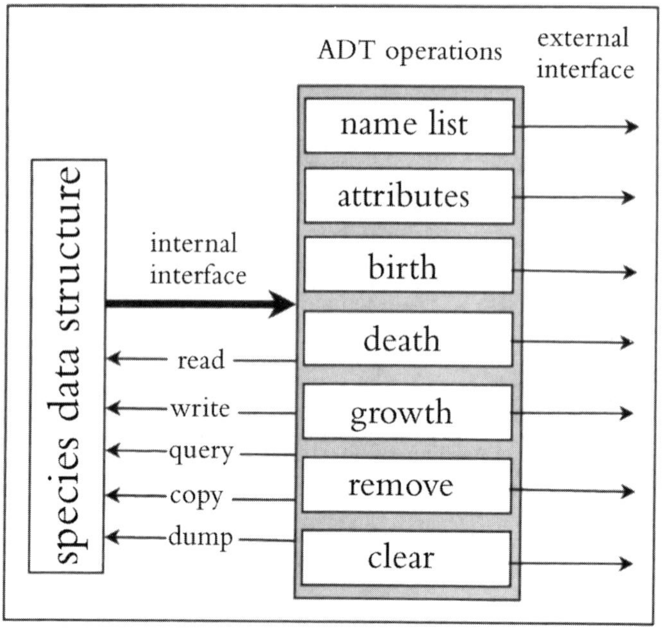

as the mean fire return interval on the site (encapsulated by *landtype*), the time since last fire, current fuel accumulation regime, and species fire-tolerance class. The *seeding* routine queries for the species list, current age cohorts, effective seeding and maximum seeding distance, species shade-tolerance class, species vegetative reproduction probability, and species establishment coefficients on the site. The *site* object uses its internal interface to work with its internal data structure to respond to each external query. The internal interface includes read, write (update), query, copy, and dump functions that are also standard for other objects. These internal operations and the internal data structure of the object are described elsewhere (He *et al.*, 1999b).

Species object

Species is a spatially constant, temporally explicit object (Fig. 6.3). Operations defined for *species* include "query" (for species name and attributes), "birth", "death", "grow", "remove", and "clear" (Fig. 6.5). The "query" operation is site specific since each site may contain a unique species list. The "birth" operation simulates either a new species seeding in, or on-site species regeneration. The latter usually applies to species with high shade tolerance. For some species with vegetative reproduction ability, "birth" simulates vegetative reproduction. The "death" operation typically simulates species reaching maximum longevity. It applies only to the particular age cohort which reaches longevity. "Grow" simulates the species age-class increment during each model iteration. "Remove" simulates one or more

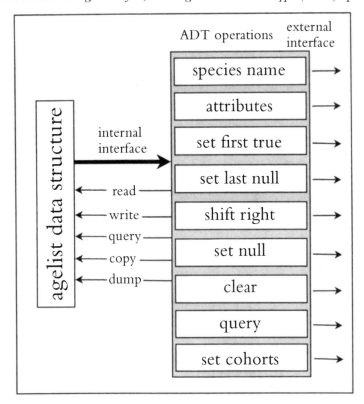

Fig. 6.6. *LANDIS agelist object, showing abstracted data type (ADT) operations.*

age-cohorts of a species removed from the site due to various causes. Disturbances, harvest, and background mortality can all result in removal of certain species age-cohorts. For example, wind disturbance tends to remove older age cohorts, while fire disturbance tends to remove younger age-cohorts. The "remove" operation differs from "death", and can remove any age-cohort of the species. "Clear" simulates the removal of entire age-cohorts of a given species on the site, and usually accompanies severe fire disturbances or clearcut harvest.

Agelist object

Agelist is also a spatially constant, temporally explicit object in the model (Fig. 6.3). *Agelist* can be considered a lower level *species* operation (Fig. 6.6), containing a bit-level data structure of species age-cohorts (He *et al.* 1999b). In programming, *agelist* is the base class of *species*, and *species* inherits the features defined for *ageclass*. "Set first true" simulates "birth", when a new age-cohort is set present at 10-year-old. "Set last null" simulates "death", when the last age-cohort of a species is set to null. "Shift right" simulates "grow", when all age-cohorts increase 10 years by moving rightward in the bit-level data structure. "Set null" simulates "remove" of certain species age-cohorts. "Clear" simulates the removal of entire species

Fig. 6.7. LANDIS landtype object, showing abstracted data type (ADT) operations.

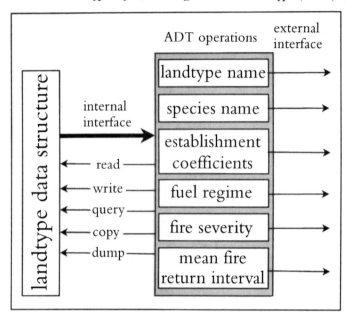

age-cohorts. Other *agelist* operations include "query age-cohort", a query frequently made by many processes, and "set age-cohorts", used to set the initial age information. With data *abstraction*, *agelist* is only accessible through *species*.

Landtype object

Landtype is a static, spatial object (Fig. 6.3). Several parameters are contained within the landtype object (Mladenoff *et al.*, 1996), including: (i) species establishment coefficients, (ii) disturbance characteristics such as mean fire return interval, and (iii) fuel accumulation and decomposition features (Fig. 6.7). These can vary among landtypes, but are homogeneous within a landtype. As previously discussed, landtype is a spatially scaleable object. More explicit simulations can be conducted if more differentiable environmental information is available in finer-grained *landtypes*. For example, if lightning is a primary driver of fire, terrain units which are more likely to have lightning can be processed from high resolution digital elevation models (DEMs). These units can be assigned shorter mean fire return intervals than less susceptible landtypes. Operations requested by other objects to *landtype* include establishment coefficients queried by *seeding*, and fuel regime, fire severity, and mean fire return interval queried by the *fire* object (Fig. 6.7).

Fire object

Fire is a spatially explicit and stochastic object (Fig. 6.3). *fire* simulates disturbance size, probability, ignition, spread, and severity using mathematically defined distri-

Fig. 6.8. LANDIS fire object, showing abstracted data type (ADT) operations.

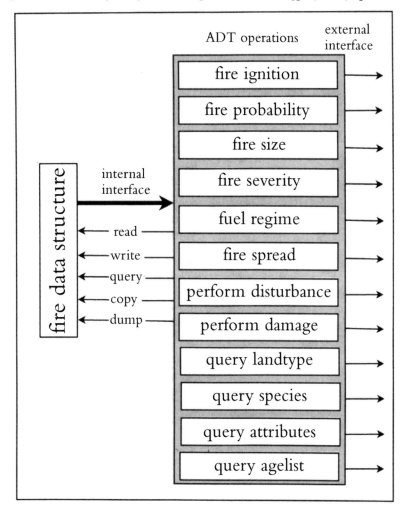

butions in combination with various algorithms. The fire disturbance module and its interaction with succession is described in detail in He and Mladenoff 1999).

Fire size (S, Fig. 6.8), is a function of mean disturbance size (MS) based on the following distribution:

$$S = a\,(10.0)^r \times MS \qquad (1)$$

MS is the estimate of the mean disturbance patch sizes; r is a normalized random number. The parameter a is called the *fire disturbance size coefficient*, and is used for model calibration (He and Mladenoff, 1999). With $r \sim N(0,\sigma^2)$, S follows the log-normal distribution with small disturbances being more frequent than large disturbances. Since r is randomly generated, S bears stochastic features.

Fire probability (P, Fig. 6.8) follows a negative exponential distribution (Eq. 2) based on the mean fire return interval (*MI*), and is a distribution used in other similar studies (Johnson, 1992; Johnson & Gutsell, 1994; Baker *et al.*, 1991; Turner *et al.*, 1993).

$$P = b \times \mathit{lf} \times MI^{-(e+2)} \qquad (2)$$

MI is the mean number of years for fire to recur on a certain site. A smaller *MI* has higher fire probability. The parameter *lf*, denoting the year since last fire, linearly modifies *P*. With larger *lf*, there is a higher fire probability. The parameter *b* is the *fire probability coefficient* used for model calibration (He and Mladenoff, 1999).

Fire ignition (Fig. 6.8) involves an algorithm to randomly locate sites for starting a fire disturbance. The ignition coefficient, a function of map size, determines the number of sites checked during each model iteration. In LANDIS ignition does not necessarily result in a fire. Once a site is chosen for ignition, the fire probability *P* is computed (Eq. 2), and compared to a uniform random number *p* (0–1). Ignition results in a fire if *P*>*p*.

Once ignited, the fire spread algorithm (Fig. 6.8) is activated. Fire can spread randomly to adjacent sites based on susceptibility, but is more likely to spread in the wind direction that was randomly determined at the time of ignition. Fire spread is more computationally intensive, involving: (i) computing *P* of any new site to which the fire spreads, (ii) generating a random number *p*, and (iii) comparing *P* with *p* to determine if the site can burn. If disturbed, species data and the fuel regime of the site will drive the extent of fire damage. Fire-related mortality is determined from species-specific fire tolerance classes (retrieved from species attributes), fire susceptibility (retrieved from agelist), and fire severity class (He and Mladenoff, 1999). Thus, there can be cases of low severity fires where no species are killed, or high severity fires where all species are killed. If damage (species killed) occurs, the time since last fire on the site is set to 0.

The parameter *lf* (Eq. 2) defines the fuel regime on the site (Fig. 6.8). A larger *lf* implies more fuel accumulation and corresponding higher fire severity class. The relationship between *lf* and severity classes is defined for each landtype with the assumption that the fuel accumulation and decomposition rates are homogeneous within a landtype. Windthrow occurring within the previous decade or last several decades can increase fire severity class by adding more fuel to the *site*. On mesic landtypes for example, as time since last windthrow increases, fire severity class returns to a lower level, assuming greater fuel decomposition (Mladenoff *et al.*, 1996).

Wind object

The wind object is similar in design to the fire object (Fig. 6.3) and will not be fully discussed here. Species wind susceptibility classes are approximated based on species' life-spans. Species life-span is divided into five classes according to its respective life-span proportion: 20%, 50%, 70%, 85%, 100%. Thus, susceptibility class one (with age <20% of species life span) is the youngest age class and least susceptible; susceptibility class five (with age >85% of species life span) is the oldest age class and most susceptible. Wind disturbance severity class is categorical from 1 to 5, corresponding to the susceptibility class. Differential windthrow susceptibility by landtype is not currently included in the model.

Program and hardware characteristics

Data format and hardware requirements

LANDIS currently inputs and outputs popular raster GIS file formats, that can be converted either to or from various GIS platforms. LANDIS is a platform-independent application implemented for MS-Win32 families and UNIX operating systems (He et al., 1996). An important advantage of LANDIS is the ability to simulate fairly large landscapes in a reasonably short time. LANDIS can simulate a 500 by 500 pixel landscape for 500 years in minutes. Processing time will vary depending on machine configuration and the complexity of the simulation task (He et al., 1996).

Graphic interfaces

LANDIS has two graphic interfaces, the LANDIS *viewer* and an Arcview™ based graphic user interface (GUI). The LANDIS viewer is a stand-alone program that can display and analyze GIS files created by LANDIS. It can be used to display species' spatial information, fire or wind disturbance patterns, species age-class groups, and forest type maps at each time-step of a simulation. The LANDIS viewer is easy to use, fast, and requires less memory than GIS interfaces. However, the tradeoff is that it cannot perform complex spatial operations, query-by-example, or view large maps (e.g., larger than 500 × 500 pixels). The other optional LANDIS graphic interface, Arcview™ GUI, uses Avenue scripts from Arcview™. By incorporating all of the functions of the LANDIS viewer, this interface is able to integrate maps, tables, and charts. It also has more sophisticated GIS functions.

Spatial analysis program

APACK is a statistical analysis software package useful for calculating various landscape indices. It was initially designed to analyze LANDIS output maps, but it now can be applied to any raster GIS map with various formats (Mladenoff and Dezonia, 1997). APACK can be run under both PC and UNIX platforms. Numerous landscape indices can be individually selected for calculation, including fractal dimension, perimeter/area ratio, average polygon area, diversity, total area by type, average polygon perimeter, dominance, contagion, edge electivity index, electivity for overlays, percolation ratio, angular second moment, inverse different moment, and lacunarity (Mladenoff and Dezonia, 1997).

Model behavior and sensitivity analysis

Methods such as sensitivity analysis or error analysis are commonly used to evaluate ecological models (e.g., Dale *et al.*, 1988; Woodward & Rochefort, 1991; Botkin & Nisbet, 1992). These methods attempt to analyze the model behavior by ranking the parameters according to their contribution to the overall model response. Such an evaluation is useful for a model like LANDIS since it assists evaluation of model results and provides feedback to model design.

Methods

We created a random landscape of 100 × 100 cells (sites), each cell 30 m × 30 m size. Sugar maple (*Acer saccharum*), quaking aspen (*Populus tremuloides*), northern red oak (*Quercus borealis*) and open space were created as four categories randomly distributed on the landscape in approximately equal proportions. This design ensures that seed sources are not limiting anywhere on the landscape for purposes of the tests. In this case, seed of the three species can disperse to any site regardless of which seed dispersal routine is selected. To simplify the simulation, the three species were assigned as 10-year-old age cohorts, and included only one active landtype across the entire landscape. The designed mean fire return interval is 800 years and mean fire disturbance size is about 1 000 000 m^2 (100 ha or 1111 cells).

Model calibration and result verification

Sensitivity tests were performed by adjusting individual parameters ± 10 and 20% in separate model runs, and the results compared to the standard run with unmodified parameter values. These were done for most LANDIS parameters, but our discussion here will focus on two fire-related variables, mean fire return interval (standard run value of 800 years) and mean fire disturbance size (standard run value of 100 ha); and the species establishment coefficient (standard run value of 0.5), which responds to landtype and disturbance. Our discussion will also be limited to one measurement, species areal abundance on the landscape, as the most important response variable in the simulations.

Various stochastic components are built into LANDIS including disturbance, seed dispersal, and seedling establishment. Simulation results from independent runs are presumably different from each other unless a fixed random number seed is used (He *et al.*, 1996). In simulating disturbance, LANDIS uses either empirical or assumed means according to the defined distributions. As reported elsewhere, the variation of the simulated mean fire disturbance size for 20 different runs can be as high as ± 50% at the 85% confidence interval (He and Mladenoff, 1999). Therefore, model calibration and verification are important to ensure that either the empirical or the assumed means are correctly simulated. Conducting multiple, replicate simulations is often not feasible due to the large spatial data sets and long time spans involved. Furthermore, algorithms creating new maps from the multiple, replicate maps are not yet available. Thus, verification and calibration of a single run is important for stochastic models as discussed elsewhere for LANDIS (He and Mladenoff, 1999).

Simulation verification and model calibration were performed following the routine discussed in He and Mladenoff (1999). In this procedure, sensitivity analysis is based on comparisions of output from a model run with a single modified parameter against a base model run. A fixed random number seed is used for both simulations, so that change between the two runs is attributable to the degree of parameter modification and not stochastic effects.

Results

Standard run

A 500-year LANDIS run was conducted with the standard parameters, generating fire and windthrow disturbances based on the assigned return interval parameters. The model randomly generated 19 fires and 20 windthrow disturbances with varied sizes (Fig. 6.9(a)). Fires at years 100, 190, 270, 320, and 330 were relatively large. A very large windthrow event occurred at year 170, with several significant wind disturbances at years 250, 340, 450.

The disturbances interact with other processes such as establishment, and background mortality (when maximum longevity is reached), to affect trajectories of species abundance. Dominant influences on succession for different factors can be observed from the trajectories (Fig. 6.9(b)). At year 0, all species have equal abundance at about one-quarter of the total area. At year 10, aspen reaches its sexual maturity first (20 yr) and begins to seed into the quarter open area. Aspen is not able to seed into cells where oak or maple reside due to its lower shade tolerance. At year 10, aspen abundance doubles, but oak and maple remain unchanged since neither are mature. Red oak is less shade tolerant than sugar maple but more than aspen. At year 20 oak is mature and seeds into the aspen dominated area. Oak abundance then reaches three-quarters of the landscape by year 30. Sugar maple reaches sexual maturity (40 yr) at year 30 of the simulation, and is able to seed into other areas at year 40. Fuel has accumulated and small fires occur at years 70 to 90. These fires reduce the sugar maple, which is most susceptible to fire of the three species. The area of the other species remains relatively unchanged, since red oak and aspen either survive or resprout under these low severity fires. Fire at year 100 removes a significant amount of sugar maple and creates open space. Red oak and aspen both benefit from the fire as shown by the increase in their trajectories for a few decades at year 100.

From year 110 to 160, sugar maple gradually recovers from the major fire, but does not reach its former levels due to small fires elsewhere on the landscape. Red oak and aspen remain relatively stable with oak increasing slightly over aspen. At year 170, the average age of the second generation aspen has reached 60 years old, and first generation red oak reaches 170 years. Oak and aspen are now older and more susceptible to windthrow disturbance. The large windthrow at year 170

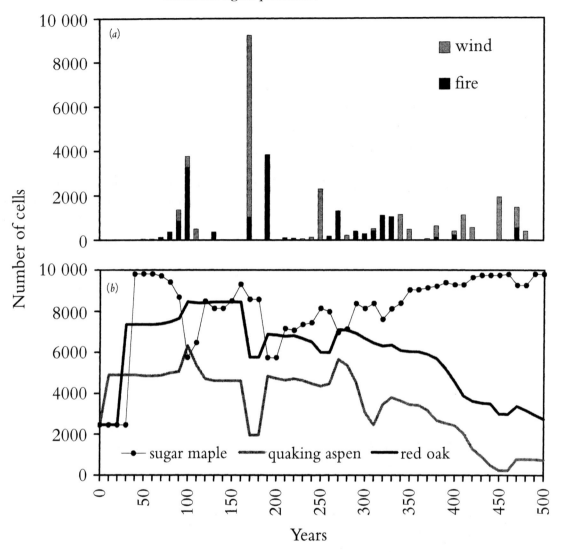

Fig. 6.9. (a) Year and size of fire and wind disturbance under assigned standard parameters. (b) The abundance of sugar maple, quaking aspen and red oak under the assigned standard disturbance regime parameters.

removes a significant amount of aspen and oak from the landscape. Sugar maple in the understory is younger and not greatly affected by the wind. Shortly after the large windthrow, fire susceptibility is enhanced by the greater level of fuels present and a large fire significantly reduces sugar maple at year 190. Aspen benefits most from the fire disturbance and quickly occupies the open space. For the remainder of the simulation several small fires occur, but sugar maple largely out competes red oak and aspen with this disturbance regime, fully occupying the landscape. Red oak substantially declines in abundance, and aspen is nearly removed from the landscape (Fig. 6.9(b)).

Altered mean disturbance size

Responses of sugar maple and aspen are examined for the test runs with altered parameters, although red oak also was included in the simulations. As in the standard run, the response of red oak to disturbance generally remains intermediate to the other two species. Compared with the standard run, fire disturbance size fluctuated as mean disturbance size was adjusted ±10% (Fig. 6.10(a), Eq. 1). In these tests, the dominant roles of each factor (except fire) discussed in the standard run still play approximately the same roles at their respective time steps. To avoid confusion, we will describe the parameter changes as raising and lowering, and the species abundance responses as increases and decreases.

Lowering mean disturbance size by 10% results in an average 1.5% increase in sugar maple abundance, since it is the most shade tolerant and fire sensitive among the three species (Fig. 6.10(b), Table 6.2). Compared with the standard run, quaking aspen decreases in abundance by an average 5.7% (Fig. 6.10(c), Table 6.2). In the second test run, lowering mean disturbance size by 20% (Fig. 6.11(a)), both sugar maple and quaking aspen respond more strongly compared to the +10% scenario. Sugar maple abundance increases an average 3.4% (Fig. 6.11(b)), 1.9% more than the +10% scenario, and quaking aspen decreases 11.9% (Fig. 6.11(c)), a 6.2% greater response than the +10% scenario (Table 6.2).

Raising mean disturbance size by 10% (Fig. 6.10(a)) results in sugar maple abundance decreasing on average 5.0% (Fig. 6.10(b)). Quaking aspen benefits from this decrease, with its abundance increasing an average 8.5%. But the change of abundance of aspen is more obvious than that of maple (Fig. 6.10(c)). Raising mean disturbance size by 20% (Fig. 6.11(a)) results in sugar maple abundance decreasing an average 6.6% (Fig. 6.11(b)), or 1.6% greater decrease than the +10% scenario. Quaking aspen benefited from this decrease, its abundance increasing an average 13.4% (Fig. 6.11(c)), 4.9% more than +10% scenario (Table 6.2).

Altered mean fire return interval

Raising the fire disturbance interval decreases fire probability, and lowering fire disturbance interval increases fire disturbance probability. The results of the changes are not as direct as found in changing mean disturbance sizes (Table 6.3). Mean return interval is used to calculate disturbance probability P, according to Eq. 2, which is site specific.

There is little obvious change between the standard run and the simulation with the mean fire return interval raised 10% (Fig. 6.12(a)). There is one fewer fire over the simulation, with a fire dropped at year 340. The effect of that drop can be observed by the slight increase in sugar maple abundance at year 340 (Fig. 6.12(b)) and a small decrease in aspen abundance at year 340 (Fig. 6.12(c)). Overall, these changes reflect an average abundance increase of 0.9% for sugar maple and 3.5% decrease for aspen (Table 6.2). Raising mean fire return interval by 20% resulted

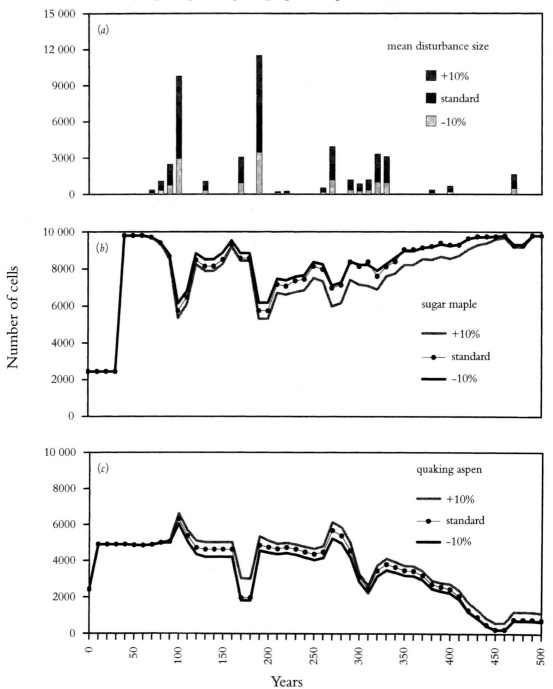

Fig. 6.10. (a) *Simulated disturbances with mean disturbance size ±10%.* (b) *and* (c) *are abundance of sugar maple and quaking aspen in response to* (a).

Table 6.2. *Species abundance changes between standard run and simulations with altered fire parameters at the end of 500-year simulations*

Parameter altered	Sugar maple		Quaking aspen	
	Cells	Change (%)	Cells	Change (%)
MS−10%	8158	+1.5	3335	−5.7
MS+10%	7636	−5.0	3834	+8.5
MI−10%	7337	−8.7	4378	+23.8
MI+10%	8107	+0.9	3412	−3.5
MS−20%	8311	+3.4	3115	−11.9
MS+20%	7503	−6.6	4009	+13.4
MI−20%	7330	−8.8	4378	+23.8
MI+20%	8805	+9.6	2310	−34.7

MS − mean fire size, MI − mean fire return interval. Change (%) is based in comparison of simulations with modified parameters with the standard parameter value run.

in four fires dropped at years 190, 260, 340, and 470 respectively (three more fires dropped comparing with the +10% scenario) (Fig. 6.13(*a*)). Sugar maple largely responds to these changes, especially from the drop of a relatively large fire at year 190 (Fig. 6.13(*b*)). Sugar maple mean abundance increases substantially by 9.6%, 8.7% greater compared with the +10% scenario (Table 6.2). Aspen, on the other hand, responds more abruptly at year 190 (Fig. 6.13(*c*)). Its average abundance decreases significantly by 34.7%, 31.2% more than the +10% scenario.

For the −10% return interval scenario, there are four more fires generated at years 30, 210, 420, and 450 (Fig. 6.12(*a*)). These four extra fires remove substantial amounts of sugar maple (Fig. 6.12(*b*)), resulting in the average abundance of sugar maple decreasing about 8.7% (Table 6.2). Again quaking aspen substantially benefits from the fires with its average abundance increasing 23.8% (Fig. 6.12(*c*), Table 6.2). For the −20% scenario, only a small fire at year 30 was added over the −10% scenario. Both sugar maple and aspen abundance trajectories change little in response to this single disturbance increase (Fig. 6.12(*b*), (*c*) and Fig. 6.13(*b*),(*c*)).

Altered species establishment coefficient

Average species abundance responses reflect the same directional trend expected from increasing or decreasing their establishment coefficients. Raising the species establishment coefficients by 10% results in 1.5, 4.3, and 11.5% increases of sugar maple, red oak, and quaking aspen, respectively (Fig. 6.14(*a*), (*b*), (*c*)). Raising them by 20% results in 3.0, 6.9, and 13.9% greater increase for these three species respectively (Fig. 6.15(*a*), (*b*), (*c*)). A 10% lowering of species establishment coefficients results in 4.7, 4.3, and 7.8% decrease of sugar maple, red oak, and quaking aspen respectively (Fig. 6.14(*a*), (*b*), (*c*)), while greater magnitudes of change were observed for all species with the −20% alteration of establishment coefficients (Fig. 6.15(*a*)–(*c*)).

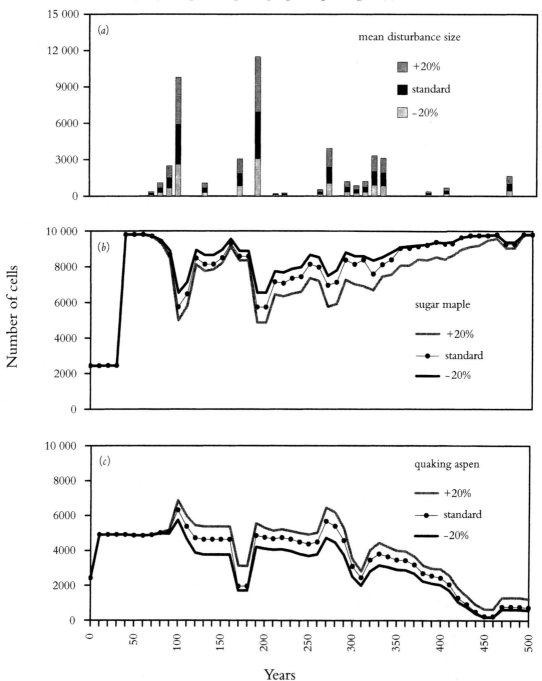

Fig. 6.11. (a) Simulated disturbances with mean disturbance size ±20%. (b) and (c) are abundance of sugar maple and quaking aspen responding to (a).

Table 6.3. *Species abundance changes corresponding to altered establishment coefficient (e) sensitivity at the end of 500-year simulations*

Establishment coefficient	Sugar maple (%)		Red oak (%)		Quaking aspen (%)	
	Abundance	Change	Abundance	Change	Abundance	Change
e+10%	71.5	+1.5	59.3	+4.3	25.2	+11.5
e−10%	67.2	−4.7	54.5	−4.3	20.8	−7.8
e+20%	72.6	+3.0	60.8	+6.9	25.8	+13.9
e−20%	64.9	−8.0	50.5	−11.3	17.9	−20.6

Summary

Model behavior observed for all scenarios indicate that the model responds reasonably to the parameter changes. The raising and lowering of mean fire return intervals do not produce proportional results in the increase and decrease of species abundance compared with the standard run (Table 6.2). The reasons include both stochastic factors and the mechanism of the fire probability equation (Eq. 2). As discussed in *fire object*, when $P > p$ (a LANDIS generated random probability) disturbance is initiated. But even when P is lowered due to the +10 or 20% altered mean fire return intervals (MI) scenarios, P is still generally larger than the random probability p, and therefore most disturbances are still performed. Secondly, linearly changing MI does not result in a linear response of P, since P follows a negative exponential distribution (Eq. 2). Generally, raising fire return interval will produce smaller changes than lowering the return interval (He and Mladenoff, 1999). For the same reason, mean disturbance size (MS) does not respond linearly to the adjustments. This behavior can be modified by altering the distributions used in the algorithm.

The raising and lowering of species establishment coefficients do not always result in an equal response in species abundance (Table 6.3) due to the differences in species life history characteristics and competition on occupied landtypes. For dominant shade-tolerant species, such as sugar maple, the +10% scenario did not create a significant increase. The response of red oak under +10% falls in the middle. Aspen, the least shade-tolerant species, responds the strongest in all cases. The most shade-tolerant species (sugar maple) is less affected by such changes, and the least competitive but most opportunistic species (aspen) responds the greatest.

Application

A fuller demonstration of the model is provided here by simulating a large, heterogeneous landscape with multiple landtypes and fuller complement of species and age classes. A large portion of northern Wisconsin was simulated, an area within the northern US Great Lakes States. This region has been highly altered by human activity in the last century, and is dominated by young, second- and third-growth

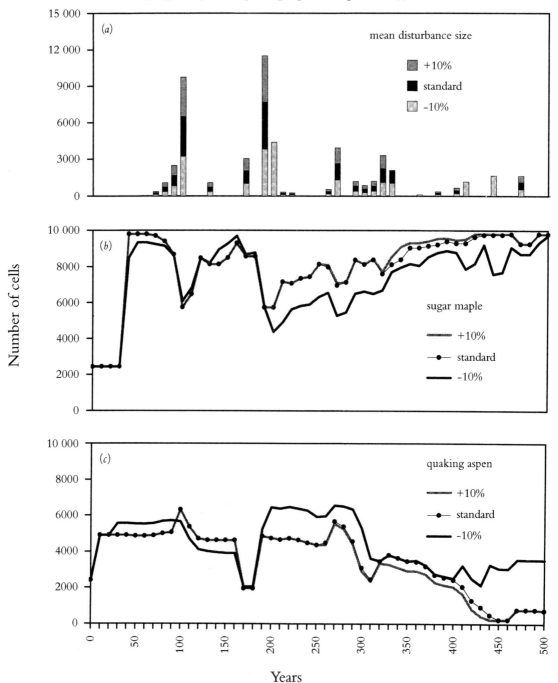

Fig. 6.12. (a) Simulated disturbances with mean fire return interval ±10%. (b) and (c) are abundance of sugar maple and quaking aspen in response to (a).

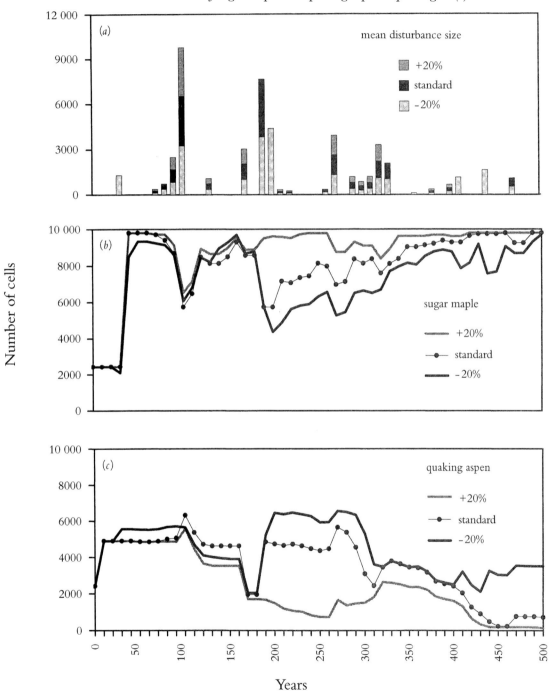

Fig. 6.13. (a) *Simulated disturbances with mean fire return interval ±20%.* (b) *and* (c) *are abundance of sugar maple and quaking aspen responding to* (a).

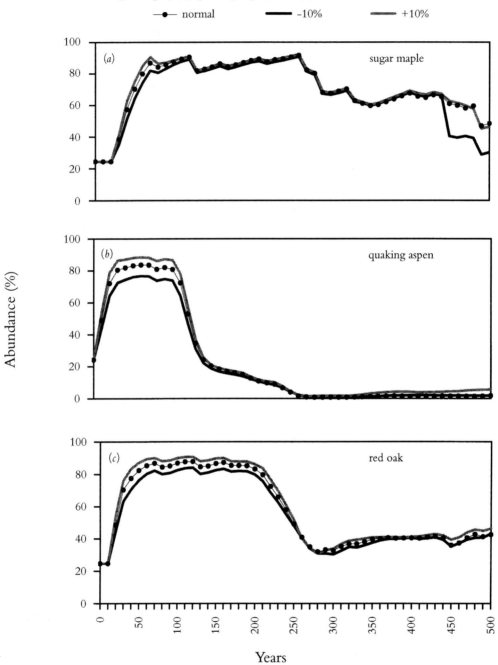

Fig. 6.14. Species abundance response to establishment coefficient changes of ±10% for (a) sugar maple, (b) quaking aspen, and (c) red oak.

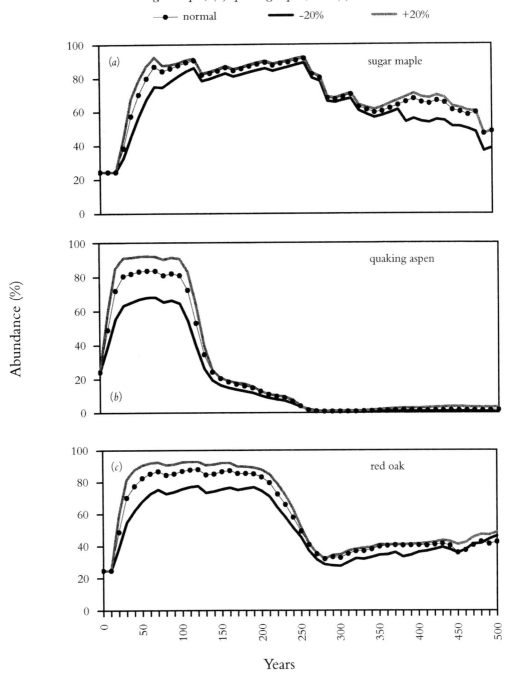

Fig. 6.15. Species abundance response to establishment coefficient changes of ±20% for (a) sugar maple, (b) quaking aspen, and (c) red oak.

forests (Mladenoff and Pastor, 1993). This application of the model addresses the question of how the regional landscape would recover from the current condition if natural successional processes operated, both with and without fire and wind disturbance. This simulation provides a baseline projection of what could occur without continued forest harvesting.

Methods and input data

Our study area comprises about 1.5 million hectares in northwestern Wisconsin. Ecologically, the area is in the transitional zone between boreal forest to the north and temperate forests to the south (Curtis, 1959; Pastor and Mladenoff, 1992). Twenty-three forest species were incorporated in the simulation, with dominant species spatial distributions derived from a species-level classification of Landsat TM imagery (Wolter *et al.*, 1995). This layer maintains dominant species patch structure on the landscape. Secondary species were assigned primarily from their association with the canopy dominants and importance in the US Forest Service Eastwide Forest Inventory and Analysis (FIA) database for the study area (Hansen *et al.*, 1992). Species age class information and associated species spatial distributions were derived by combining TM imagery, FIA, and ecoregions, and assigned probabilistically (He *et al.*, 1998, 1999). Landtype data for this area were available from a quantitative ecoregion classification (Host *et al.*, 1996). The entire study area comprises about 850×550 cells with the cell size of $200 \text{ m} \times 200 \text{ m}$, 10 ecoregions (landtypes), and 194 map input classes.

LANDIS runs were conducted with no disturbance (both wind and fire disturbance turned off), and with both disturbances on. When simulating disturbances, moderate disturbance regimes were assumed for both fire and wind. The mean disturbance size (MI) for fire is set to about 1.5% of the total area and maximum fire size is <12% of the total area. Mean return intervals (MI) for fire vary among landtypes from 200 to 1000 years based on the literature (e.g., Canham and Loucks, 1984; Frelich and Lorimer, 1991; Heinselman, 1973, 1981). Wind disturbance is set more diffuse than fire. Mean disturbance size (MS) for wind is set to about 1.0% of the study area, and maximum wind disturbance size is about 4% of the study area. The mean disturbance interval (MI) for wind is set at 1000 years.

Model calibration and evaluation was carried out according the procedure described above (see *Model behavior and sensitivity analysis* and *Model calibration and evaluation*) where single model runs are compared using paired simulations with a fixed random number seed (detailed by He and Mladenoff, 1999). For a given parameter, the simulated mean (M') is assessed for the degree to which it approximates the known mean (M) from historical or empirical data of the study area. M' can be described as a proportion (of M) (Guertin and Ramm 1996) where Accuracy = $M'/M \times 100$. For example, disturbance size coefficient a (Eq. 1) and disturbance probability coefficient b (Eq. 2) can be adjusted accordingly to ensure acceptable model accuracy within a predetermined range, and that the model assumptions are being correctly simulated. Percent accuracy (Guertin and Ramm,

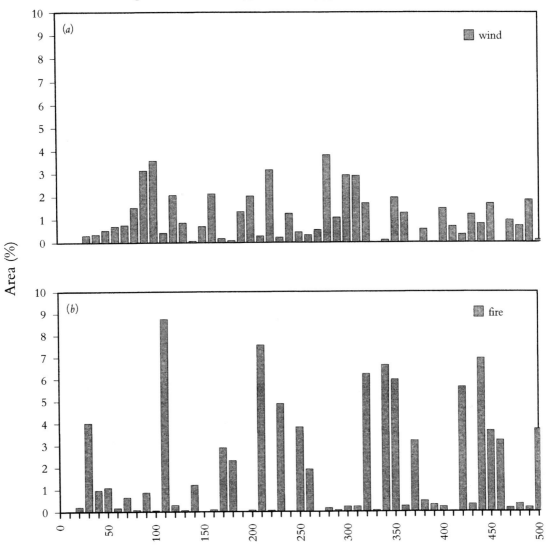

Fig. 6.16. Simulated (a) wind and (b) fire disturbances after calibration.

1996) assesses the similarity of the modified run to the base run (He and Mladenoff, 1999).

Results and analysis

Wind

Various windthrow events occurred during the 500-year LANDIS run (Fig. 6.16(a)). The simulated MS is 4034 cells, which is approximately an 85% accuracy based on the known MS set in the simulation. Simulated wind MI is 910 years, which is 91% accuracy of the MI designed for this area. Windthrow has a stronger effect on altering species age structure than it does on species composition, but at some locations stronger windthrow events, such as severity classes 4 and 5, may create gaps in even aged stands. The shapes of windthrow events are not deterministic. Rather, they are the result of interactions of site condition, species age information, and wind severity classes. For example, there are several patches of windthrow at year 160 with different shapes and severity classes (Fig. 6.17(a): see color section). The cumulative windthrow map (Fig. 6.17(b)) illustrates that windthrow creates a more diffuse, salt-and-pepper pattern throughout the landscape than does fire, which generally spreads contiguously.

Fire

Fire is more difficult to simulate than wind since it has greater variation in disturbance sizes (Fig. 6.16(b)) and mean return interval among landtypes. On landtypes with 200-year MI, such as landtype 5, the simulated MI is 271 years (78% accuracy); on landtypes with 500-year MI, including landtypes 4,7, 8, and 11, the simulated MI is 542 years (91% accuracy) on landtypes with 800 year MI including landtypes 3 and 7, simulated MI is 670 years (84% accuracy); on landtypes with 1000-year return interval including landtypes 2 and 6, simulated MI is 1053 years (95% accuracy). The spatially explicit descriptions of fires can be shown as a map of examples (Fig. 6.17(c),(d): see color section).

Less severe fires can also alter species age structures by removing younger age-cohorts. Stronger severity fires have greater effects on the landscape, which often result in new patches of different species composition. For example, the 420-year fire occurs in several patches randomly located on the landscape, with fire severity classes 2, 3, and 5 (Fig. 6.17(c)). Impacts of these fires can be examined at the individual species level. Both red pine and jack pine are fire-intolerant species and common on landtype 5, the pine barrens. Red pine is substantially killed by the severity 3 fire spreading on to the barrens (Fig. 6.18(a): see color section). The open space created makes it possible for jack pine, an early successional species, to become established (Fig. 6.18(b): see color section). The fire primarily on landtype 9 is a severity class 3 fire (Fig. 6.17(c)). The open space created by this fire allows quaking aspen, an early successional species favored on this landtype, to establish itself (Fig. 6.18(c)). Northern red oak is also able to seed into the open area due to the surrounding seed source availability (Fig. 6.18(d): see color section).

More general fire impacts can be examined at the community level. Before year 420, pine is dominant on the barrens (landtype 5), and maple is dominant on landtype 6 (Fig. 6.19(a),(b): see color section). At year 420 after the fire occurred, the patch on landtype 5 is still dominated by pine since other species have low establishment probabilities on this landtype (Fig. 6.19(c): see color section). On landtype 6, aspen and other hardwoods benefit from the severity 5 fire and are able to invade (Fig. 6.19(c)).

Succession dynamics

As discussed in the sensitivity analysis, species abundance trajectories are a result of seeding and establishment, mortality, wind and fire disturbance, and competition. The dominant role of each factor can be observed at different stages of forest succession from the individual species trajectories. In the following, some of the most common deciduous and conifer species in the area are focused on.

The magnitude of the impact of disturbance on different species varies. The most shade-tolerant species such as sugar maple, hemlock, and balsam fir are not generally favored by fire disturbance (Fig. 6.20(a),(b),(c)). Sugar maple remains dominant on landtypes other than 5 and 9, since the fire return interval on the other landtypes is 500 years and above. At year 500, sugar maple is able to spread to 35% of the landscape without fire, and 25% with fire (Fig. 6.20(a)). Hemlock abundance is also usually negatively affected by fires. It generally increases, reaching about 11% of the landscape at year 500, about 10% less than without fire disturbance (Fig. 6.20(b)). The low abundance percentage of hemlock at year 0 indicates the historical cutting and current human impacts on this species (Mladenoff and Pastor, 1993; Mladenoff and Stearns, 1993). Our simulation suggests that, even by restoring pre-European disturbance regimes, former dominant species in our region such as hemlock, yellow birch, oak, and pine require 100–500 years to recover their former landscape equilibrium proportions. Human alteration of these landscapes to a degree that limits seed sources contributes to slow species recovery, along with altered disturbance regimes.

White pine, a relatively widespread species, needs fire to become established on most landtypes. With fire disturbance its abundance reaches about 9% of the landscape at year 500 (Fig. 6.20(d)). Its low occurrence at year 0 also reflects previous harvesting of this species. Jack pine, a species primarily on pine barrens, responds positively to fires (Fig. 6.20(e)). Jack pine depends on fire to remove red pine and oak, which are more shade-tolerant competitors of jack pine on the barrens. Aspen, red oak, and yellow birch all show positive responses to fire throughout the landscape with their abundances reaching around 8% (Fig. 6.20(f), (g), (h)).

Application summary

This demonstration of the LANDIS model illustrates application to a large, heterogeneous landscape, with the simulation based on current forest input. Simulating this northern Wisconsin landscape demonstrates landscape recovery from current conditions, which is a landscape highly altered by human use during the past century (Mladenoff and Pastor, 1993). We simulated the interaction of fire and windthrow disturbances, which are the two dominant disturbance modes operating in this region. Fire and windthrow interact with species distribution, abundance and seed sources to drive successional change on the landscape. The spatial pattern of the current, human-dominated landscape is shown to alter natural dynamics by dramatically reducing and increasing different tree species, their age-class distribu-

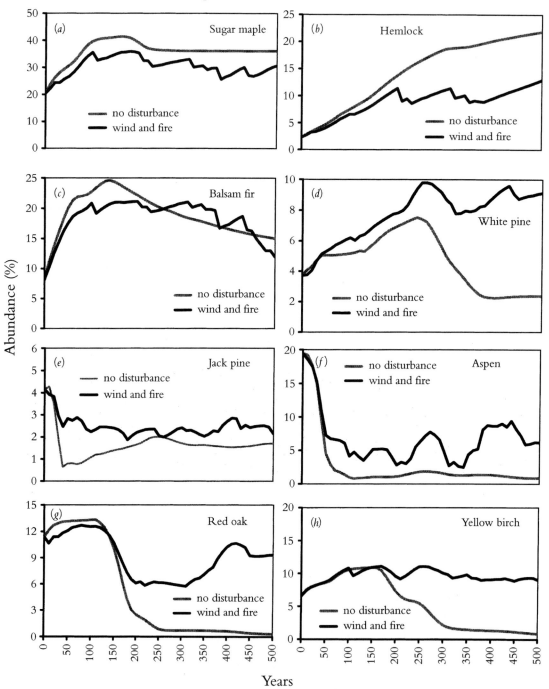

Fig. 6.20. Comparison with and without wind and fire disturbances for (a) *sugar maple,* (b) *hemlock,* (c) *balsam-fir,* (d) *white pine,* (e) *jack pine,* (f) *aspen,* (g) *red oak, and* (h) *yellow birch.*

tion, and seed sources. The simulation suggests that hundreds of years would be required for the forest landscape to return to approximate pre-settlement conditions, even without continued harvesting.

Conclusions

The design and behavior of LANDIS has been described, illustrating the model ability to integrate fire and wind disturbance with forest succession, and simulate forest landscape change at the tree species level. With object-oriented design, the model provides flexibility of upgrading as our knowledge of ecological processes increases. New modules such as timber harvest, which is completed (E. J. Gustafson *et al.*, unpublished data), insect defoliation, or forest disease can be added to the model without affecting model integrity. The model responded reasonably to the parameter changes in the sensitivity analysis. Changes of ±20% show greater responses than the corresponding ±10% scenario. The model does not produce equal or proportional responses with raising or lowering of some parameters due to both stochastic and non-linear mechanisms built into the model.

With built-in stochastic components and the semi-quantitative method employed to record species age cohort information, LANDIS is not designed to predict the occurrence of a given event or change on a single real location. The model is best viewed as tool for projecting plausible landscape patterns resulting from different simulated assumptions and scenarios. Such models are useful for increasing understanding of the complex interactions that occur on landscapes (Dale, 1998). If realistically parameterized and properly calibrated, LANDIS can be used to examine the trend and pattern of change over long time periods. It can be effectively used to examine forest landscape change under a set of assumptions, such as severe fire disturbance, climate warming, or various management scenarios. The model preserves the flexibility of coping with various input data at a variety of scales. This can be an important feature where not all desired data are available for a study region.

The model application to a real landscape shown above with only natural processes operating, can serve as a useful baseline against which to assess various landscape management or other change scenarios. Current or projected forest harvesting can be simulated to project landscape-scale consequences of forest management scenarios (Gustafson *et al.*, 1999). The model can also be used to assess effects of climate warming on the forest landscape, by linking LANDIS with an ecosystem process model that can directly accept climate variables, producing species establishment coefficients for current and changed climate conditions (He *et al.* (1999a).

References

Baker, W. L., Egbert, S. L. and Frazier, G. F. (1991). A spatial model for studying the effects of climatic change on the structure of landscapes subject to large disturbances. *Ecological Modelling*, **56**, 109–25.

Botkin, D. B. (1993). *Forest dynamics: An ecological model.* Oxford, UK: Oxford University Press.

Botkin, D. B. and Nisbet, R. A. (1992). Forest response to climatic change – effects of parameter estimation and choice of weather patterns on the reliability of projections. *Climatic Change,* **20**, 87–111.

Botkin, D. B., Janek, J. F. and Wallis, J. R. (1972). Some ecological consequences of a computer model of forest growth. *Journal of Ecology,* **60**, 849–72.

Bugmann, H. K. M. (1996). A simplified forest model to study species composition along climate gradients. *Ecology,* **77**, 2055–74.

Canham, C. D. and Loucks. O. L. (1984). Catastrophic windthrow in the presettlement forests of Wisconsin. *Ecology,* **65**, 803–9.

Carrano, F. M. (1995). *Data abstraction and problem solving with C++—walls and mirrors.* Redwood City, CA: The Benjamin/Cummings Publishing Company, Inc.

Coad, P. and Yourdon, E. (1991). *Object-oriented design.* Englewood Cliffs, NJ: Yourdon Press.

Curtis, J. T. (1959). *The vegetation of Wisconsin: An ordination of plant communities.* Madison, WI: University of Wisconsin Press.

Dale, V. H. (1998). Models provide understanding, not belief. *Bulletin Ecological Society of America,* **79**, 169–70.

Dale, V. H., Jager, H. L., Gardner, R. H. and Rosen, A. R. (1988). Using sensitivity and uncertainty analyses to improve predictions of broad-scale forest development. *Ecological Modeling,* **42**, 165–78.

ERDAS (1994). *ERDAS field guide, Version 7.5.* Atlanta, GA: ERDAS Inc.

ESRI (1996). *Arc/info grid commands.* Redlands, CA: ESRI Inc.

Frelich, L. E. and Lorimer, C. G. (1991). Natural disturbance regimes in hemlock hardwood forests of the upper Great Lakes region. *Ecological Monographs,* **61**, 145–64.

Gao, P., Zhan, C. and Menon, S. (1996). An overview of cell-based modeling with GIS. In *GIS and environmental modeling: progress and research issues,* ed. M. F. Goodchild., L. T. Steyaert, and B. O. Parks, pp. 315–24. Fort Collins, CO: GIS World Books.

Gardner, R. H., Hargrove, W. W., Turner, M. G. and Romme, W. H. (1996). Climate change, disturbances and landscape dynamics. In *Global change and terrestrial ecosystems,* ed. B. H. Walker and W. L. Steffen, pp. 149–72. IGBP Book Series No. 2. Cambridge, UK: Cambridge University Press.

Green, D. G. (1989). Simulated effects of fire, dispersal and spatial pattern on competition within forest mosaics. *Vegetatio,* **82**, 139–53.

Guertin, P. J. and Ramm, C. W. (1996). Testing Lake States TWIGS: five-year growth projections for upland hardwoods in northern lower Michigan. *Northern Journal of Applied Forestry,* **13**, 182–8.

Hansen, M. H., Frieswyk, T., Glover, J. F. and Kelly, J. F. (1992). The eastwide forest inventory data base: user's manual. USDA Forest Service, General Technical Report NC 151.

He, H. S., Mladenoff, D. J. and Boeder, J. (1996). LANDIS, a spatially explicit model of forest landscape disturbance, management, and succession – LANDIS 2.0 users' guide. Department of Forest Ecology and Management, University of Wisconsin–Madison, Madison, WI, USA.

He, H. S. and Mladenoff, D. J. (1999). Dynamics of fire disturbance and succession on a heterogeneous forest landscape: a spatially explicit and stochastic simulation approach. *Ecology*, **80**, 80–99.

He, H. S., Mladenoff, D. J., Radeloff, V. C. and Crow, T. R. (1998). Integration of GIS data and classified satellite imagery for regional forest assessment and landscape planning. *Ecological Applications*, **8**, 1072–83.

He, H. S., Mladenoff, D. J. and Crow, T. R. (1999a). Linking an ecosystem model and a landscape model to study forest species response to climate warming. *Ecological Modelling*, 112: 213–33.

He, H. S., Mladenoff, D. J. and Boeder, J. (1999b). Object-oriented design of LANDIS, a spatially explicit and stochastic forest landscape model, *Ecological Modelling*.

Heinselman, M. L. (1973). Fire in the virgin forests of the boundary waters canoe area, Minnesota. *Quaternary Research*, **3**, 329–82.

Heinselman, M. L. (1981). Fire intensity and frequency as factors in the distribution and structure of northern ecosystems. In *Fire regimes and ecosystem properties*, ed. H. A. Mooney, T. M. Bonnicksen, N. C. Christensen, pp. 7–57. USDA Forest Service, General Technical Report WO-26.

Host, G. E., Polzer, P. L., Mladenoff, D. J., and Crow, T. R. (1996). A quantitative approach to developing regional ecosystem classifications. *Ecological Applications*, **6**, 608–18.

Johnson, E. A. (1992). *Fire and vegetation dynamics: studies from the North American boreal forest*. Cambridge, UK: Cambridge University Press.

Johnson, E. A. and Gutsell, S. L. (1994). Fire frequency models, methods and interpretations. *Advances in Ecological Research*, **25**, 239–86.

Keane, R. E., Morgan, P. and Running, S. W. (1996). Fire-BGC – a mechanistic ecological process model for simulating fire succession on coniferous forest landscapes of the northern Rocky Mountains. USDA Forest Service, Research Paper INT-RP-484.

Keane, R. E., Arno, S. F. and Brown, K. J. (1989). FIRESUM – an ecological process model for fire succession in western conifer forests. USDA Forest Service, General Technical Report INT-266.

Mladenoff, D. J. and Dezonia, B. (1997). *APACK 2.0 user's guide*. Department of Forest Ecology and Management, University of Wisconsin-Madison, Madison, WI, USA.

Mladenoff, D. J., Host, G. E., Boeder, J. and Crow, T. R. (1996). LANDIS: a spatial model of forest landscape disturbance, succession, and management. In *GIS and environmental modeling: Progress and research issues*, ed. M. F. Goodchild., L. T. Steyaert and B. O. Parks, pp. 175–80. Fort Collins, CO: GIS World Books.

Mladenoff, D. J. and Stearns, F. (1993). Eastern hemlock regeneration and deer browsing in the northern great lakes region – a re-examination and model simulation. *Conservation Biology*, **7**, 889–900.

Mladenoff, D. J. and Pastor, J. (1993). Sustainable forest ecosystems in the northern hardwood and conifer forest region: concepts and management. In *Defining sustainable forestry*, ed. G. H. Aplet, N. Johnson, J. T. Olson and V. A. Sample, pp. 145–80. Washington DC: Island Press.

Paepacke, A., ed. (1993). Object-oriented programming – the CLOS perspective. Cambridge, MA: The MIT Press.

Pastor, J. and Mladenoff, D. J. (1992). The southern boreal-northern hardwood forest border. In *A system analysis of the global boreal forest*, ed. R. L. Shugart and G. B. Bonan, pp. 216–40. Cambridge, UK: Cambridge University Press.

Roberts, D. W. (1996). Modeling forest dynamics with vital attributes and fuzzy systems theory. *Ecological Modeling*, **90**, 161–73.

Rumbaugh, J., Blaha, M., Premerlani, W., Eddy, F. and Lorensen, W. (1991). *Object-orientated modeling and design*. New York, NY: Prentice-Hall Inc.

Shugart, H. H. (1984). *A theory of forest dynamics: the ecological implications of forest succession models*. New York: Springer-Verlag.

Sigfried, S. (1996). *Understanding object-oriented software engineering*. New York, NY: IEEE Press.

Turner, M. G., Romme, W. H. and Gardner, R. H. (1994). Landscape disturbance models and the long-term dynamics of natural areas. *Natural Areas Journal*, **14**, 3–11.

Turner, M. G., Romme, W. H., Gardner, R. H., O'Neill, R. V. and Kratz, T. K. (1993). A revised concept of landscape equilibrium: Disturbance and stability on scaled landscapes. *Landscape Ecology*, **8**, 213–27.

Varhol, P. D. (1992). *Object-oriented programming – the software development revolution*. Charleston, SC: Computer Technology Research Corp.

Wolter, P. T., Mladenoff, D. J., Host, G. E. and Crow, T. R. (1995). Improved forest classification in the Northern Lake States using multi-temporal Landsat imagery. *Photogrammetric Engineering and Remote Sensing*, **61**, 1129–43.

Woodward, F. I. and Rochefort, L. (1991). Sensitivity analysis of vegetation diversity to environmental change. Research Letter, *Global Ecology and Biogeography Letters*, **1**, 7–23.

Yourdon, E. and Argilar, C. (1996). *Case studies in object oriented analysis and design*. Upper Saddle River, NJ: Yourdon Press.

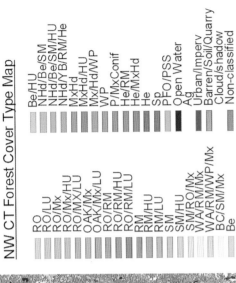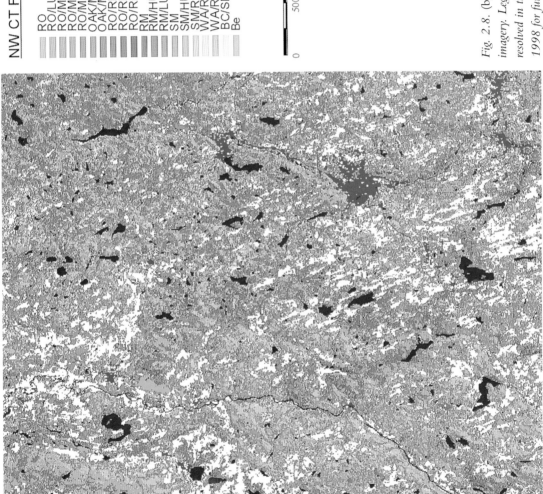

Fig. 2.8. (b) Forest classification map derived from Landsat TM imagery. Legend shows the set of coverages, color coded, that were resolved in the image classification analysis (see Mickelson et al., 1998 for further details). See p.33 for legend key.

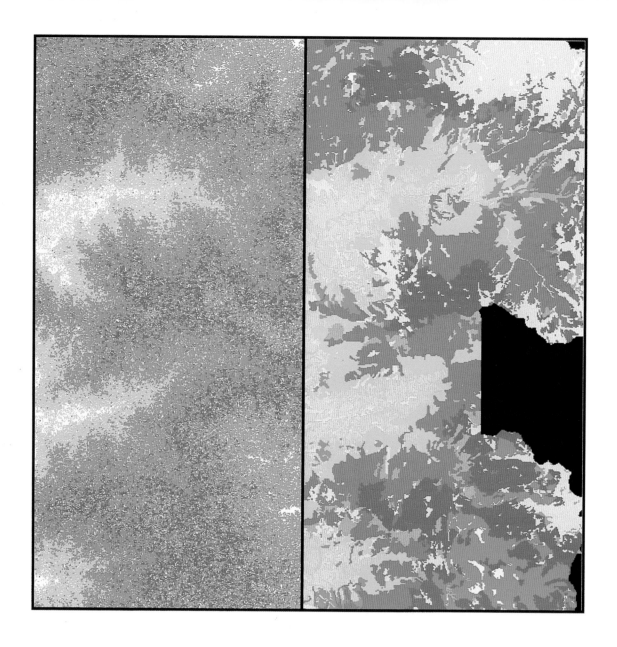

Fig. 4.4 Sample of Sequoia – Kings Canyon National Park as simulated at 30-m resolution by MetaFor (left) and as a vegetation map created by the Park (right). Cover types do not correspond between maps, but brown and orange shades are foothills chaparral and hardwoods; bright green, Ponderosa pine types; darker greens, white and red fir types; teal green, high-elevation conifers; white and beige, barren ground. The black area in the map on right is missing data. The striking point of this comparison is that neither image is a true depiction of the Park's vegetation.

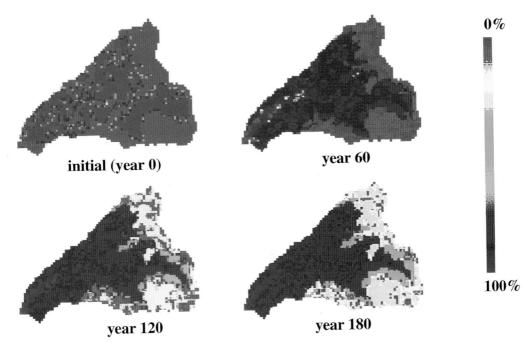

Fig. 4.7. Example simulations of the H.J. Andrews Forest, a ~6300-ha watershed in the central Cascades of Oregon, as simulated with the Mosaic model (from Ablan, 1997).

Fig. 4.10 Simulated conditions 60 years from the present, for the Horse Creek Watershed located in the central Oregon Coast Range. Ten, 20, and 20% of the harvest units were treated at years 1, 25, and 50, respectively, using a dispersed harvest strategy. Stand-level treatments consisted of removing 20% (left panel) or 60% (right) of the overstory basal area. Arrows indicate the tendency of different removal levels to favor conifers (left, at 20%) or hardwoods (right, at 60% removal).

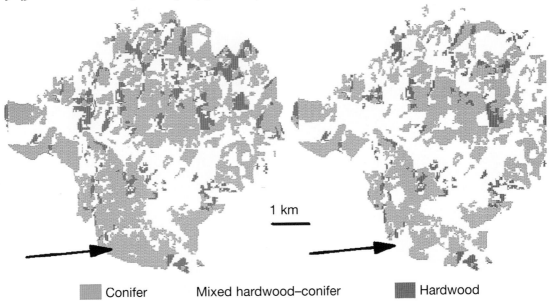

Fig. 5.7. Landscape alpha diversity for four different fire return intervals at year 100: (a) *fire return interval = 1×,* (b) *(FRI) fire return interval = 2×,* (c) *fire return interval = 4×,* (d) *current conditions.*

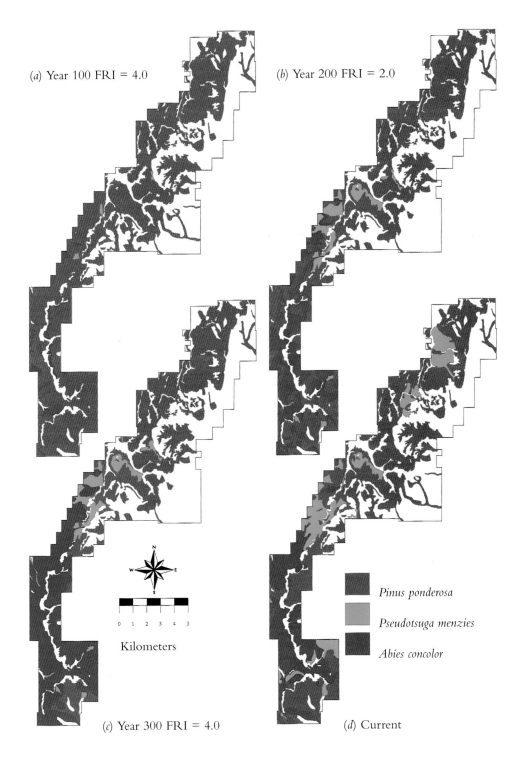

Fig. 5.8. Distributions of dominant species at three points in simulation from historic conditions with a 4× fire regime, and current conditions: (a) *100 years,* (b) *200 years,* (c) *300 years,* (d) *current conditions.*

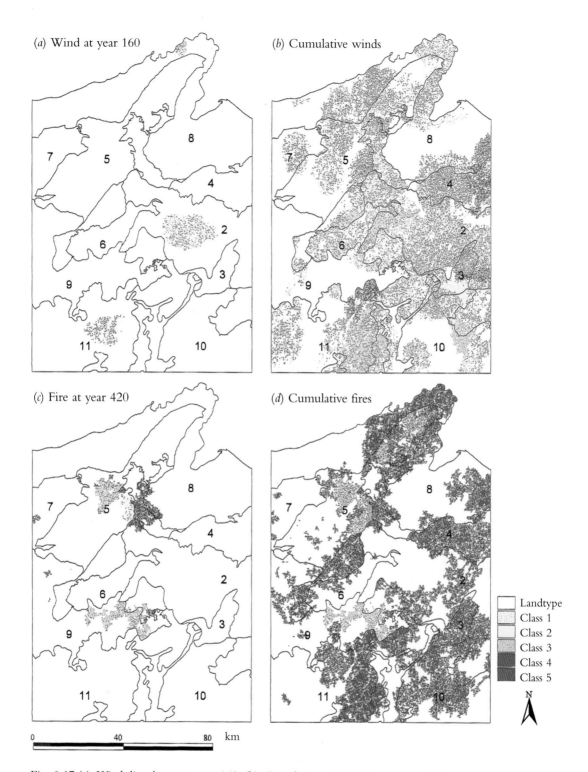

Fig. 6.17 (a) Wind disturbances at year 160. (b) Cumulative wind disturbances over 500-year simulation. (c) Fire disturbances at year 420. (d) Cumulative fire disturbances over 500-year simulation.

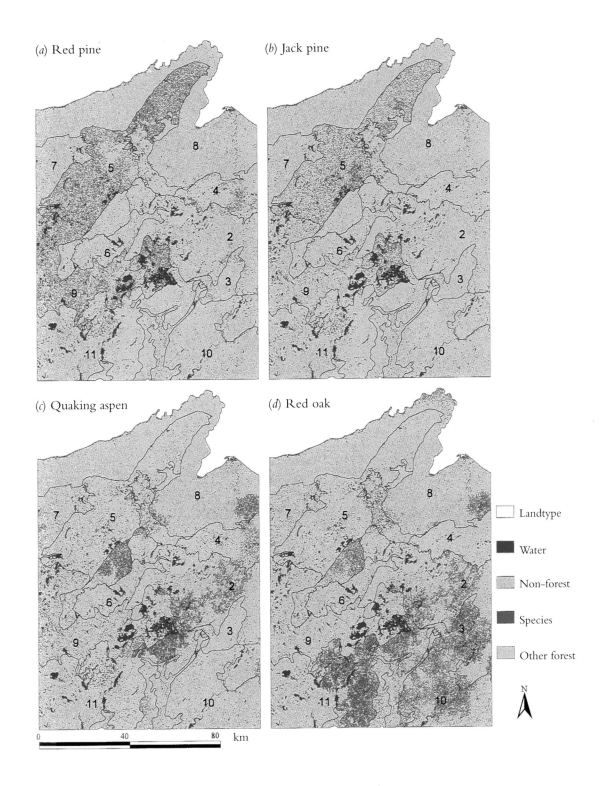

Fig. 6.18 (a) Red pine, (b) jack pine, (c) quaking aspen, and (d) red oak responses at year 420 after fires.

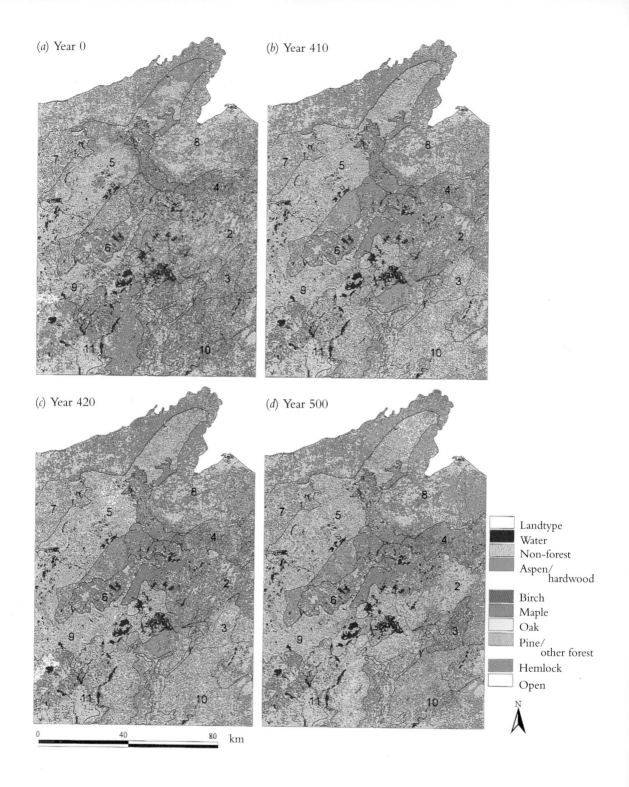

Fig. 6.19. Dominant forest types at (a) year 0, (b) year 410 before year 420 fires, (c) year 420 after fires, and (d) year 500, respectively.

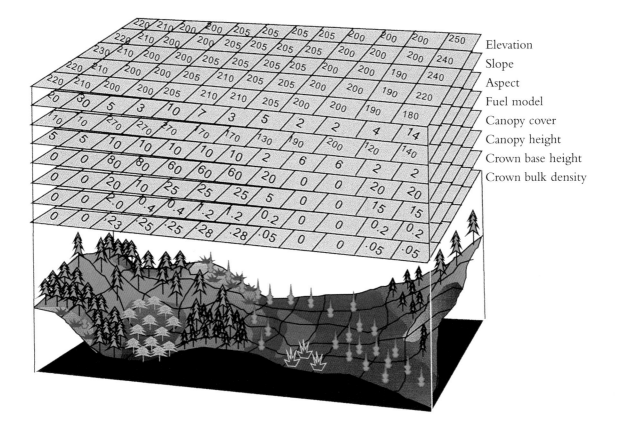

Fig. 8.3. Raster GIS themes used for spatial input data to FARSITE.

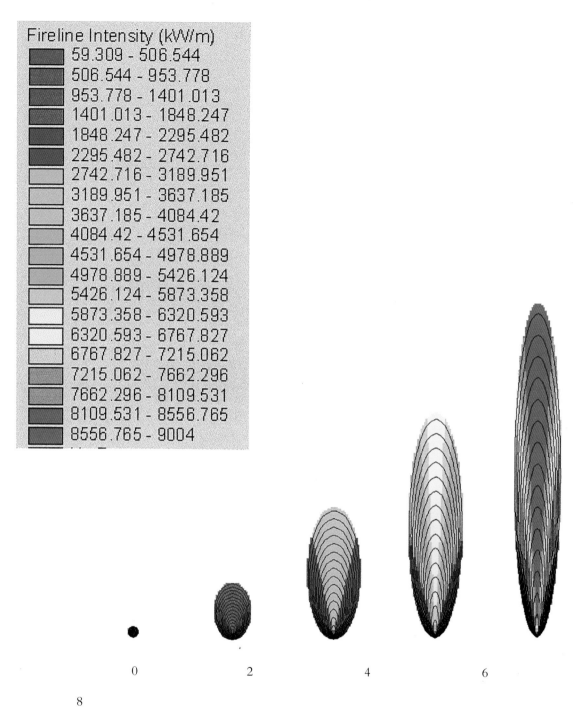

Fig. 8.4 Fire shapes and intensity patterns for constant wind and fuel moisture condition for flat terrain.

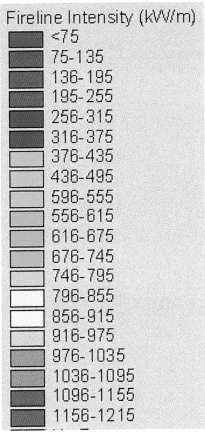
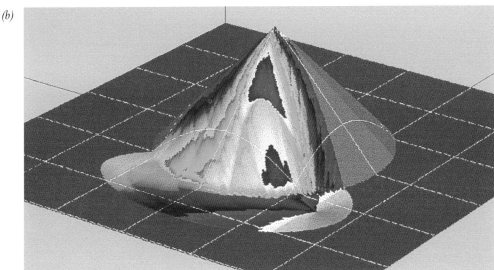

Fig. 8.5. Fire growth and intensity when burning onto a conical hill covered by grass fuel. Wind was constant from the south at 5 m s^{-1}. Fire started at 12 00 and progressed for 5 hours. Changes in fireline intensity are produced by changing fuel moisture during the afternoon and vectoring of wind and slope on different faces of the cone. (a) Horizontal view, (b) perspective view.

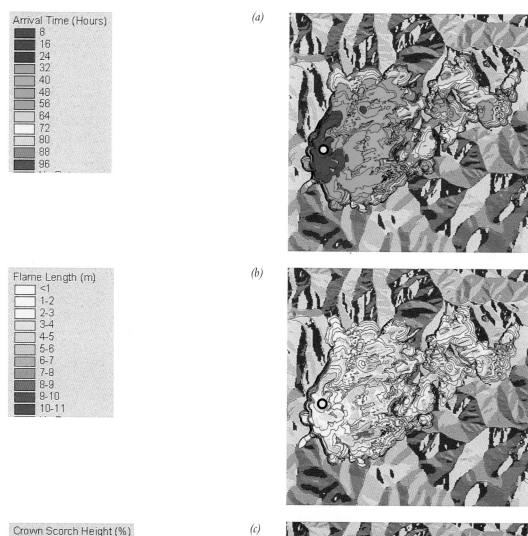

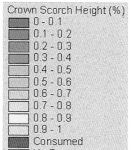

Fig. 8.6. Four-day fire simulation in the Selway-Bitterroot Wilderness, Idaho. (a) Fire progression, (b) flame length, and (c) percentage crown scorch. Higher intensities and crown scorch occurred in the afternoons, during passage of a cold front, and with uphill fire runs. Lower intensities occurred at night and from backing and flanking spread.

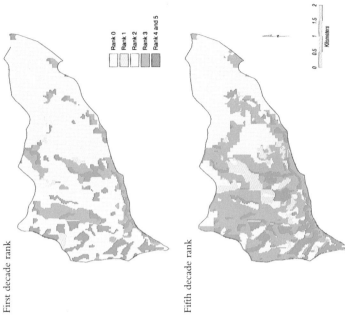

First decade rank

Fifth decade rank

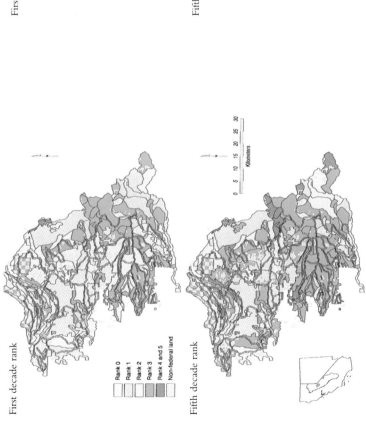

First decade rank

Fifth decade rank

Fig. 9.16. Late successional/old growth (LS/OG) rank of the simulation olygons in period one and period five from a typical simulation of SAFE FORESTS, with a period being 10 years in length (rank 0 = lowest level of structural complexity and contribution to late forest function, rank 5 = highest level). Between periods one and five in this simulation, the proportion of the forest in LS/OG ranks three and four increased indicating an increase the amount of structurally complex forest (Johnson et al., 1996).

Fig. 9.17. Details of late successional/old growth (LS/OG) rank in an individual simulation polygon in period one and period five from a typical simulation of SAFE FORESTS. Each stratum within the simulation polygon can have a different rank, depending on its stand conditions (rank 0 = lowest level of structural complexity and contribution to late forest function, rank 5 = highest level).

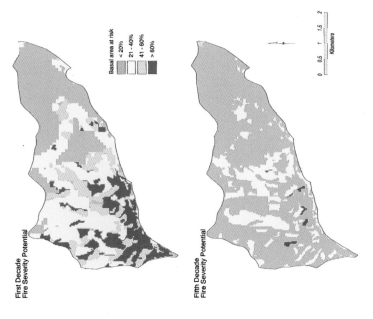

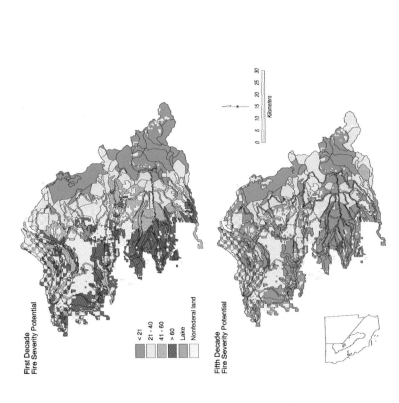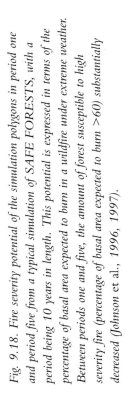

Fig. 9.18. Fire severity potential of the simulation polygons in period one and period five from a typical simulation of SAFE FORESTS, with a period being 10 years in length. This potential is expressed in terms of the percentage of basal area expected to burn in a wildfire under extreme weather. Between periods one and five, the amount of forest susceptible to high severity fire (percentage of basal area expected to burn >60) substantially decreased (Johnson et al., 1996, 1997).

Fig. 9.19. Details of the fire severity potential of an individual simulation polygon in period one and period five from a typical simulation of SAFE FORESTS. Each stratum within the simulation polygon can have a different fire severity potential, depending on its stand conditions. Between periods one and five, the amount of forest within the polygon susceptible to high severity fire (percentage of basal area expected to burn >60) substantially decreased.

*Figures 9.18 and 9.19 are reprinted from the Journal of Forestry, **96** (1): 44-5 published by the Society of American Foresters, 5400 Grosvenor Lane, Bethesda, MD 20814-2198. Not for further reproduction.*

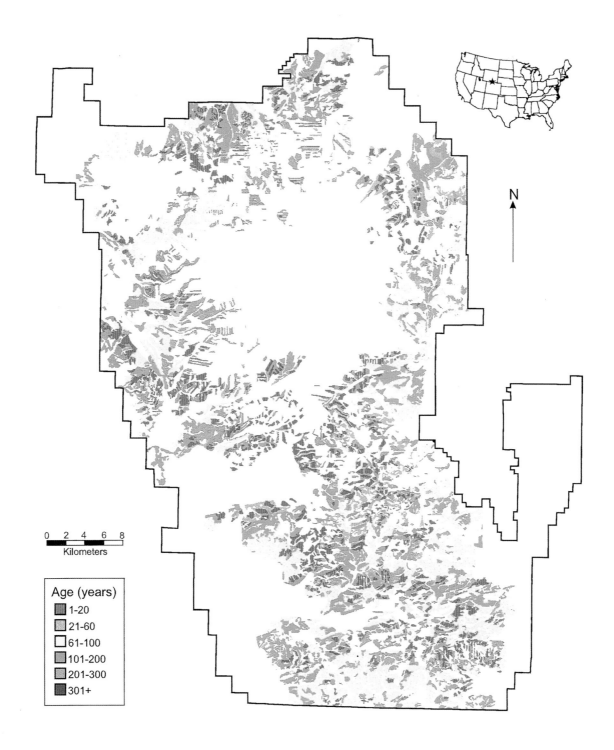

Fig. 11.4. The Snowy Range study area used in the validation/verification work and in the simulation experiment. This is the initial map of the age of lodgepole pine forest area on the MBNF that is also considered suitable for timber harvesting. The MBNF border is also shown.

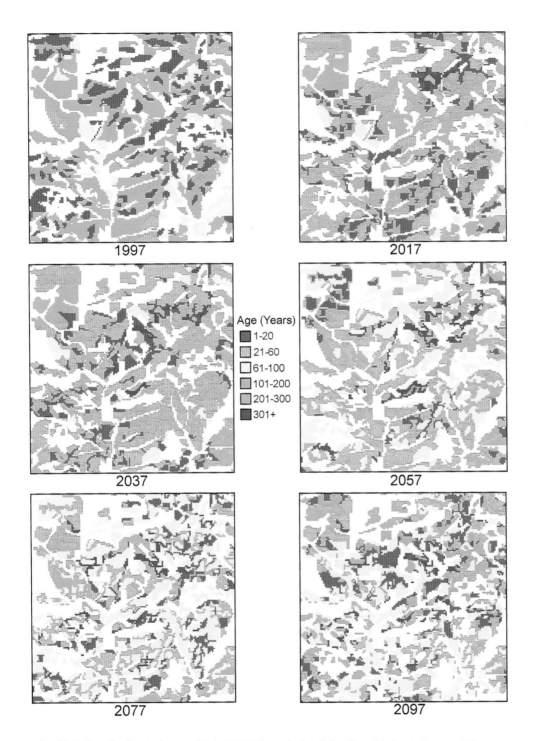

Fig. 11.8. Result of a single run of LANDLOG on the Medicine Bow National Forest at 20-year intervals over the next century. The model was run on the whole study area shown in Fig. 11.4, but only a small part (7 km × 7 km) of that area is illustrated here. The general trend toward replacement of large patches of old forest by smaller patches of younger forest is evident over the time period of this simulation run.

7

Predicting forest fire effects at landscape scales
Robert H. Gardner, William H. Romme and Monica G. Turner

Introduction

The sudden changes in terrestrial and aquatic ecosystems induced by large forest fires are a significant factor shaping ecosystem dynamics and vegetation patterns at landscape scales (Bonan *et al.*, 1990; Moreno and Oechel, 1994; Overpeck *et al.*, 1991; Romme and Turner, 1991; Swetnam, 1993; Whelan, 1995). It is difficult to predict the landscape-scale effects of forest fires because the relationships between weather, fire, and the dynamics of terrestrial ecosystems require an understanding of fine-grained details of fire ignition and spread, the patterns of vegetation and their response to fire, and the long-term trends and potential shifts in weather and climate (Bessie and Johnson, 1995). In addition, changes in land use patterns in many regions of the world are further exacerbating the problem of prediction as clearing and land abandonment produces a heterogeneous pattern of fuel accumulation, and subsequent changes in the risk of large fires (Ales *et al.*, 1992; Davis and Burrows, 1993). These combined factors have increased the possibility that large areas of forested land may be permanently altered due to shifts in the pattern of vegetation regrowth and succession (Neilsen, 1993). The effects of shifts in disturbance regimes must be anticipated if management plans are to prevent reductions in biodiversity and ensure ecosystem productivity (Hunter, 1993; Gauthier *et al.*, 1996).

Our current understanding of the ecological effect of forest fires is based on a variety of empirical and theoretical studies. Analysis of relationships between fire and weather (for instance, Balling *et al.*, 1992; Johnson and Wowchuk, 1993; Moritz,1997; Renkin and Despain, 1992) have allowed effects to be evaluated over the duration of contemporary records (*ca.* 100 yr), while studies based on intensive sampling of stand histories and dendocronology (for instance, Agee, 1996; Arno, 1980; Romme, 1982; Romme and Despain, 1989; Swetnam and Betancourt, 1990; Swetnam and Baisan, 1996) and stratigraphic charcoal data (Clark, 1990; Millspaugh, 1997) have created records dating back 800 yr or longer. Study of fires within stands of the giant sequoia (*Sequoiadendron giganteum*) has produced very long records which have allowed the reconstruction of fire history over the past 2000 years (Swetnam, 1993). As impressive and important as these studies are for establishing relationships between weather, fuels, and fire, simulation models are necessary to evaluate changes in fire regimes due to shifts in climate and the landscape patterns of forested areas.

A variety of simulation methods have been used to establish relationships between the fire regime (e.g., the frequency, severity and extent of disturbances) and the effects that these fires have on the recovery of disturbed areas. It is the purpose of this paper to present an overview of models which provide insight into the prediction of fire effects at landscape scales. The interest is specifically on models which predict spatial patterns of disturbance and recovery. Non-spatial models of fire effects (see Li and Apps, 1995) are not specifically considered unless they have contributed directly to the development of spatially explicit modeling approaches. Therefore, models developed for fire danger rating systems and prescribed burns are not considered in this review. This overview identifies: the important features of landscape-scale models; similarities and differences in predictions; relationships between model complexity and predictability; and an evaluation of the accuracy and reliability of model results. It concludes with a discussion of features that are missing from most landscape models and a comment on the current limits of predictability.

Models of fire effects at landscape scales

A large variety of models have been developed for simulating landscape-scale patterns of fire effects. These models range from simple formulations for investigating theoretical properties of spatial systems to complex formulations requiring extensive computer time needed to simulate the fire ignition and spread, the effect of fire on vegetation, and the eventual recovery of the disturbed area. Comparison among these diverse models is difficult because of the large number of model formulations that have been published, the variety of purposes considered by the different models, and the uniqueness of the landscapes to which they have been applied. One way to organize this task is to define categories of models (and model applications) that can assist in identifying similarities and differences among the different models. It is suggested that a classification scheme with six categories provides a useful perspective (Table 7.1). The classes are: *theoretical models*, which are simple formulations designed to explore fundamental concepts such as critical thresholds and fractal structures (e.g., MacKay and Jan, 1984; Duarte *et al.*, 1994), and self-organized criticality (e.g., Bak *et al.*, 1990); *exploratory models* which are similar to the theoretical models, but are more complex and provide greater details concerning landscape effects (e.g., von Niessen and Blumen, 1988; Ratz, 1995; Green, 1989); *physical models* based on principles of thermodynamics of fire spread as a function of moisture and structure of the fuels being burned that are then extended to landscape scales (e.g., Davis and Burrows, 1993; Keane *et al.*, 1996;

Table 7.1. *Classification scheme for models of fire effects on forested landscapes*

Model types	Characteristics	Applications
A. Theoretical	• Very simple model formulations • Parameters, spatial and temporal scales are often arbitrary	• Explore and characterize behavior of spatial systems • Identify attributes of critical phenomena
B. Exploratory	• Increased complexity with models simulating realistic fire scenarios • Parameters and landscapes may be arbitrary	• Explore dynamics of complex systems, including heterogeneity of fuels • Identify relative importance of factors affecting spatial and temporal dynamics of fires
C. Physical	• Mathematics based on detailed thermodynamics of fire behavior • Parameter values based on experimental measurements • Simulations with realistic landscapes	• Use information on fine-grained dynamics of fire behavior to extrapolate to broad spatial and temporal scales
D. Probabilistic	• Probability rules used to explain local dynamics of ignition and spread • Simulation techniques are extensions of methods of cellular automata	• Hypothesized that simple formulations of fine-grained dynamics allows improved predictions of landscape effects while lowering computational burden
E. Shape	• Use of elliptical geometry of fire shapes to characterize fire behavior	• Hypothesized that elliptical geometry is an accurate representation of fire behavior and can be used to simulate fire effects
F. Statistical	• Use of historical data to establish statistical characteristics of disturbance regimes • Sampling from measured or hypothesized frequency distributions to simulate fire effects	• Use of historical information on fire regimes to characterize and simulate dynamics at broad temporal and spatial scales

Kessell, 1979; Vasconcelos and Guertin, 1992); *probabilistic models* developed as extensions of cellular automata models to simulate fires in heterogeneous landscapes (e.g., Antonovski *et al.*,1992; Clarke *et al.*, 1994; Gardner *et al.*, 1997); *shape models* based on the geometry of fire patterns resulting from variable spread rates produced by wind and topography (e.g., Catchpole *et al.*, 1992; Cheney *et al.*, 1993; Wallace, 1993); and *statistical models* using estimated probability density functions to characterize and/or simulate fire regimes (e.g., Johnson and van Wagner, 1985; Baker *et al.*, 1991).

The proposed classification scheme is based on the observation that there are distinct sets of perspectives, objectives and solution techniques used to study the relationships between forests and disturbances due to fire. For instance, theoretical models usually ask general questions regarding disturbance and self-organization, often relying on simple representations both of forests and fires. Statistical models, on the other hand, rely more on empirical studies as a basis for extrapolation and prediction. These categories were developed in an attempt to characterize as broad a range of models types as possible. The question of interest, then, is how do these different types, and the perspectives they represent, help us to understand and predict fire effects at landscape scales.

The classification scheme suggested here is both arbitrary and simplistic. This is evident, in part, because the categories for different models are not completely exclusive, i.e., some models may belong to more than one category. In addition, the categories are not independent: models (and model parameters) in one category may have been developed from models in another. Nevertheless, the scheme presented here is heuristically useful because it organizes the diversity of model approaches, facilitates comparison between models, and assists in the synthesis of results. The classification is not intended to indicate that one approach is better than another but rather to demonstrate that all approaches are interesting and informative, helping to advance studies of this important area.

One of the challenges of simulating the landscape-scale effect of fires is the great range of temporal and spatial scales which must be considered to adequately represent both fires and the recovery of landscape from these fires. For instance, significant changes in the moisture of herbaceous and fine woody fuels can occur within a single day and will likely change over a few hectares as forest age and composition changes with space. Similarly, the structure of coarse woody fuels provided by the forest varies over hectares but changes little from year to year. Variations in topography (slope, elevation, and aspect) have well-known effects on fuel moisture and the rate of spread and intensity of fires, varying greatly from hectare to hectare but changing imperceptibly from decade to decade. The inclusion of wind is a difficult problem for most models because it varies from minute to minute, shifts with changes in topography, and can be directly altered by the fire itself. Fire ignition events are often separated in time by days or even years, but precise knowledge of the spatial locations of fire starts may be required to adequately predict landscape scale patterns of fire effects. Because temperature affects relative humidity, diurnal

Table 7.2. *The models of different types used as examples in the text*

Model Type[a]	Model Name[b]	Citation
A. Theoretical	BCT90	Bak *et al.*, 1990
	MJ84	MacKay and Jan, 1984
B. Exploratory	VNB88	von Neissen and Blumen, 1988
	RTZ95	Ratz, 1995
C. Physical	FIREMAP	Vasconcelos and Guertin, 1993
	FIRE-BGC	Keane *et al.*, 1996
D. Probabilistic	EMBYR	Gardner *et al.*, 1997
	CLK94	Clarke *et al.*, 1994
E. Shape	CTH92	Catchpole *et al.*, 1992
	WALL93	Wallace, 1993
F. Statistical	JW85	Johnson and Van Wagner, 1985
	DISPATCH	Baker *et al.*, 1991; Baker, 1992

[a] Model types are defined in Table 7.1.
[b] FIREMAP, FIRE-BGC, EMBYR and DISPATCH are names given to the models by the authors in the papers cited. Other names are arbitrary for purposes of identification in the text.

changes in temperature can be important in explaining the spatial pattern of fire effects. Fire spotting is difficult to simulate because it depends on factors causing variation in fuels, forest structure, fire intensity and the location of new fire starts due to landing fire brands. Although the prediction of changes in community structure with time remains an important challenge, the temporal and spatial scales of forest regrowth are easier to include because regrowth is slow (order of decades) and resolutions need not be finer than hectares to adequately describe spatial variation in landscape patterns of forest regrowth. Thus, the simulation of patterns in fire-prone landscapes requires reliable methods for extrapolating processes which occur at many different temporal and spatial scales. In fact, the model categories represented in Table 7.2 are, to a large extent, due to different approaches used to extrapolate effects through time and space. The problem of temporal and spatial extrapolation is a popular theme in ecology (Levin, 1992; Schneider, 1994), but firm principles have yet to emerge allowing the resolution of this dilemma. Therefore, in the comparisons which follow, the problem of predicting across scale will be one basis for comparisons within and between the categories listed in Table 7.1. Finally, detailed analyses of two models are presented for each category (Table 7.2) that illustrate the approach and results. These models were chosen because they illustrate the approach and results typical to each category. Reference to other models which lend insight into the landscape-scale effects of changes in fire regime may be found within the text.

Model comparisons

Theoretical models

It is true that all models are in some sense theoretical. What are referred to here as 'theoretical models' are the subset of models with extremely simple formulations designed to explore the behavior of spatial systems with no intention of predicting patterns of fire effects for *actual* landscapes. Thus, in these models, fire spread is often represented as a random process of spread to neighboring sites and spatial and/or temporal dimensions may be arbitrary (Table 7.3). The results of these models can be quite interesting because they reveal fundamental characteristics of fire-dominated landscapes. Although these models have been often ignored by fire ecologists, they are responsible for stimulating new approaches to modeling fires at landscape scales (e.g., Clarke *et al.*, 1994; Gardner *et al.*, 1997; Green, 1989; von Niessen and Blumen, 1988).

Perhaps the simplest imaginable model of fire in forested landscapes is that proposed by MacKay and Jan (1984; referred to as MJ84 in Tables 7.2 and 7.3). The landscape is represented in MJ84 as a two-dimensional lattice with all sites occupied by identical trees. Fire is initiated in the center of the lattice and then spreads to neighboring trees with probability p. The purpose of the MJ84 is to demonstrate the behavior of fire with changes in p and estimate scaling parameters (i.e., the fractal dimension of the number of trees burned with time).

A similar model formulation was studied by Bak *et al.* (1990; BCT90 in Tables 7.2 and 7.3). The stated purpose of the BCT90 model is "...to demonstrate in a toy model the emergence of scaling and fractal energy dissipation." To accomplish this, a two-dimensional lattice was formulated (*note:* higher dimensions can be simulated with this model!) with trees occupying individual sites on the resulting grid. Each site may burn in a single time step and can "grow back" with probability p. The simulation results indicate that the system remains critical (i.e., remains close to the point where fires can propagate across the landscapes) as long as trees continue to grow. The implication is that because forests regrow after fires, landscapes will constantly exist near this critical threshold, with resulting patterns "... reflective of turbulent dynamics of growth and fire" (Bak *et al.*, 1990).

The development of simple models is instructive because they allow one to fully characterize the spatial dynamics of these systems. What is useful for the study of fire-prone landscapes is that dynamics observed in simple models can be expected to apply to more complex landscape models. For instance, it is known that fuels accumulate with time and therefore fire suppression may actually force the system beyond the critical threshold, i.e., the landscape may reach a "hypercritical" state where small changes in conditions (i.e., wind and weather) may result in catastrophic fires. The challenge for fire ecologists is to transform the estimated thresholds of simple theoretical models to measurable thresholds for actual landscapes.

Table 7.3. Comparison of fire and landscape processes represented by the different models

	Models[b]									
	Theoretical	Exploratory		Physical		Probabilistic		Shape		Statistical
Model variable or process[a]	BCT90 MJ84	VNB88	RTZ95	FIREMAP	FIRE-BCG	EMBYR	CLK94	CTH92	WALL93	JW85	DISPATCH
Fuel structure	x			x	x	x	c	c	x		x
Fuel moisture				x	x	x	c	c			x
Ignition			r	f	pdf	pdf	f				r
Wind				t	t	t	c	c	x		
Fire intensity				x	x	x	x	x	x	x	
Fire spotting						x	x				
Topography				x	x	x	x	c	x		
t resolution	a		dc	hr	y	e	min	na	min	y	d
t extent	a		ml	hr	cen	ml	hr-d	na	hr	cen	cen-mi
S resolution (ha)	~0.05 ~4	~0.05		04	~10	0.25	0.06	na	0.001	na	25
S extent (ha)	26 ~40k	25	25	~1.6k	50k	90k	3.3k	na	30	na	1000k
Forest regrowth	p				e	m					dt

[a] A blank entry indicates that the model does not consider the variable or process indicated. An x, or other symbol, indicates the process is represented by the model. A c indicates that the process is considered by the model, but is held constant during any single simulation. An na indicates that this process or variable does not apply to the methods used by the model. The model processes and special symbols are:

Fuel structure: spatially variable combustion characteristics of fuel.
Fuel moisture: dynamic changes in moisture affecting combustion of fuels.
Ignition: f is a fixed ignition site, r is randomly selected ignitions sites, and pdf indicates that a statistical probability distribution was used to calculate the location of fire starts.
Wind: t is temporal variation in wind velocity and direction and c is constant wind. No model included in this table had both spatial and temporal variation in wind.
Fire intensity: fire temperature and/or duration to estimate disturbance intensity.
Fire spotting: fire propagation beyond the fire line.
Topography: spatial variability of fire effects due to variations in slope and aspect.
t resolution: temporal resolution of simulation: a arbitrary, e is event-driven simulation, s seconds, d days, y years, dc decades, cen centuries, and ml millennia
t extent: the length of time simulated. Units same as for temporal resolution.
S resolution: spatial resolution in hectares.
S extent: spatial extent of simulation (e.g., ~40k refers to spatial extent of approximately 40 000 ha)
Forest regrowth: reestablishment and regrowth of the forest following fires, p is a simple probability, dt is deterministic forest regrowth, m is a Markov model, and e is a ecosystem dynamic model for forest succession.

[b] Model names are the same as those listed in Table 7.2.

Exploratory models

Exploratory models are distinguished from theoretical models by an increased level of realism in model processes and parameters and an attempt to identify the importance of these factors as they affect the spatial and temporal dynamics of fire-prone landscapes. While theoretical models concentrate on reaching a fundamental understanding of spatial dynamics, exploratory models also emphasize the practical application of results.

The model presented by von Niessen and Blumen (1988, VNB88 in Tables 7.2 and 7.3) provides a good example of this distinction. The purpose of the VNB88 model was to continue to explore the critical character of forest fires (i.e., small variations of parameters which may produce huge differences in results), but to do so within a framework more representative of fires within forested landscapes. The additional features of VNB88 developed to accomplish this objective include: the use of a double lattice to represent both surface and crown fires; individual sites that can represent a single tree or a group of trees; and the inclusion of anisotropic effects of wind on fire spread. The model's time step remains somewhat arbitrary (i.e., it is equal to the average time it takes a tree(s) on a site to burn) and fire spread is restricted to the ignition of four nearest neighbor sites. The fuel layer is homogeneous and the recovery of the forest from fire is not modeled. Thus, VNB88 incrementally builds on methods used by theoretical models, but with additional relevant detail to simulate dynamics of surface and crown fires and wind.

The results of numerous simulations by von Niessen and Blumen (1986, 1988) with variants of the VNB88 model showed that the critical threshold (i.e., the conditions which may produce fires which burn across the landscape) was independent of the form of the model, but the value of the parameters describing critical behavior vary from model to model depending on difference in parameters and assumptions. Hence, the VNB88 model demonstrates that quantitative methods used to measure critical values for actual landscapes will be model dependent.

A second example of an exploratory model, also based on methods derived from cellular automata methods (Wolfram, 1984; Hogeweg, 1988; Caswell and Etter, 1993) as was VNB88, is given by Ratz (1995, RTZ95 in Tables 7.2 and 7.3). The purpose of the RTZ95 model was to test the effect of long-term patterns of disturbance by fire on the spatial structure of forests. The key factor of this model is the capability of testing the alternative effects of change (or no change) in flammability of forest stands with age. The RTZ95 model simulates fires on a 200 × 200 grid of forest stands with each grid site equal to 4 ha. Fire spreads isotropically to nearest neighbor stands and the recovery of the stand after fire is deterministically modeled. Although age-dependent changes in flammability have been discussed by several authors (e.g., Johnson and Van Wagner, 1995; Johnson, 1992; Renkin and Despain, 1992; Romme, 1982), the effects of changes in flammability on the spatial structure of fire-prone landscapes remains poorly understood (Ratz, 1995).

One interesting result of the RTZ95 model is that a variety of spatial patterns can be attributed to fire disturbances (Ratz, 1995), with more complex patterns produced with the assumption of age-dependent flammability of forest stands. Because of data limitations, only a restricted set of comparisons between model simulations and data were possible. (It is worth noting that this is the first example presented here of a model attempting comparison of results with patterns of fire effects in landscapes.) Even though the comparison was limited by the lack of sufficient data, the level of agreement between model results and data was rather good (Ratz, 1995) indicating that the model formulation was relevant for simulating landscape-scale dynamics of fire disturbance. In spite of the different assumptions regarding age-dependent flammability, all simulations produced large fires (i.e., the critical thresholds were always exceeded).

Green (1989) developed an interesting exploratory model which simulated fire, seed dispersal, life history characteristics of individual "species", environmental gradients, and successional dynamics. The spatial scale of the lattice used to model plant growth and fire spread is arbitrary with time steps of a single year. The attributes of each cell include the vegetation type and time since the last fire. Fires were ignited at random and the elliptical pattern of fire shape simulated via frequency distributions (see the following discussion of *Shape Models*). The results showed a wide range of patterns produced by the interaction of species life history characteristics, environmental gradients, and disturbance. For instance, vegetation "clumping" occurs even when environmental gradients are absent because short-range dispersal of seeds provides local competitive advantages. Fires cause patches to coalesce into vegetation zones separated by sharper borders than those produced by the gradients alone. Also, local patterns of seed production allowed established populations to exclude invaders.

None of the exploratory models is intended to be a realistic representation of actual landscapes. Never the less, the contribution these models have made to our understanding of fire-prone landscapes is significant. Results of a variety of models show that the fire regime dominates the landscape patterns of vegetation, that the spatial arrangement of fuels (and not simply the total amount) and forest regrowth are important variables affecting the probability of future disturbances. Because these models are not as simple as the theoretical models, analytical solutions are either elusive or impossible. Fortunately, the high speed of current computers and the simplicity of the model formulations allows spatial dynamics over large areas and long time spans to be conveniently simulated and analyzed.

Physical models

The physical models of fire spread are direct descendants of the pioneering work of Rothermel (1972) that produced a well-used set of computer models of fire behavior (Albini, 1976; Andrews, 1986; Cohen, 1986). The Rothermel model has been implemented in BEHAVE (Andrews, 1986) as a set of equations to estimate the spread rate of fires as a function of heat released and the physical and chemical characteristics of the fuels. The advantage of this approach is that these equations (along with other related variables and parameters) have all been determined under precisely controlled experimental conditions. Knowledge of the rate of spread allows the geometry of fires in homogeneous fuels to be predicted and estimates of the perimeter length, area, and intensity of fires to be made. The motivating factor in the development of BEHAVE, and other related models of fire behavior, is that information about time-dependent changes in fires as a result of changes in fuels, wind and weather is important for allocating resources and protecting lives while fighting fires.

Reasons to adapt this detailed model to landscape-scale problems of fire effects include:

(i) the successful representation of experimental systems by the Rothermel equations. Other models either do not estimate changes in the rate of spread and intensity of fire as a function of the physical characteristics of fuels and weather or use some adaptation of the Rothermel formulation to do so;

(ii) the extensive experience applying models such as BEHAVE for fire control and management has produced confidence in the methods as a valid representation of fire behavior; and

(iii) the use of highly detailed, fine-grained models as the basis for estimating changes over broad areas and long time spans continues to have wide appeal.

However, using models such as BEHAVE for predicting fire effects at landscape scales is a daunting task. Changes in fuel moisture, packing and geometry of fuels as well as changes in wind speed and direction, and humidity must all be estimated across the entire fire front. Often these variables are either unknown or unmeasurable at landscape scales. The experience with the real-time application of BEHAVE for estimating worst-case scenarios during the 1988 Yellowstone fires is informative in this regard (Rothermel, 1991). Because weather conditions were so extreme, and the effects of fire spotting (i.e., the initiation of new fires by transport of burning material beyond the fire line) on the ignition of new crown fires was so important, alterations in assumptions of fuel type were necessary to simulate the rapid spread of fires experienced in Yellowstone in 1988 (Rothermel, 1991). Furthermore, the nature of the fuels and weather forecasts were uncertain (i.e., the weather conditions in Yellowstone in 1988 were extreme) causing prediction of fire behavior to depend on expert experience as much as on the model simulations (Rothermel, 1991).

The Rothermel equations for simulating landscape effects of fire have been adapted for chaparral ecosystems by Malanson (1984) and Davis and Burrows (1993), demonstrating a strong linkage between vegetation fragmentation, frequency of fire ignition and landscape patterns of post-fire vegetation. An adaptation of these equations for forested landscapes has been proposed by Vasconcelos and Guertin (1992; FIREMAP in Tables 7.2 and 7.3). The FIREMAP model involves the conversion of the continuous form of the equations for fire spread of BEHAVE (Andrews, 1986) to a discrete form which allows gridded landscape data to be used to characterize the spatial heterogeneity of fuel types. This discrete transformation also permits digital elevation data to be used to estimate changes in fire behavior with changes in topography. During a simulated fire the computer program reads in the landscape data and calculates fire characteristics at each time step for each cell. The process is repeated for each successive update in weather conditions and new output maps are then generated. The analysis of map layers allows the time-dependent changes in fire spread and intensity to be analyzed. Although the temporal and spatial extent of the simulations performed with FIREMAP are limited (Table 7.3), the spatial characterization of fire spread and the effects of the spatial variability in fire intensity are important for estimating landscape-scale changes in ecosystem dynamics as the result of shifting fire regimes. In spite of the limitations of FIREMAP, the comparison of model simulations with patterns recorded during a single fire event indicates that this representation may satisfactorily mimic observed patterns (Vasconcelos & Guertin, 1992).

The most extensive application of the physical-based approach has been recently published by Keane *et al.* (1996*b*; FIRE-BGC in Tables 7.2 and 7.3). FIRE-BGC was designed to integrate a series of fire disturbance models for fire ignition (FIRESTART) and spread (FARSITE; Finney, 1998) with a forest succession models to replace disturbed forests (FIRESUM; Keane *et al.*, 1989) and estimate ecosystem attributes of the affected areas (FOREST-BCG; Running and Coughlan, 1988). Other models are used to generate weather scenarios and evaluate the historical patterns of fire occurrence (see Keane *et al.*, 1996*a* for additional details). The simultaneous simulation of multiple models is effected by communication among UNIX workstations executing each model within this complex system. The intriguing aspect of linkage of models operating over different temporal and spatial scales is that it is possible to track multiple ecosystem properties (e.g., evapotranspiration, net primary production, nitrogen availability, standing crop biomass of forest, etc.) simultaneously as they are altered by fire and as they, in turn, alter the fire regime. However, this complexity makes it difficult to identify the importance of model assumptions, verify the validity of the computer formulation and linkage, and test the sensitivity of model response to alternative sets of parameters and input values. Of course, these issues are not unique to this model – only the dynamics of the simple theoretical models can be throughly analyzed. The problem of complexity and the propagation of error is an important issue in extrapolating results to landscape scales (Gardner *et al.*, 1982; Milne, 1992; Turner *et al.*, 1989).

The results of a series of simulations with FIRE-BGC for a 200-year period of a 50 000 ha region within Glacier National Park (Keane *et al.*, 1996*a,b*) showed increases in net primary productivity and available nitrogen when fires were included in the simulations, with standing crop biomass and evapotranspiration showing an expected decrease when fires were included. The interesting result is that tree increment data generated by model simulations agreed well with available tree-ring data from Glacier National Park. This type of comparison is unique among fire models because it uses information that is independent of the data used to develop and calibrate the model.

Probabilistic models

Probabilistic models are ones which simulate the spread of fire on a gridded landscape via probabilistic rules that are similar to those used in cellular automata models (Wolfram, 1984). Because probabilistic models are similar to the theoretical and exploratory classes of models, it might seem more logical to present probabilistic models before discussing the physical models. However, the probabilistic models are also directly related to the physical models. The model of Clarke et al. (1994; CLK94 in Tables 7.2 and 7.3) begins by a discussion of the landscape-scale application of the Rothermel equations and a justification of a simpler representation of the processes modeled by BEHAVE. The model of Gardner et al. (1997; EMBYR in Tables 7.2 and 7.3) also relies on the experimental results of models based on the Rothermel equations to estimate parameters of the probability, rate of spread, and fire intensity.

The purpose of the CLK94 model (Table 7.1) is to provide a simple and reliable representation of fire intensity and spread using methods based on percolation theory that can efficiently simulate the spatial pattern of fires at landscape scales. The CLK94 model considers a diverse set of characteristics (Table 7.3) that are relevant to assessing landscape effects, including the representation of spatial variability in fuel structure, temporal variability in wind speed and direction, and the effect of topography on fire spread and intensity. It is worth noting that only the two probabilistic models (i.e., CLK94 and EMBYR, Table 7.3) and the FARSITE model explicitly consider the effects of fire spotting on the spread and pattern of fire effects.

The development of the CLK94 model was motivated by the desire to use remotely sensed thermal infrared imagery to calibrate the model and view real-time estimates of the spatial patterns of fire spread. Because CLK94 was not designed for the study of landscape effects of fire, a number of factors are held constant for any given simulation (for instance, fuel structures, moisture contents, and ignitions are fixed, Table 7.3), the spatial and temporal scales are set to values appropriate for simulating single fire events, and the effect of fires on the forested ecosystems is not considered. In spite of these limitations, the model structure is of interest to landscape studies because it is possible to use Monte Carlo simulations of the model to produce spatially explicit estimates of risk (risk maps are spatial estimates of the probability of an individual site being burned: see Clarke et al., 1994). The use of risk maps is an appropriate method for evaluating fires in heterogeneous landscapes because small random changes in unmeasurable variables (e.g., sudden bursts of wind, slight changes in fuel moisture, etc.) makes the direct comparison of actual fires with a single model simulation quite misleading. What is more appropriate is to be able to state that an actual fire is a likely member of the ensemble of simulations produced by the model under a set of conditions representative of the observed fire. Significant changes in these risk maps are appropriate evidence of landscape-scale effects of a given fire regime.

Many of the features of EMBYR are quite similar to those of CLK94. EMBYR propagates fire by user-defined sets of probabilities of fire spread from cell to cell. The probabilities are dependent on the age of the forest stand and the fuel moisture conditions (i.e., probabilities of spread increase with increasing age and decreasing fuel moisture). In addition, during the simulation the probabilities are adjusted for each site depending on topography and shifts in wind speed and direction. Fire spotting is simulated by lofting burning embers downwind from the fire front (Gardner *et al.*, 1997). Weather records can be input directly into EMBYR or generated from instrumental records. Simulations with EMBYR are quite rapid because it is an event-driven model updating only the status of burning cells. Monte Carlo methods can also be applied to estimate spatial patterns of risk with EMBYR (Hargrove *et al.*, in press). Because EMBYR was specifically designed to simulate fire at landscape scales, a Markov model of forest recovery is included. The spatial and temporal scales of landscape simulations are coarser grained with broader extent than CLK94.

Monte Carlo simulations with EMBYR over 1000 years for three different weather conditions (i.e., weather conditions typical of the past 30 years, slightly wetter conditions, and slightly drier conditions) has demonstrated the sensitivity of landscape pattern to changes in the fire regime (Gardner *et al.*, 1997). Fewer but more severe fires occur when conditions were wetter than normal, and more frequent, smaller fires resulting in greater landscape fragmentation occurred when conditions were drier than normal. These results are consistent with both empirical and simulation studies that show that small changes in climate can produce dramatic differences in the fire regime (e.g., Antonovski *et al.*, 1992; Clark, 1989, 1990; Baker, 1992; Suffling *et al.*, 1988; Romme and Turner, 1991; Davis and Burrows, 1993; Swetnam, 1993).

Shape models

Catchpole *et al.*, (1992) have noted that the "... simple ellipse has been a most useful model of the perimeter of a free-burning wildland fire." These authors maintain that only three inputs are required to simulate the shape of a fire: the intensity of the fire front, the elliptical shape factor, and the backfire spread rate. Although this approach has been criticized when fuels are non-uniform (e.g., Green, 1989; Clarke *et al.*, 1994), other studies have shown that the ellipse is a satisfactory model of fire spread under uniform fuel conditions (e.g., Anderson, 1983; Alexander, 1985). The shape models are, in fact, easily derived from the equations developed by Rothermel for the physical models assuming that fuels are homogeneous and continuous and weather conditions are constant because the "... elliptical fire shape implies a variation in the spread rate at different angles to the prevailing wind direction."

Catchpole *et al.* (1992) suggest that the elliptical model could be used to generate tables of fire size and effect for different fuel types and burning conditions, allowing rapid simulations of fires at landscape scales. Because the elliptical model most closely matches reality in a homogeneous fire environment, with steady-state weather conditions, model output provides risk estimates for areas were fuel loadings and topography are known (Catchpole *et al.*, 1992). Wallace (1993; WALL93 in Tables 7.2 and 7.3) developed a more general model using elliptical geometry to simulate fire patterns in landscapes with variability topography and fuels. Unfortunately, errors in WALL93 simulations tend to produce unacceptably large errors with time and space – indicating a property which may be related to the geometric methods used by the shape models.

Statistical models

The statistical description of the past history of fires has allowed the estimation of frequency distributions describing the average fire interval, fire frequency, average age of the vegetation, and the renewal rate of disturbed forests. Johnson and Van Wagner (1985) in a much cited paper have demonstrated that the Weibull distribution is the most suitable form for representing the fire regime, because the Weibull allows the hazard function (or the estimate of the instantaneous rate of burning) to vary as a function of the age of the vegetation. (The exponential distribution assumes a constant hazard function and is a special case of the Weibull distribution.) The Weibull distribution allows estimates to be made of the distribution of times since last fire or the probability of having fires within any given time span. The application of methods proposed by Johnson and Van Wagner (1985) result in projections of the effects of the current disturbance regime over time and space. Because the method assumes a constant regime, the parameters for the Weibull must be re-estimated for different landscapes and/or climatic regimes or for different time periods within the same landscape. In addition, the statistical method does not provide spatially explicit information of fire size and effect, seriously limiting its application to landscape studies.

Although the statistical approach developed by Johnson and Van Wagner (1985) lacks a spatial component, the method has been adapted and applied at landscape scales by modifications allowing the estimation of the effect of spatial patterns of disturbance. One example of this adaptation is the DISPATCH model (Baker *et al.*, 1991; Baker, 1992), which was designed to address the effect of changing climate (i.e., shifts in the disturbance regime) on vegetation patterns at landscape scales. Consequently, DISPATCH is constructed so that it is possible to simulate landscape change from hundreds to thousands of years over very large areas (Table 7.3), allowing the effect of heterogeneity of fuels, temporal changes in moisture, and changes in topography on disturbances to be considered (Baker *et al.*, 1991; Baker, 1992).

The simulation procedure used by DISPATCH involves random selection of weather conditions from pre-specified synoptic climate types, random selection of a disturbance with sizes set by unique negative exponential distributions associated with each synoptic climate, and random propagation of disturbance as a function of vegetation type, age, elevation, and topography until the specified disturbance size is reached. Because disturbance spreads randomly from cell to cell, the DISPATCH model resembles a combination of a statistical model and a theoretical (or cellular automata) model.

The results with DISPATCH have shown that landscape structure fluctuates widely even if a disturbance regime remains constant for over a 250-year period (Baker et al., 1991). Shifting patterns of land use and climate indicate that management techniques are unlikely to be able to reconstruct pre-settlement conditions of forested landscapes (Baker, 1992). Obviously, these simulations represent a "test of hypothesis" of the nature of interactions at landscape scales and should not be expected to be precise predictions of the future states of the landscape (Baker, 1992). This caveat is true of all landscape models – simulations over long temporal and spatial scales will always be difficult to verify. In many ways the purposes of landscape models are similar to the theoretical models that explore the dynamics of interacting processes. The only difference might be that the parameters for the landscape models have been developed from a stronger empirical base and are intended to represent actual landscape and weather conditions.

Discussion

Fires are important events shaping the dynamics of terrestrial ecosystems in many regions of the world (Turner et al., 1994; Whelan, 1995; Goldammer and Furyaev, 1996). Forest fires directly affect community composition and structure (Heinselman, 1981; Noble and Gitay, 1996) and result in significant releases of greenhouse gases (Auclair and Carter, 1993; Cofer et al., 1996). Ecosystem effects due to forest fire are significant with the consumption of biomass resulting in modifications of soil temperature and chemistry (Goldammer and Furyaev, 1996), the removal of seeds and seed sources in high intensity fires (but also the removal of seed predators), significant changes in cycling of nutrients and some loss of soils due to subsequent erosion (Meyer et al., 1992; Trabaud, 1994), and the redistribution and eventual sequestration of elemental carbon (Clark and Richard 1996; Kasischke et al., 1995). The combined effects of community and ecosystem alteration due to fires often result in conditions which favor the continued propagation of new fire regimes (Bergeron, 1991). The long-term consequences of these changes remain difficult to anticipate because changes in climate and land-use (including fire suppression) are causing continual shifts in fire regimes (e.g., Johnson and Larsen, 1991) with the result that landscape patterns will be in a constant state of disequilibrium (Ales et al., 1992; Boychuk et al., 1997; Johnson et al., 1990; Turner and Romme, 1994). Because current methods of analysis and prediction have emphasized the fine-grained dynamics of fire behavior and the corresponding management practices that reduce fire hazard, the long-term interactions of forest ecosystems with different fire regimes has been neglected.

Empirical and theoretical studies of the effects of fire in forested landscapes are beginning to reveal interesting and unexpected relationships, including the following:

(i) Theoretical models have demonstrated that the pattern of recovery and regrowth of forests experiencing frequent fires results in the self-organization of these landscapes to a critical threshold where the connectance of fuels is sufficient to sustain very large fires (Bak et al., 1990; Green, 1989; Hargrove et al., in press; von Niessen and Blumen, 1986). The exact value of the critical threshold (i.e., the degree of connectance of fuels) depends on both climate and the rate of recovery of the forest (Gardner et al., 1997; Hargrove et al., in press; Swetnam and Betancourt, 1990). Because landscape patterns of systems near the critical threshold reflect the effects of turbulent dynamics (Bak et al., 1990), the variability of forest structure at fine scales can be expected to be very high.

(ii) One useful property of systems near the critical threshold is that scaling rules can be developed to extrapolate fine-grained interactions to much larger areas (Beer and Entin, 1991; Orbach, 1986; Stauffer and Aharony, 1992). Scaling rules have been applied to different formulations of probabilistic models simulating fires within forested landscapes (MacKay and Jan, 1984) and the results found to be robust across different formulations (von Niessen and Blumen, 1986). This result is quite important because the study of forested areas requires that dynamics be predicted for very large regions. However, forest ecologists have yet to fully explore this result for other model types with the goal of developing simpler, more efficient models.

(iii) It is generally accepted, as noted previously, that forests subject to frequent fires are unlikely to be in state of equilibrium – either in terms of landscape-scale patterns of disturbance and regrowth, species composition and community structure, or ecosystem dynamics. What is equally important, especially for long-term projections of effects, is that this disequilibrium requires precise information regarding initial conditions (Keane et al., 1996a) and that shifting regimes will confound management practices intended to maintain specific sets of conditions (Baker, 1992).

(iv) Coupling ecosystem models with models of fire disturbance has produced several interesting and unanticipated results including increased levels of available nitrogen, increased net primary productivity and reduced levels of biomass and evapotranspiration (Keane et al., 1996a, b). Further efforts to link ecosystem models with models of fire behavior of other forested landscapes is especially needed.

(v) Small shifts in climate can produce dramatic shifts in fire regime, with subsequent changes in the age structure and the spatial arrangement of mature forest (Gardner et al., 1997).

(vi) The high level of uncertainty associated with the dynamics of disturbed systems, and the lack of understanding of processes operating at landscape scales (e.g., fire spotting, changes in synoptic weather conditions, local effects of fire on wind, etc.) requires that model projections be expressed in terms of risk of change (Clark *et al.*, 1994; Hargrove *et al.*, in press; Fosberg *et al.*, 1996).

The categories developed for this review were intended, in part, to illustrate the broad range of purposes and assumptions characteristic of each model type. Although other classification schemes have been offered (e.g., Li and Apps, 1995), all such schemes are confounded by fundamental differences in the structure of the models and differences in the data used for their development and application. Only when simultaneous simulations of models are performed can the importance of specific differences in processes considered and structure of models be evaluated. What is interesting is that the models within our six categories share three distinct "threads" of development from: (i) the fire behavior methods based on the Rothermel model of fire spread (Rothermel, 1972); (ii) probabilistic models using a gridded landscape and methods adapted from the methods of cellular automata; (iii) and statistical models based on sampling of fire frequency and size from empirically derived frequency distributions. The shared features among apparently different models makes differences in model structure and performance even more difficult to evaluate.

It is equally clear that current models do not fully represent all factors important in predicting landscape-scale patterns of fire disturbance. For instance, the process of fire spotting remains poorly understood and infrequently included in fire models; the dynamics of winds during fire events are difficult to quantify in space and time making the scale-dependence of their effects difficult to evaluate; the description of the spatial heterogeneity of fuels, fuel moisture and weather remains coarse and uncertain; and factors contribution to land-use change are usually neglected. The inclusion of realistic dynamics of forest regrowth and succession, as well as the coupling of community and ecosystem dynamics, has only recently been attempted (Keane *et al.*, 1996a). These methods need to be expanded to a broader spectrum of forest types and climatic zones.

Clearly the inclusion of these additional variables would serve to greatly increase the complexity of current models. However, there are clear dangers to increases in model complexity, in part because the uncertainties associated with these details may overwhelm the reliability of model predictions. The additional complexity makes the process of model verification more difficult and the addition of unknowns may increase the bias associated with predictions. There are, however, corresponding dangers in the exclusive use of simple models. Key processes that affect both fine-grained dynamics and broad-scale trends may be absent from simple models unless systematic comparisons between models and data are performed. Although this problem is an old and familiar one (Gardner *et al.*, 1982;

Bartell *et al.*, 1988), these issues are only now being considered in the adequacy and reliability of disturbance models which must make predictions over extremely broad temporal and spatial scales.

Three suggestions are offered here to resolve some of these difficulties. First, it is important to remember that the results of model simulations are simply the logical consequence of a series of hypotheses and assumptions. The relevant question then is, how different would the projections be if one or more assumptions were relaxed, different processes included or disregarded, and an alternative set of hypotheses tested? This important question can best be evaluated by cross-comparison of model sensitivities with relevant empirical studies and the quantification of differences in response between models. Examples of such comparisons can be found in other modeling studies (e.g., Melillo *et al.*, 1995; Rose *et al.*, 1991*a*, *b*). Inter-model comparisons, which use similar data sets and disturbance scenarios to assess model sensitivities, will allow the relative importance of specific processes to be quantified. Second, there must be a renewed effort to develop simple yet reliable models for extrapolation of results across scales. Comparisons of differences between models depending on their temporal and spatial resolution can be inferred from Table 7.3. However, rigorous sets of model experiments are necessary to estimate errors and uncertainties associated with the cross-scale predictions. If scaling rules can be developed for complex models then the effects of climate change (which acts at broad scales) might be revealed in changes in disturbance regimes and corresponding shifts in forest structure. Finally, the study of fire in natural systems must be expanded to include systems altered by changes in land use. In many regions of the world changes in agricultural practices are increasing the risk of fire, while in other regions forest fragmentation is reducing the risk of large fires. These shifts in land use confound the estimates of net carbon storage and nutrient release from forest disturbance.

Acknowledgments

Research for this paper was supported by funding from the National Science Foundation (BSR-9016281 and BSR-9018381). This paper is contribution number 3156 of the Appalachian Laboratory, University of Maryland Center for Environmental Science.

References

Agee, J. K. (1996). Fire regimes and approaches for determining fire history. In *The use of fire in restoration*, ed. C. Hardy and S. F. Arno. USDA Forest Service General Technical Report INT-GTR-341.

Albini, F. A. (1976). *Computer-base models of wildland fire behavior: A user's manual.* USDA

Forest Service unnumbered publication, Intermountain Forest and Range Experiment Station, Ogden, UT.

Ales, R. F., Martin, A., Ortega, F. and Ales, E. E. (1992). Recent changes in landscape structure and function in a mediterranean region of SW Spain (1950–1984). *Landscape Ecology*, **7**, 3–18.

Alexander, M. E. (1985). Estimating the length-to-breadth ratio of elliptical forest fire patterns. In *Proceedings of the 8th Conference on Fire and Forest Meteorology*, 29 April–2 May 1985, Detroit, Michigan Society of American Foresters, Bethesda, MD. SAF Publ. 85-04, pp. 287–304.

Anderson, H. E. (1983). Predicting wind-driven wild land fire size and shape. USDA Forest Service, Intermountain Forest and Range Experiment Station, Research Paper, INT-305, Ogden, UT.

Andrews, P. L. (1986). *BEHAVE: fire behavior prediction and fuel modeling system–BURN subsystem*, Part 1. USDA Forest Service Intermountain Research Station Ogden, UT, General Technical Report INT-194.

Antonovski, M. Ya., Ter-Mikaelian, M. T. and Furyaev, V. V. (1992). A spatial model of long-term forest fire dynamics and its application to forests in western Siberia. In *A systems analysis for the global boreal forest*, ed. H. H. Shugart, R. Leemans and G. B. Bonan, pp. 373–403. Cambridge: Cambridge University Press.

Arno, S. F. (1980). Forest fire history in the northern Rockies. *Journal of Forestry*, **78**, 460–5.

Auclair, A. N. D. and Carter, T. B. (1993). Forest wildfires as recent source of CO_2 at northern latitudes. *Canadian Journal of Forest Research*, **23**, 1528–36.

Bak, P., Chen, K. and Tang, C. (1990). A forest-fire model and some thoughts on turbulence. *Physics Letters A*, **147**, 297–300.

Baker, W. L. (1992). Effects of settlement and fire suppression on landscape structure. *Ecology*, **73**, 1879–87.

Baker, W. L., Egbert, S. L. and Frazier, G. F. (1991). A spatial model for studying the effects of climatic change on the structure of landscapes subject to large disturbances. *Ecological Modelling*, **56**, 109–25.

Balling, Jr, R. C., Meyer, G. A. and Wells, S. G. (1992). Climate change in Yellowstone National Park: is the drought-related risk of wildfires increasing? *Climatic change*, **22**, 35–45.

Bartell, S. M., Cale, W. G., O'Neill, R. V. and Gardner, R. H. (1988). Aggregation error, research objectives, and relevant community structure. *Ecological Modelling*, **41**, 157–68.

Beer, T. and Entin, I. G. (1991). Fractals, lattice models, and environmental systems. *Environmental International*, **17**, 519–53.

Bergeron, Y. (1991). The influence of island and mainland lakeshore landscape on boreal forest fire regimes. *Ecology*, **72**, 1980–92.

Bessie, W. C. and Johnson, E. A. (1995). The relative importance of fuels and weather on fire behavior in subalpine forests. *Ecology*, **76**, 747–76.

Bonan, G. B., Shugart, H. H. and Urban, D. L. (1990). The sensitivity of some high-latitude boreal forests to climatic parameters. *Climatic Change*, **16**, 9–29.

Boychuk, D., Perera, A. H., Ter-Mikaelian, M. T., Martell, D. L. and Li, C. (1997). Modelling the effect of spatial scale and correalted fire disturbances on forest age distribution. *Ecological Modelling*, **95**, 145–64.

Caswell, H. and Etter, R. J. (1993). Ecological interactions in patchy environments: from patch-occupancy models to cellular automata. In *Patch dynamics*, ed. S. A. Levin, T. M. Powell, J. H. Steele, New York: Springer-Verlag.

Catchpole, A., Alexander, M. E. and Gill, A. M. (1992). Elliptical-fire perimeter- and area-intensity distributions. *Canadian Journal of Forest Research*, **22**, 968–72.

Cheney, N. P., Gould, J. S. and Catchpole, W. R. (1993). The influence of fuel, weather and fire shape variables on fire-spread in grasslands. *International Journal of Wildland Fire*, **3**, 31–44.

Clark, J. S. (1989). Effects of long-term water balances on fire regime, north-western Minnesota. *Journal of Ecology*, **77**, 989–1004.

Clark, J. S. (1990). Fire and climate change during the last 750 yr in northwestern Minnesota. *Ecological Monographs*, **60**, 135–59.

Clark, J. S. and Richard, P. J. H. (1996). The role of paleofire in boreal and other cool-coniferous forests. In *Fire in ecosystems of boreal Eurasia*, ed. J. G. Goldammer and V. V. Furyaev, pp. 65–89. The Netherlands: Kluwer Academic Publishers.

Clarke, K. C., Brass, J. A. and Riggan, P. J. (1994). A cellular automaton model of wildfire propagation and extinction. *Photogrammetric Engineering and Remote Sensing*, **60**, 1355–67.

Cofer, W. R., III, Winstead, E. L., Stocks, B. J., Cahoon, D. R., Goldammer, J. G. and Levine, J. S. (1996). Composition of smoke from North American boreal forest fires. In *Fire in ecosystems of boreal Eurasia*, ed. J. G. Goldammer and V. V. Furyaev, pp. 465–75. The Netherlands: Kluwer Academic Publishers.

Cohen, J. D. (1986). *Estimating fire behavior with FIRECAST: User's manual.* United States Department of Agriculture, Forest Service, Pacific Southwest Forest and Range Experiment Station, General Technical Report PSW-90.

Davis, F. W. and Burrows, D. A. (1993). Modeling fire regime in Mediterranean landscape. In *Patch dynamics*, ed. S. A. Levin, T. M. Powell and J. H. Steele, pp. 247–59. New York: Springer-Verlag.

Duarte, J. A. M. S., Marques Carvalho, J. and Ruskin, H. J. (1992). The direction of maximum spread in anisotropic forest fires and its critical properties. *Physica A*, **183**, 411–21.

Finney, M. A. (1998). *FARSITE: Fire area simulation – model development and evaluation.* USDA Forest Service, Rocky Moutain Research Station, Research Paper RMRS-RP-4.

Fosberg, M. A., Stocks, B. J. and Lyham, T. J. (1996). Risk analysis in strategic planning. In *Fire in ecosystems of boreal Eurasia*, ed. J. G. Goldammer and V. V. Furyaev, pp. 495–504. The Netherlands: Kluwer Academic Publishers.

Gardner, R. H., Cale, W. G. and O'Neill, R. V. (1982). Robust analysis of aggregation error. *Ecology*, **63**, 1771–9.

Gardner, R. H., Hargrove, W. W., Turner, M. G. and Romme, W. H. (1997). Climate change, disturbances and landscape dynamics. In *Global change and terrestrial ecosystems*, ed. B. H. Walker and W. L. Steffen, pp. 149–72. IGBP Book Series No. 2, Cambridge: Cambridge University Press.

Gauthier, S., Leduc, A. and Bergeron, Y. (1996). Forest dynamics modelling under natural fire cycles: a tool to define natural mosaic diversity for forest management. *Environmental Monitoring and Assessment*, **39**, 417–34.

Goldammer, J. G. and V. V. Furyaev (1996). Fire in ecosystems of boreal Eurasia: Ecological impacts and links to the global system. In *Fire in ecosystems of boreal*

Eurasia, ed. J. G. Goldammer and V. V. Furyaev, pp. 1–20. The Netherlands: Kluwer Academic Publishers.

Green, D. G. (1989). Simulated effects of fire, dispersal, and spatial pattern on competition within forest mosaics. *Vegetatio*, **82**, 139–53.

Hargrove, W. W., Gardner, R. H., Turner, M. G., Romme, W. H. and Despain, D. G. (1999). Simulating fire patterns in heterogeneous landscapes. *Ecological Modelling*, in press.

Heinselman, M. L. (1981). Fire intensity and frequency as factors in the distribution and structure of northern ecosystems. In *Fire regimes and ecosystem properties*, ed. H. A. Mooney, T. M. Bonnicksen, N. L. Christensen, J. E. Lotan and W. A. Reiners, pp. 7–57. USDA Forest Service Gen. Tech. Rep. WO-26.

Hoganson, H. H. and Rose, D. W. (1984). A simulation approach for optimal management scheduling. *Forest Science*, **30**, 220–38.

Hogeweg, P. (1988). Cellular automata as a paradigm for ecological modeling. *Applications in Mathematics and Computation*, **27**, 81–100.

Hunter, M. L., Jr. (1993). Natural fire regimes as spatial models for managing boreal forests. *Biological Conservation*, **65**, 115–20.

Johnson, E. A. (1992). *Fire and vegetation dynamics: studies from the North American boreal forest*. New York, NY: Cambridge University Press.

Johnson, E. A. and Larsen, C. P. S. (1991). Climatically induced change in fire frequency in the southern Canadian Rockies. *Ecology*, **72**, 194–201.

Johnson, E. A. and Van Wagner, C. E. (1985). The theory and use of two fire history models. *Canadian Journal of Forest Research*, **15**, 214–20.

Johnson, E. A., Fryer, G. I. and Heathcott, M. J. (1990). The influence of man and climate on frequency of fire in the interior wet belt forest, British Columbia. *Journal of Ecology*, **78**, 403–12.

Johnson, E. A. and Wowchuk, D. R. (1993). Wildfires in the southern Canadian Rocky Mountains and their relationship to mid-tropospheric anomalies. *Canadian Journal of Forest Research*, **23**, 1213–22.

Kasischke, E. S., Christensen, N. L. Jr. and Stocks, B. J. (1995). Fire, global warming, and the carbon balance of boreal forests. *Ecological Applications*, **5**, 437–51.

Keane, R. E., Arno, S. F. and Brown, J. K. (1989). FIRESUM – An ecological process model for fire succession in Western conifer forests. USDA Forest Service, Intermount. Forest Range Experimental Station, Ogden, UT, General Technical Report INT-266, 76pp.

Keane, R. E., Ryan, K. C. and Running, S. W. (1996*a*). Simulating effects of fire on northern Rocky Mountain landscapes with the ecological process model FIRE-BGC. *Tree Physiology*, **16**, 319–31.

Keane, R. E., Morgan, P. and Running, S. W. (1996*b*). FIRE-BGC – A mechanistic ecological process model for simulating fire succession on coniferous forest landscapes of the Northern Rocky Mountains. Intermountain Research Station Research Paper INT-RP-484.

Kessell, S. R. (1979). *Gradient modeling, resource and fire management*. New York: Springer-Verlag.

Levin, S. A. (1992). The problem of pattern and scale in ecology. *Ecology*, **73**, 1943–67.

Li, C. and Apps, M. J. (1995). Disturbance impacts on forest temporal dynamics. *Water, Air and Soil Pollution*, **82**, 429–36.

MacKay, G. and Jan, N. (1984). Forest fires as critical phenomena. *Journal of Physics A: Mathematics General*, **17**, L757–60.

Malanson, G. P. (1984). Fire history and patterns of California coastal sage scrub. *Vegetatio*, **57**, 121–8.

Melillo, J. M. and VMAP participants. (1995). Vegetation/ecosystem modeling and analysis project: comparing biogeography and biogeochemistry models in a continental-scale study of terrestrial ecosystem responses to climate change and CO_2 doubling. *Global Biogeochemical Cycles*, **9**, 407–37.

Meyer, G. A., Wells, S. G., Balling, R. C. Jr. and Jull, A. J. T. (1992). Response of alluvial systems to fire and climate change in Yellowstone National Park. *Nature*, **357**, 147–50.

Millspaugh, S. H. (1997). Late-glacial and holocene variations in fire frequency in the Central Plateau and Yellowstone-Lamar Province of Yellowstone National Park. Dissertation. University of Oregon, Eugene, OR.

Milne, B. T. (1992). Spatial aggregation and neutral models in fractal landscape. *American Naturalist*, **139**, 32–57.

Moreno, J. M. and Oechel, W. C., eds. (1994). *The role of fire in Mediterranean-type ecosystems*. New York: Springer-Verlag.

Moritz, M. (1997). Analyzing extreme distriburbance events: Fire in Los Padres National Forest. *Ecological Applications*, **7**, 1252–62.

Neilson, R. P. (1993). Transient ecotone response to climatic change: some conceptual and modeling approaches. *Ecological Applications*, **2**, 385–95.

Noble, I. R. and Gitay, H. (1996). A functional classification for predicting the dynamics of landscapes. *Journal of Vegetation Science*, **7**, 329–36.

Orbach, R. (1986). Dynamics of fractal networks. *Science*, **231**, 814–19.

Overpeck, J. T., Bartlein, P. J. and Webb, T. III. (1991). Potential magnitude of future vegetation change in Eastern North America: Comparison with the past. *Science*, **254**, 692–5.

Ratz, A. (1995). Long-term spatial patterns created by fire: a model oriented towards boreal forests. *International Journal of Wildland Fire*, **5**, 25–34.

Renkin, R. A. and Despain, D. G. (1992). Fuel moisture, forest type, and lightning-caused fire in Yellowstone National Park. *Canadian Journal of Forestry Research*, **22**, 37–45.

Romme, W. H. (1982). Fire and landscape diversity in subalpine forests of Yellowstone National Park. *Ecological Monographs*, **52**, 199–221.

Romme, W. H. and Despain, D. G. (1989). Historical perspectives on the Yellowstone fires of 1988. *BioScience*, **39**, 695–9.

Romme, W. H. and Turner, M. G. (1991). Implications of global climate change for biogeographic patterns in the greater Yellowstone ecosystem. *Conservation Biology*, **5**, 373–86.

Rose, K. A., Brenkert, A. L., Cook, R. B., Gardner, R. H. and Hettelingh, J. P. (1991a). Systematic comparison of ILWAS, MAGIC, and ETD watershed acidification models: 1. Mapping among model inputs and deterministic results. *Water Resources Research*, **27**, 2577–98.

Rose, K. A., Brenkert, A. L., Cook, R. B., Gardner, R. H. and Hettelingh, J. P. (1991b). Systematic comparison of ILWAS, MAGIC, and ETD watershed acidification models: 2. Monte Carlo analysis under regional variability. *Water Resources Research*, **27**, 2591–603.

Rothermel, R. C. (1972). A mathematical model for predicting fire spread in wild land fuels. USDA Forest Service, Intermountain Forest and Range Experiment Station Research Paper, INT-115, Ogden UT.

Rothermel, R. C. (1991). Predicting behavior of the 1988 Yellowstone fires: Projections versus reality. *International Journal of Wildland Fire*, **1**, 1–10.

Running, S. W. and Coughlan, J. C. (1988). A general model of forest ecosystem processes for regional applications. I. Hydrologic balance, canopy gas exchange and primary production processes. *Ecological Modelling*, **42**, 125–54.

Schneider, D. C. (1994). *Quantitative ecology: spatial and temporal scaling*. San Diego, CA: Academic Press.

Stauffer, D. and Aharony, A. (1992). *Introduction to percolation theory*, 2nd edn. London: Taylor & Francis.

Suffling, R., Lihou, C. and Morand, Y. (1988). Control of landscape diversity by catastrophic disturbance: a theory and a case study of fire in a Canadian boreal forest. *Environmental Management*, **12**, 73–8.

Swetnam, T. W. (1993). Fire history and climate change in giant sequoia groves. *Science*, **262**, 885–9.

Swetnam, T. W. and Baisan, C. H. (1996). Historical fire regime patterns in the southwestern United States since AD 1200. In *Fire effects in southwestern forests: proceedings of the second La Mesa fire symposium*, ed. C. D. Allen, pp. 11–32. Los Alamos, 1994. USDA General Technical Report RM-GTR-286.

Swetnam, T. W. and Betancourt, J. L. (1990). Fire-southern oscillation relations in southwestern United States. *Science*, **249**, 1017–20.

Trabaud, L. (1994). The effect of fire on nutrien losses and cycling in a *Quercus coccifera* garrigue (southern France). *Oecologia*, **99**, 379–86.

Turner, M. G., O'Neill, R. V., Gardner, R. H. and Milne, B. T. (1989). Effects of changing spatial scale on the analysis of landscape pattern. *Landscape Ecology*, **3**, 153–62.

Turner, M. G. and Romme, W. H. (1994). Landscape dynamics in crown fire ecosystems. *Landscape Ecology*, **9**, 59–77.

Turner, M. G., Romme, W. H. and Gardner, R. H. (1994). Landscape disturbance models and the long-term dynamics of natural areas. *Natural Areas Journal*, **13**, 3–11.

Vasconcelos, M. J. and Guertin, D. P. (1992). FIREMAP – simulation of fire growth with a geographic information system. *International Journal of Wildland Fire*, **2**, 87–96.

von Niessen, W. and Blumen, A. (1986). Dynamics of forest fires as a directed percolation model. *Journal of Physics A: Mathematics General*, **19**, L289–93.

von Niessen, W. and Blumen, A. (1988). Dynamic simulation of forest fires. *Canadian Journal of Forestry Research*, **18**, 805–12.

Wallace, G. (1993). A numerical fire simulation model. *International Journal of Wildland Fire*, **3**, 111–16.

Whelan, R. J. (1995). *The ecology of fire*. New York: Cambridge University Press.

Wolfram, S. (1984). Cellular automata as models of complexity. *Nature*, **311**, 419–24.

8

Mechanistic modeling of landscape fire patterns
Mark A. Finney

Introduction

Fire as a landscape process is of broad interest to ecologists and land managers. Fires alter forest age-distributions (Heinselman, 1973; Van Wagner, 1978), are sensitive to climate (Balling *et al.*, 1992, Swetnam and Bettancourt, 1990; Swetnam, 1993; Timoney and Wein, 1991), can be manipulated by fire suppression (Baker, 1992; Barrett, 1994), and affect directions for land management policy (Hunter, 1993; Lesica, 1996; Huff *et al.*, 1995; Johnson *et al.*, 1995; Omi, 1996). Fire models are used for ecological research into spatial disturbance and recovery patterns (Turner *et al.*, 1989; Green, 1989; Ratz, 1996; Boychuk *et al.*, 1997), forest landscape dynamics (Keane *et al.*, 1996; Mladenoff *et al.*, 1996; Boychuk and Perera, 1997; Li *et al.*, 1997), and fire planning (Kessell, 1976; Methven and Feuenkes, 1988; Beer, 1990). For ecological modeling, fire or disturbance patterns have usually been simulated by directly applying stochastic algorithms to modify spread directions and rates (Turner *et al.*, 1989; Green, 1989; Baker *et al.*, 1991; Mladenoff *et al.*, 1996; Gardner *et al.*, 1996), or to burn a proportion of the landscape area (Ratz, 1996; Li *et al.*, 1997). Another approach is to focus on simulating the fire processes so that the cause-and-effect relationships for a given pattern can be studied. There is great interest in analyzing landscape patterns that result from fire to determine how those patterns relate to ecological theories (Romme, 1982; Baker, 1992; Suffling *et al.*, 1988). Because many landscape patterns are produced by variation in fire behavior, a mechanistic simulation of fire behavior and fire growth is useful for explaining how, why, and when such patterns can form.

Mechanistic simulation models (e.g., process models) try to represent a system as a set of fundamental processes that each describe cause and effect relationships between physical variables. Often, empirical relationships must substitute for individual processes that are not understood well enough for a more detailed description. The general mechanistic approach is useful for studying fire patterns because it allows an evaluation of: (i) the role of specific environmental factors in creating patterns of fire behavior and effects, (ii) the effects of each component process on the simulated fire pattern, and (iii) how spatial and temporal dependencies affect fire patterns.

A mechanistic simulation of fire growth must contain components that describe specifically how fuels, weather, and topography affect fire behavior. Wildland fire research over the past several decades has led to the development of numerous models for different fire behaviors (Rothermel, 1972; Albini, 1976, 1979; Van Wagner, 1977; Forestry Canada Fire Danger Group, 1992). Systems such as BEHAVE in the US (Burgan and Rothermel, 1984; Andrews, 1986) and the Canadian Fire Behavior Prediction System (Forestry Canada Fire Danger Group, 1992) have incorporated these models as tools for fire management applications. For mechanistically simulating fire as an ecological process, these models represent the crucial link between the largely independent environmental variables and the fire behavior that produces those ecological effects.

Recent advances in mathematics (Richards, 1990, 1995) as well as computing have provided a means of linking separate models of fire behavior into a practical spatial simulation of two-dimensional wildland fire growth (Finney, 1998). In the US, the simulation model *FARSITE* (Fire Area Simulator) integrates component models for surface fire, crown fire, fire acceleration, spotting, and fuel moisture (Finney, 1994, 1998). *FARSITE* was originally developed as a tool for making long-range projections of prescribed natural fires in large wilderness areas of the western US (Finney, 1994; Finney and Ryan, 1995). In has since been applied to other problems including planning for fire management activities (Van Wagtendonk, 1996) and ecological modeling of landscape fire patterns as a component of spatial forest succession models (Keane *et al.*, 1996*a*, *b*).

The mechanistic structure of *FARSITE* has allowed some insight into the causes of variable fire behavior across a landscape. Simulated fire behavior patterns can be related to their causative factors that change both spatially and temporally. As the fire front expands across a landscape, it encounters different fuels and topography under particular weather conditions that may be unique to that place and time. Spatial heterogeneity in fire behavior results because of the interdependent combinations of variables that drive fire behavior. Weather is obviously the most variable influence on fire behavior in space and time. Changing temperature and humidity affects fuel moisture throughout the day, and differentially by elevation, slope, and aspect. Winds change speed and direction and strongly influence the fire spread rate, direction, and intensity. Fuel structure and topography vary with space but are constant in time (at least during a single fire). The ignition location establishes the context for relative fire spread direction on that landscape (backing, flanking, or heading) that strongly affects fire behavior for a given set of environmental conditions. The ignition location also establishes the possible routes that fire can travel to other points on the landscape.

Variable fire behavior causes variable fire effects. This variation occurs at all spatial scales but is especially noticeable, for example, within large burns in forests affected by crown fire. Here, the wide range in potential fire behavior makes differences in fire effects more obvious. The different behavior of surface fires and crown fires causes a wide range in tree mortality and crown damage that is highly visible (Agee and Huff, 1980; Despain et al., 1989; Morrison and Swanson, 1990; Turner and Romme, 1994). Causes of this variation can be difficult to interpret after the fire, sometimes prompting descriptions of these patterns as "random" or "stochastic", especially if the fire's progress was not observed. Explanations of such spatial patterns can often be found, however, once the time domain for fire travel has been established and the pre-fire landscape conditions have been mapped. Such analysis requires detailed temporal data on weather and winds as well as spatial information on fuels, vegetation, topography, and fire growth (Wade and Ward, 1973; Anderson, 1968; Simard et al., 1983; Alexander, 1991; Rothermel, 1993; Butler and Reynolds, 1997). Although fire will not be completely predictable, it is likely to be understandable in mechanistic terms as a time- and space-dependent physical and ecological phenomenon.

This chapter presents a review of fire growth simulation, describes the constituent fire behavior models incorporated into *FARSITE*, and demonstrates the spatial consequences of this approach to fire behavior variation across the landscape.

Fire growth modeling

Fire growth models originated with the need to calculate fire size and perimeter length for fire-fighting operations. The earliest research efforts were directed toward determining the shape of fires burning under relatively uniform environmental conditions (Hornby, 1936; Fons, 1946). Using a constant fire shape, the relative location of the ignition point, and an estimate of the forward spread rate, changes in fire size and perimeter could be calculated as a function of time. The fire shape most commonly used has been the ellipse (Van Wagner, 1969; Alexander, 1985; Andrews, 1986). It is mathematically simple and apparently fits well to most empirical data on fire shapes (Green et al., 1983). Fire shapes vary from circular without wind on flat terrain to highly elongated or eccentric ellipses produced by high winds and steep slopes (Alexander, 1985). Some evidence suggests that fires may be better described as egg-shaped, ovoid, or double ellipsed (Peet, 1967; Albini, 1976; Anderson, 1983). The practical importance of using one of these more exotic shapes over a simple ellipse may be negligible, however. The differences occur largely in the backing or rearward flanking directions that constitute a low proportion of spread and intensity compared to the forward flanks and head. Furthermore, if environmental conditions are constant, the simple ellipse can be easily used for all fire growth modeling without a computer (Van Wagner, 1969), even for producing spatially explicit intensities and spread rates (Catchpole et al., 1982, 1992).

Environmental conditions do not remain constant throughout the duration of most fires. As the fire gets larger, it encounters different topography and fuel types. The longer it burns, the more likely it will be subjected to changing weather.

This environmental heterogeneity requires more complex simulation methods to produce the proper effect on spatial patterns of fire growth and behavior.

Simulation of two-dimensional fire growth since the 1970s can be classified into one of two approaches: cellular or vector. The difference lies in how space and time variables are used. Cellular models use the fixed distances between regularly spaced grid cells to solve for fire's arrival time from one cell to the next. Vector models use a specified time interval to solve for the distance fire would travel in a calculated direction. Although the differences appear as simple inverses, their ramifications are far-reaching, and have limited the ability of cellular models to simulate expected fire shapes under heterogeneous environmental conditions (French, 1992). This has compromised their utility as operational tools in fire management and their accuracy in implementing fire behavior models in two dimensions.

The cellular approach involves a discrete process of "ignitions" within the regular structure of a gridded landscape. The earliest implementation of this by Kourtz and O'Regan (1972) showed how fire could travel along a fixed number of radii between cells under homogenous conditions of fuels, weather, and topography. Model iterations update the arrival time from burning cells to each unburned cell connected to it within some radial distance. The radius determines the number of cells involved in each iteration and consequently the number of angular sides acquired by the fire (O'Regan *et al.*, 1976; Feunekes, 1991; French, 1992). This angular distortion to fire shape is a serious problem for practical uses. Distortion results when fire travel is constrained to a fixed set of pathways between cells when more direct routes, and shorter arrival times, are possible. The distortion can be minimized under homogeneous environmental conditions by increasing the radius for each iteration (O'Regan *et al.*, 1976; French, 1992; Xu and Lathrop, 1994). While increasing the demand on computing power, this also produces a legacy or holdover effect that influences fire growth long after a temporal change occurs (i.e., wind direction or speed). Cells not ignited before the change still contain arrival times that were reduced during earlier conditions; the legacy arrival times in these cells continue to influence the sequence of ignitions long after new conditions begin affecting the fire. Resetting all unburned cells to an initial state merely removes the benefit of greater precision intended originally by increasing the radius. Many workers have experimented with this and related cellular techniques (see Green, 1983; Feunekes and Methven, 1987; Vasconcelos *et al.*, 1990; Ball and Guertin, 1992). Other techniques for modeling fire growth as cellular automata include the transfer of fractional burned area (Richards, 1988; Karafyllidis and Thanailakis, 1997), probability-driven models (Von Niessen and Blumen, 1988; Beer and Enting, 1990; French, 1992; Gardner *et al.*, 1996; Ratz, 1996), or fractal models (Clarke *et al.*, 1994). Under uniform conditions, almost any technique can probably reproduce idealized fire shapes (ellipsoids). They can also be used to create spatial patterns of burned cells. Cellular models in general, however, have not been able to produce the expected responses under test conditions that intro-

Fig. 8.1. Schematic of Huygens' principle for fire front expansion. (a) The fire front is defined by vertices, (b) Uniform conditions at each vertex allow constant shapes and sizes of elliptical wavelets and produces elliptical fire growth (gray) over a finite time step, and (c) non-uniform conditions where the local fuels, weather and topography at each vertex determine different shapes, sizes, and orientation of wavelets, resulting in complex fire growth patterns.

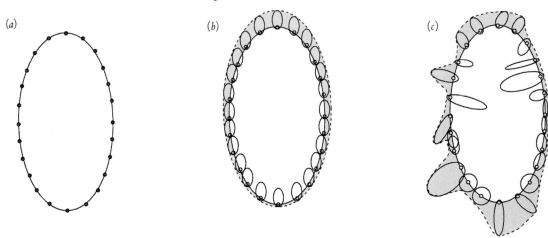

duce spatial and temporal heterogeneity (French, 1992). For this reason, the vector approach was chosen for use in developing the *FARSITE* simulation.

The vector or wave-type models avoid the problems encountered by cellular models in dealing with spatial and temporal heterogeneity. With vector models, both the direction and distance of fire travel are determined independently of the resolution of the spatial input data. Here, the fire front is represented as a series of vertices (Fig. 8.1a) that collectively define the edge of a spreading fire at a particular instant in time (Sanderlin and Sunderson, 1975; Anderson et al., 1982). The environmental conditions local to each vertex are used to compute the forward fire spread rate and its direction. The fire is propagated from each vertex assuming "Huygens' principle" applies to a fire front as it was originally intended for light waves. Huygens' principle states that a wave front can be propagated using any point on its edge as an independent source of a new "wavelet" (Anderson et al., 1982). The wavelets refer to elliptical fires of a size determined by a fixed time step and the fire spread rate local to each vertex. The orientation of these elliptical wavelets is determined by the maximum fire spread direction θ, calculated as the resultant vector of local wind and slope (Finney, 1998). The shape of each wavelet is a function of the midflame wind-slope vector (U, m s^{-1}, expressed as an effective windspeed) that determines the eccentricity of an ellipse (length-to-breadth ratio *LB*) under locally uniform conditions. Several *LB* equations have been developed (Alexander, 1985; Andrews, 1986; Rothermel, 1991). *FARSITE* uses one modified from Anderson (1983):

Fig. 8.2. Elliptical dimensions and parameters used by Richards (1990) equations for fire growth (Eqs. 3 and 4).

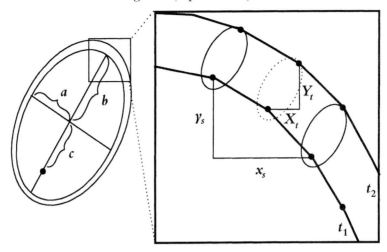

$$LB = 0.936 \; e^{(0.2566U)} + 0.461 \; e^{(-0.1548U)} - 0.397 \quad (1)$$

For surface fires, the windspeed used is reduced for canopy cover (%) and tree height (*m*) (Albini and Baughman, 1979) but for crown fires the overstory wind conditions are used. The rear focus of the ellipse is assumed to be the ignition point (Alexander, 1985), such that the heading-to-backing ratio *HB* is:

$$HB = (LB + (LB^2 - 1)^{0.5})/(LB - (LB^2 - 1)^{0.5}) \quad (2)$$

Conceptually, the fire front is expanded over each time step by aggregating the individual wavelets into an "envelope" around the previous fire front (Fig. 8.1(*b*)). Because the conditions at each vertex produce independent elliptical wavelets of potentially different shapes and sizes, this technique is flexible in representing highly heterogeneous conditions encountered by a fire in both space and time (Fig. 8.1(*c*)).

A number of mathematical methods have been developed for propagating wavelets with this technique (Richards, 1990, 1995; Knight and Coleman, 1993; Wallace, 1993; and Dorrer, 1993). *FARSITE* uses equations from Richards (1990) with modifications for sloping terrain:

$$X_t = \frac{a^2 \cos\theta \, (x_s \sin\theta + y_s \cos\theta) - b^2 \sin\theta \, (x_s \cos\theta - y_s \sin\theta)}{(b^2 \, (x_s \cos\theta + y_s \sin\theta)^2 - a^2 \, (x_s \sin\theta - y_s \cos\theta)^2)^{1/2}} + c \sin\theta \quad (3)$$

$$Y_t = \frac{-a^2 \sin\theta \, (x_s \sin\theta + y_s \cos\theta) - b^2 \cos\theta \, (x_s \cos\theta - y_s \sin\theta)}{(b^2 \, (x_s \cos\theta + y_s \sin\theta)^2 - a^2 \, (x_s \sin\theta - y_s \cos\theta)^2)^{1/2}} + c \cos\theta \quad (4)$$

where X_t and Y_t are the spread rate components of fire growth at each vertex, the *a*, *b*, and *c* parameters describe the elliptical dimensions (Fig. 8.2), x_s and y_s are the directional components that determine the orientation angle of the vertex on the fire front ($x_{i-1}-x_{i+1}$, $y_{i-1}-y_{i+1}$), and θ is the direction of maximum fire spread (resultant wind slope vector). It is critical to recognize that on sloping topography, all inputs and outputs from these equations relate to the local surface plane not to the horizontal plane. Storage and display of fire growth, however, must be on the horizontal plane. This requires all input parameters to be transformed first to terrain-following coordinates for use in Eqns. (3) and (4) and the outputs then transformed back to horizontal coordinates (see Finney, 1998).

FARSITE description

FARSITE (Fire Area Simulator) is a stand-alone fire growth model that incorporates fire behavior models for surface fire, crown fire, fire acceleration, spotting, and fuel moisture using the vector modeling technique of Huygens' principle. The data inputs to *FARSITE* consist of eight raster GIS themes (Table 8.1) that describe the terrain, surface fuels, and crown fuels (Fig. 8.3: see color section), and two data streams for wind and weather (Table 8.2 and Table 8.3). Other inputs are fuel-specific initial fuel moistures and spread rate adjustment factors used for calibrating the model's output with observed fire progression.

The weather stream contains precipitation, temperature, and humidity patterns on a daily basis (Table 8.2). The temperature and humidity values are maxima and minima that allow cosine interpolation of their values for any time of the day. The wind stream (Table 8.3) contains event-driven temporal changes in horizontal wind speed and direction at the US standard reference height (6.1m or 20 ft). It also specifies cloud cover (percentages) that decrease solar radiation reaching the top of the vegetation.

Fire behavior is assumed to follow a typical sequence of activity. First, the behavior or a surface fire is calculated (Rothermel, 1972). If the environmental conditions permit, this fire may transition to some form of crown fire (Van Wagner, 1977, 1993) that can initiate spotting (Albini, 1979). When environmental conditions change to produce faster spread rates at each time step, the fire is accelerated toward the new spread rate (Forest Canada Fire Danger Group, 1992) rather than jumping immediately to the faster rate.

Fuel moisture

Dead and live fuel moistures greatly affect fire behavior. Moisture content of fine dead fuels varies throughout the day according to temperature, humidity, solar irradiance, wind speed, and fuel size. The user provides an initial suite of fuel moisture conditions by surface fuel model for dead and live fuels. *FARSITE* then calculates moisture content (percentage of dry weight) of dead woody fuels in the

---

[1] *FARSITE* is available free of charge (www.montana.com/sem) and requires an IBM compatible computer (Pentium-class CPU or better) with Microsoft Windows 95 (16MB+RAM) or Windows NT (32MB).

Table 8.1. *Raster inputs to FARSITE and their usage in the simulation*

Raster theme	Units	Usage
Elevation	m, ft	Adiabatic adjustment of temperature and humidity from the reference elevation input with the weather stream
Slope	%, deg	Used for computing direct effects on fire spread, and along with Aspect, for determining the angle of incident solar radiation (along with latitude, date, and time of day) and transforming spread rates and directions from the surface to horizontal coordinates
Aspect	deg Az	(see Slope)
Fuel model		Provides the physical description of the surface fuel complex that is used to determine surface fire behavior (see Anderson, 1982). Included here are loadings (weight) by size class and dead or live categories, ratios of surface area to volume, and bulk depth
Canopy cover	%	Used to determine an average shading of the surface fuels (Rothermel et al., 1986) that affects fuel moisture calculations. It also helps determine the wind reduction factor that decreases windspeed from the reference velocity of the input stream (6.1 m above the vegetation) to a level that affects the surface fire (Albini and Baughman, 1979)
Crown height	m, ft	Affects the relative positioning of a logarithmic wind profile that is extended above the terrain. Along with canopy cover, this influences the wind reduction factor (Albini and Baughman, 1979), the starting position of embers lofted by torching trees, and the trajectory of embers descending through the wind profile (Albini, 1979)
Crown base Height	m, ft	Used along with the surface fire intensity and foliar moisture content to determine the threshold for transition to crown fire (Van Wagner, 1977; Alexander, 1988)
Crown bulk density	kg m^{-3} lb ft^{-3}	Used to determine the threshold for achieving active crown fire (Van Wagner, 1977, 1993)

1 h and 10 h timelag categories using the models from BEHAVE (Rothermel *et al.*, 1986; Hartford and Rothermel, 1991). The 100-h timelag fuel moisture is computed using the National Fire Danger Rating System (Bradshaw *et al.*, 1984). Moisture content of live fuels (shrubs, green grass, etc.) are input by the user but are not modified by the simulation. The environmental inputs required by the dead fuel moisture models are provided by the weather stream, wind stream, and

Table 8.2. *Sample of weather stream format*

Month	Day	Precip. (mm)	Hour (am)	Hour (pm)	Temp. (°C min)	Temp. (°C max)	Humid. (% max)	Humid. (% min)	Elevation (m)
7	31	0	0500	1500	7	29	65	25	915
8	1	0	0500	1500	7	29	65	25	915
8	2	0	0500	1500	13	28	55	18	915
8	3	2	0500	1500	13	25	55	18	915
8	4	0	0500	1500	7	26	66	25	915
8	5	0	0500	1500	14	31	56	25	915
8	6	0	0500	1500	16	32	45	24	915
8	7	0	0500	1500	16	32	45	24	915

The weather stream specifies precipitation, and maximum and minimum temperature, and humidity for each day at a particular reference elevation.

the spatial GIS data on terrain and forest cover. Solar irradiance influences the rate of fuel drying and is computed for a given pixel from the latitude, time of day, cloud cover, slope, aspect, and canopy structure (Rothermel et al., 1986). The air temperature, relative humidity, and windspeed are then used to compute moisture contents on an hourly basis for fuels in the 1 h and 10 h categories and daily for the 100-h fuels. For each fire behavior calculation, the current fuel moisture contents at a given time are computed from their initial conditions. This technique produces moisture data for only those locations that are involved in a calculation. It has proven faster than progressively calculating moisture contents for all cells across the landscape at each time step, regardless of their involvement in subsequent computations.

Surface fire

A *surface fire* burns in the grass, shrubs, or downed woody material lying in contact with the ground surface. *FARSITE* uses the Rothermel (1972) fire spread equation to compute the steady-state spread rate R (m min^{-1}) and fireline intensity I_b (kW m^{-1}) of a surface fire. Surface fuels are described by their loading (dry weight per unit area) by size class and live/dead category, the surface-area-to-volume ratios for each size class, and the bulk depth of the fuel complex. These parameters are combined to form a fuel model (Anderson, 1982). To calculate fire behavior for a given fuel model, the Rothermel (1972) equation requires data on the environmental conditions, including moisture content by size class for live and dead fuels (% dry weight), midflame wind speed, and topographic slope.

Crown fire

Crown fire describes fire burning in the foliage and fine branches of trees. *FARSITE* uses the crown fire criteria developed by Van Wagner (1977, 1993) to determine

Table 8.3. *Sample portion of the wind stream format*

Month	Day	Hour	Wind speed (6.1m, km/h)	Wind dir (Azimuth)	Cloud cover (%)
7	31	2000	20	234	0
7	31	2200	8	90	0
8	1	0000	6	90	0
8	1	0200	17	258	0
8	1	0400	18	271	0
8	1	0610	15	267	0
8	1	0800	13	260	0
8	1	1000	16	275	0
8	1	1200	14	230	0
8	1	1400	12	181	0
8	1	1600	9	182	0
8	1	1800	3	164	0
8	1	2000	2	174	0
8	1	2200	6	176	0
8	2	0000	5	189	0
8	2	0200	11	181	0
8	2	0400	5	176	0
8	2	0600	6	250	0
8	2	0800	9	250	0
8	2	1000	11	260	0
8	2	1200	31	270	0
8	2	1400	39	270	0
8	2	1600	42	270	0
8	2	1800	40	260	0
8	2	2000	29	200	0
8	2	2200	13	211	0
8	3	0000	5	190	0
8	3	0200	2	195	0
8	3	0400	5	200	0
8	3	0600	6	196	0
8	3	0800	9	200	0
8	3	1000	5	234	0
8	3	1200	6	240	0
8	3	1400	7	220	0

The wind stream contains wind speed and direction changes to the nearest minute along with cloud cover.

if a surface fire makes the transition to some form of crown fire and then if that crown fire achieves a faster spread rate typical of "active" crown fires. Van Wagner (1977) suggested that the crown is ignited if the surface fire intensity I_b exceeded a threshold value I_o determined by the availability of crown fuels (e.g., proximity to the surface fire) and the ignition energy required to ignite them:

$$I_o = (0.010 \ CBH \ (460 + 25.9M))^{3/2} \qquad (5)$$

where CBH is the crown base height (m), M is moisture content (%). If I_b meets or exceeds I_o, then at least some of the crown fuels become ignited. These burning crown fuels increase the intensity but not the spread rate unless a crown fire threshold (RAC m min^{-1}) is surpassed that determines the critical mass flow rate through the crown fuels:

$$RAC = 3.0/CBD \qquad (6)$$

where CBD is the crown bulk density, a stand-level crown fuel descriptor. Higher CBD facilates active crown fires. Beyond this threshold, the fire is an active crown fire and burns with a faster heading crown fire spread rate. The crown fire spread rate was based on Rothermel's (1991) correlation of 3.34 times the surface fire spread rate for a timber understory fuel model (US fire behavior fuel model 10, with wind reduction factor of 0.4). The crown fire spread rate at each vertex depends on its orientation on the fire front relative to the direction of maximum spread using the elliptical dimensions for a crown fire (Eq. 1).

Fire acceleration

Fire acceleration defines the rate of increase in fire spread rate for a given ignition source (e.g., point fire or line fire) assuming environmental conditions remain constant. Point source fires may take 20 minutes or more to accelerate to an equilibrium spread rate (McAlpine and Wakimoto, 1991) whereas than line-source fires accelerate faster (Johansen, 1987). FARSITE uses the logarithmic model developed for the CFBPS (Forestry Canada Fire Danger Group, 1992) to calculate the spread rate R (m min^{-1}) after time t (min) for both line and point source fires:

$$R = R_e(1-e^{-at}) \qquad (7)$$

where the a is the acceleration constant and R_e (m min^{-1}) is the new equilibrium spread rate. Point and line source fires are differentiated by the length of the fire perimeter set by the user. At the start of a new time step, the fire spread rate is accelerated from its previous value toward the new equilibrium spread rate calculated from current conditions. The acceleration constants can be set by fuel type to allow, for example, fire to accelerate faster in grass fuels than in timber or heavy slash fuels. Over relatively short time domains, fire acceleration is important to determining the fire spread rate and intensity where environmental conditions change rapidly.

Spotting

Spotting describes the lofting and transport of burning embers downwind of the main fire front where they may serve as new ignition sources. Once some form of crown fire is initiated ember transport from torching trees is simulated using the model of Albini (1979). This model was originally designed only for individual trees and groups of trees and will thus underestimate spotting distances for active crown fires. Spotting is modeled in terms of: (i) flame characteristics of the torching tree or group of trees, (ii) lofting of embers of different size classes, (iii) downwind travel of embers over the landscape, and (iv) ignition of new fires.

Flame structure and duration are determined for a given tree species and tree diameter based on their relationship to crown weight (Albini, 1979). The number of trees torching in a group is modeled as increasing with canopy cover and crown fraction burned. Little is known on ember production or size class distributions. Thus, a fixed number of embers between 0.1 cm and 2.5 cm are lofted from the tree top to their maximum heights determined by the flames from the torching tree.

The trajectory of each ember is then iterated during its descent and lateral movement across uneven terrain (Albini, 1979). The vertical windspeed profile is modeled as logarithmic, based on the reference velocity input at 6.1 m above the vegetation (Albini and Baughman, 1979). During its flight the combustion time of the ember is computed; small embers may burnout before they contact the ground. The objective of the spotting model is to use embers of different sizes to find areas where embers can ignite new fires. If a burning ember contacts fuel, it may start a new fire if it hasn't fallen within an existing fire front. Ignition itself is modeled stochastically because many important factors cannot be modeled at the relatively coarse scale of the spatial inputs, including the spatial distributions of receptive fuel (e.g., rotten wood) and fine-scale variability of the fuel bed.

Spatial and temporal simulation control

Three parameters are used to control the space and time resolution of the calculations made during a simulation: *time step*, *distance resolution*, and *perimeter resolution*. All three parameters are crucial to controlling the amount of data used in the simulation and thus, how much detail is present. The time step is the maximum amount of simulation time allowed between fire behavior calculations. Fuel moisture and winds are constantly changing over time, and the time step controls for the maximum time interval between accesses to temporal data used for computing fire behavior. The distance resolution is the farthest distance the fire is allowed to spread between successive fire behavior calculations. It ensures that the simulation uses a minimum density of spatial data in calculating fire behavior as it progresses across a landscape. The perimeter resolution is the maximum distance between perimeter vertices on the fire front. As convex portions of the fire front expand, the vertices become separated. The separation distances are checked at least once in a time step and new points are inserted at mid-span if the perimeter resolution is exceeded.

Fire growth is either limited by the time step or the distance resolution as set by the user. Time is limiting if the fire spread rate is slow enough that the spread distance from any vertex is less than the distance resolution at the end of the time step. The time step then forces additional data to be used for fire behavior calculation. Distance becomes limiting when fire spread is greater than the specified distance resolution within a given time step. The original time step is then partitioned into sub time steps, determined as the minimum time required for the fire to spread the length of the distance resolution. In this way, multiple steps are used to achieve fire growth for the original time step. At each sub time step, the perimeter resolution is checked and crossovers along the fire front are processed. Obviously, larger time steps, distance resolutions, and perimeter resolutions permit coarser approximations of simulated fire growth because data used in computations are more sparse.

Crossovers and mergers

The perimeter expansion technique used for fire growth modeling is not inherently capable of differentiating areas already burned from those not yet burned. As a result, fire fronts will cross over themselves along locally concave regions and, if multiple fires exist, "reburn" areas already burned by other fires. Specialized computational methods are required to eliminate the influence of crossing segments, to merge fire fronts that overlap, and to identify and preserve enclaves that are produced by these crosses and mergers. Enclaves are essentially new fire fronts that burn inward, eventually extinguishing themselves.

Richards and Bryce (1995) neutralize vertices that fall within already burned areas, leaving them in place but allowing no further activity at those points. Richards (1990), Knight and Coleman (1993), and Wallace (1993) describe techniques that eliminate the overlapping portions from the list of vertices that comprise a given fire. The algorithm developed for *FARSITE* is similar to the latter type, where vertices are removed when they fall inside existing fire polygons. The algorithm processes a given fire perimeter by first comparing every segment for intersection with every other segment, producing an ordered list of crossing segments. The outer edge of the main fire perimeter is then extracted by tracing the outside edge between intersecting segments. Subsequent processing uses the list of intersections to identify and preserve enclaves that are sometimes formed by the crossing. These enclaves must be preserved as separate fire fronts.

With more than one fire being simulated (for example, with spotting), the numbers of comparisons becomes factorial, requiring a search for overlap between each fire and every other fire. If bounding rectangles for two fires overlap, a more detailed procedure compares each segment on one fire with all segments on the other fire. The merger then identifies the main fire front as well as enclaves that have formed by the merger. Merging can occur between two outward burning fire fronts, and between an outward and an inward fire front. The latter situation occurs when spot fires ignite within an enclave.

Model performance

The *FARSITE* model was written in C++ and has been compiled to run under 32-bit Windows™ and several UNIX operating systems. The most common form of the model runs under Windows™ 95 or NT and has a graphical Windows interface. On a given computer, performance varies by the number of fires being simulated, the size of the fires (i.e., number of vertices), the time and space resolutions specified for the simulation, and the kinds of outputs selected. In general, the simulations are completed quickly, with those shown in Figs. 8.5 and 8.6 taking no more than about 5 minutes on an Intel Pentium Pro 200. The vector technique is computationally efficient because it requires calculations only for those vertices involved in the active fire front.

Model applications

The application of Huygens' principle to vector modeling of fire growth has been demonstrated for surface fires. Sanderlin and Sunderson (1975) found reasonable agreement between their predicted and observed growth of a Southern California chaparral fire. Anderson *et al.* (1982) and French (1992) used observed growth rates to parameterize their elliptical wavelets and found that grass fires with varying wind were well modeled with the technique. Finney (1994) reported several initial comparisons of *FARSITE* output with prescribed natural fires in forest and brush fuels in the Southern Sierra Nevada mountains. Coleman and Sullivan (1996) also described a comparison of predicted and observed spread patterns. One of the most interesting results of the validations is the apparently consistent overprediction of fire spread rates for surface fires (Finney, 1994; Finney and Ryan, 1995). Potential sources of overprediction include varying topographic sheltering of surface fuels to winds and inaccurate fuel maps. The causes of overprediction have not yet been determined conclusively, but an explanation may involve scale differences that are independent of any data or model inadequacy.

It is recognized that the scale of input data to the simulation is coarse compared to the frequency of variation in real environmental conditions affecting a fire. Winds are input at intervals of an hour or half hour (at best) and fuels and topography are typically resolved to a spatial resolution of about 30 meters. In nature, winds are more variable (on the order of seconds to minutes) as are fuels and topography (order of 10^{-1} m to 10^1 m). The disparity in scales means the models tend to calculate equilibrium conditions from the homogenized input data compared to variable fire spread rates that are really accelerating and decelerating over time and space (Albini, 1982*a,b*). The application of the modeled spread rate to large space and time scales of the simulation may not equal the cumulative spread produced by a variable environment because of lag times and non-linear spread rate responses to changing conditions. This does not imply that the fire spread rate and behavior calculations for a given suite of conditions are necessarily wrong (they behavior models in particular (e.g., Andrews, 1986; Forestry Canada Fire Danger Group, 1992). Fire patterns are not determined solely by spatial properties of the landscape (e.g., topography, fuels, or vegetation structure). Tremendous variation is caused by the weather and winds at the time the fire burns each part of that landscape; weather changes that follow diurnal and synoptic patterns that can be modeled, although not necessarily predicted into the future. The relative fire spread direction (e.g., heading, flanking, backing) can also play an important role in determining the fire behavior and consequent effects. The area burned by flanking and backing spread is usually small compared to the heading direction on fast moving and short-duration fires. Areas burned by the different relative spread directions are more evenly distributed on slow-moving fires or those that last for many weeks or months.

Mechanistic simulations are particularly useful for investigating or reconstructing the causes of patterns within a single fire. Visible crown damage patterns such as tree-crown streets (Haines, 1982) or stringers (Foster, 1983) can be simulated in *FARSITE* by varying wind direction and speed (Finney, 1998). The winds change fire spread rates around the fire front relative to the thresholds for crown fire activity (Eqns. 5 and 6). Other patterns, such as the formation of unburned islands within large burns (Eberhart and Woodard, 1987; Foster, 1983; Van Wagner, 1983) may also be investigated by simulation. The example fire in Fig. 8.6 showed unburned islands that began as large gaps between the main fire front and spot fires. The slow closure of these islands was afforded by locally unfavorable topography, fuels, or barriers that kept the heading portion of the main fire from burning rapidly into the gaps. The fire was slowly backing into these areas against wind or slope, and would have eventually burned the entire enclave if this simulation had continued. To have remain unburned, the fire surrounding the enclaves would need to have: (i) experienced a change in weather conditions (becoming more probable with longer burn times) and/or (ii) be slowed to a smolder by the absence of wind or slope assistance. Studies suggest that flaming spread may not be sustained under certain combinations of moisture content and fuel structure (packing ratio, surface-area-to-volume ratio) unless the fire is spreading with the wind (Beer, 1995) or up slope (Martin and Sapsis, 1987). These limits of sustainability would probably preclude flaming spread in backing and flanking spread directions but are not yet modeled for fires in general or for that matter, in *FARSITE* which uses only the Rothermel (1972) equation for surface fires (wind and slope modify spread rate that occurs under calm and flat conditions). Fire spread by smoldering is so slow (about 3 cm h^{-1}; Frandsen, 1991) that changing weather would likely extinguish the fire before larger enclaves could be entirely burned. Shifting wind directions could, however, rekindle smoldering sections and resume the burning of enclaves with heading fire spread.

Mechanistic simulations can be used to explore the repeatability of fire effects on different sites within a landscape or the equability of fire behavior and effects due to topography or productivity. For example, more variable fire effects might be expected where topographic position and productivity do not limit the fuel production, fire behavior, or direction from which fires can arrive from adjacent lands. Some topographic positions, like ridges or steep slopes, may be predisposed to the extreme fire behavior and effects (Geldenhuys, 1994; Kushla and Ripple, 1997; Minnich, 1988; Minnich and Chou, 1997) given wind patterns and limited productivity. By integrating fire models with forest simulators (e.g., Keane *et al.*, 1996*a*) the role of site productivity in determining repeated fire patterns and fire regimes on large landscapes can be further explored.

The future of mechanistic fire simulation will likely involve better component models for all processes such as fuel moisture, surface fire, crown fire, and three-dimensional winds. Coupling of fire and atmospheric models (e.g., Clark *et al.*, 1996; Linn and Harlow, 1998) offers a way to explore fire – environment interactions that are not possible with the two-dimensional techniques as used in *FARSITE*. Fire whorls, mass fires, plume-dominated fires are some of the many fire behaviors that are not well understood. Once modeled, these behaviors might also help to explain some fire patterns that remain mysterious today.

Notwithstanding, mechanistic models are known for their rapacious data requirements. As models are improved by adding more detailed component processes, the data required to run the models increases as well. *FARSITE* was developed for practical use by fire managers in simulating active fires and planning for potential fires. The remote sensing and computer technology necessary for generating and managing data for large landscapes are reasonable and attainable today but were not practical even a decade ago. Recognizing data limitations is the first step toward new efforts to gather data to run the models; making the data available then stimulates the development of new models. The result of this process is a steadily advancing ability to understand and predict phenomena that would not be possible if we remained sated by current technology and information.

References

Agee, J. K. and Huff, M. H. (1980). First year ecological effects of the Hoh Fire, Olympic Mountans, Washington. In *Proceedings of the sixth conference on fire and forest meteorology*, ed. R. E. Martin, R. L. Edmonds, D. A. Faulkner, J. B. Harrington, D. M. Fuquay, B. J. Stocks and S. Barr, pp. 175–81. Society of American Forestry.

Albini, F. A. (1976). Estimating wildfire behavior and effects. USDA Forest Service General Technical Report INT-30.

Albini, F. A. (1979). Spot fire distance from burning trees – a predictive model. USDA Forest Service General Technical Report INT-56.

Albini, F. A. (1982*a*). Reponse of free-burning fires to nonsteady wind. *Combustion Science Technology*, **29**, 225–41.

Albini, F. A. (1982*b*). The variability of wind-aided free-burning fires. *Combustion Science Technology*, **31**, 303–11.

Albini, F. A. and Baughman, R. G. (1979). Estimating windspeeds for predicting wildland fire behavior. USDA Forest Service Research Paper INT-221.

Alexander, M. E. (1985). Estimating the length-to-breadth ratio of elliptical forest fire patterns. *Proceedings of the eighth Conference on fire and forest meteorology*, pp. 287–304.

Alexander, M. E. (1991). The 1985 Butte Fire in Central Idaho: a Canadian perspective on the associated burning conditions. In *Fire and the environment: ecological and cultural perspectives*, ed. S. C. Nodvin and T. A. Waldrop, pp. 334–43. USDA Forest Service General Technical Report SE-69.

Anderson, D. G., Catchpole, E. A., DeMestre, N. J. and Parkes, T. (1982). Modeling the spread of grass fires. *Journal of the Australian Mathematics Society* (Ser. B.), **23**, 451–66.

Anderson, H. A. (1968). Sundance fire: an analysis of fire phenomena. USDA Forest Service Research Paper INT-56. 37pp.

Anderson, H. E. (1982). Aids to determining fuel models for estimating fire behavior. USDA Forest Service General Technical Report INT-122.

Anderson, H. E. (1983). Predicting wind-driven wildland fire size and shape. USDA Forest Service Research Paper INT-305.

Andrews, P. L. (1986). BEHAVE: fire behavior prediction and fuel modeling system-BURN subsystem, Part 1. USDA Forest Service General Technical Report INT-194.

Baker, W. L. (1992). Effects of settlement and fire suppression on landscape structure. *Ecology*, **73**(5), 1879–87.

Baker, W. L., Egbert, S. L. and Frazier, G. F. (1991). A spatial model for studying the effects of climatic change on the structure of landscape subject to large disturbances. *Ecological Modeling*, **56**, 109–25.

Ball, G. L. and Guertin, P. D. (1992). Improved fire growth modeling. *International Journal Wildlife Fire*, **2**(2), 47–54.

Balling, R. C., Meyer, G. A. and Wells, S. G. (1992). Relation of surface climate and burned area in Yellowstone National Park. *Agriculture Forestry and Meteorology*, **60**, 285–93.

Barrett, S. W. (1994). Fire regimes on andesitic mountain terrain in Northeastern Yellowstone National Park, Wyoming. *International Journal Wildlife Fire*, **4**(2), 65–76.

Beer, T. (1990). The Australian National Bushfire Model Project. *Mathematical Computer Modeling*, **13**(12), 49–56.

Beer, T. (1995). Fire propagation in vertical stick arrays: the effects of wind. International Journal Wildlife Fire, **5**(1), 43–9.

Beer, T. and Enting, I. T. (1990). Fire spread and percolation modelling. *Mathematical Computer Modeling*, **13**(11), 77–96.

Bilgili, E. and Methven, I. R. (1990). The simple ellipse: a basic growth model. *First international conference on forest fire research*, Coimbra 1990. pp. B.18-1–14.

Blackmarr, W. H. (1972). Moisture content influences ignitability of slash pine litter. USDA Forest Service Research Note SE-173.

Boychuk, D. and Perera, A. H. (1997). Modeling temporal variability of boreal landscape age-classes under different fire disturbance regimes and spatial scales. *Canadian Journal Forest Research*, **27**, 1083–94.

Boychuk, D., Perera, A. H., Ter-Mikaelian, M. T., Martell, D. L. and Li, C. (1997). Modelling the effect of spaital scale and correlated fire disturbaces on forest age distribution. *Ecological Modelling*, **95**, 145–64.

Bradshaw, L. S, Deeming, J. E., Burgan, R. E. and Cohen, J. D. (1984). The 1978

National Fire-Danger rating system: Technical Documentation. USDA Forest Service General Technical Report INT-169.

Burgan, R. E. and Rothermel, R. C. (1984). BEHAVE: Fire behavior prediction and fuel modeling system–FUEL subsystem. USDA Forest Service General Technical Report INT-167.

Butler, B. W. and Reynolds, T. D. (1997). Wildfire case study: Butte City Fire, southeastern Idaho, July 1, 1994. UDSA Forest Service General Technical Report INT-GTR-351.

Catchpole, E. A, DeMestre, N. J. and Gill, A. M. (1982). Intensity of fire at its perimeter. *Australian Forest Research*, **12**, 47–54.

Catchpole, E. A., Alexander, M. E. and Gill, A. M. (1992). Elliptical-fire perimeter- and area-intensity distributions. *Canadian Journal Forest Research*, **22**, 968–72.

Clark, T. L., Jenkins, M. A., Coen, J. and Packham, D. (1996). A coupled atmospheric-fire model: convective feedback on fire line dynamics. *Journal of Applied Meteorology*, **35**, 875–901.

Clarke, K. C. and Brass, J. A. and Riggan, P. J. (1994). A cellular automaton model of wildfire propagation and extinction. *Photogrammetric Engineering and Remote Sensing*, **60**(11), 1355–67.

Colemen, J. R. and Sullivan, A. L. (1996). A real-time computer application for the prediction of fire spread across the Australian landscape. *Simulation*, **67**(4), 230–40.

Despain, D., Rodman, A., Schullery, P. and Shovic, H. (1989). Burned area survey of Yellowstone National Park: the fires of 1988. Unpublished report, Division of Research, Yellowstone National Park.

Dorrer, G. A. (1993). Modelling forest fire spreading and suppression on basis of Hamilton mechanics methods. In *Scientific Siberian A: special issue Forest Fires Modelling and Simulation*, ed. G. A. Dorrer, pp 97–116.

Eberhart, K. E. and Woodard, P. M. (1987). Distribution of residual vegetation associated with large fires in Alberta. *Canadian Journal of Forestry Research*, **17**, 1207–12.

Feunekes, U. (1991). Error analysis in fire simulation models. MSc thesis, University of New Brunswick, Fredericton.

Feunekes, U. and Methven, I. R. (1987). A cellular fire growth model to predict altered landscape patterns. In *Perspectives in land modelling: workshop proceedings*, ed. R. Gelinas, D. Bond and B. Smit. Toronto, Ontario, Nov. 17–20, 1986. Montreal Canada: Polysciences Publications Inc.

Finney, M. A. (1994). Modeling the spread and behavior of prescribed natural fires. Proceedings of the twelfth conference on fire and forest meteorology, pp. 138–43.

Finney, M. A. (1998). *FARSITE*: Fire area simulator–model development and evaluation. USDA Forest Service Research Paper RMRS-RP-4. 47 p.

Finney, M. A. and Andrews, P. L. (1996). The *FARSITE* fire area simulator: fire management applications and lessons of summer 1994. *Proceedings of the international fire council meeting*. November 1994, Coer D'Alene ID. in press.

Finney, M. A. and Ryan, K. C. (1995). Use of the *FARSITE* fire growth model for fire prediction in US National Parks. *Proceedings of the International Emergency Management and Engineering Conference*, pp. 183–9. May 1995, Sofia Antipolis, France.

Fischer, W. C. and Bradley, A. F. (1987). Fire ecology of Western Montana forest habitat types. USDA Forest Service General Technical Report INT-223.

Fons, W. T. (1946). Analysis of fire spread in light forest fuels. *Journal of Agricultural Research*, **72**(3), 93–121.

Forestry Canada Fire Danger Group. (1992). Development and structure of the Canadian forest fire behavior prediction system, International Report ST-X-3.

Foster, D. R. (1983). The history and pattern of fire in the boreal forest of southeastern Labrador. *Canadian Journal of Botany*, **61**, 2459–71.

Frandsen, W. H. (1991). Burning rate of smoldering peat. *Northwest Science*, **65**(4), 166–72.

French, I. A. (1992). Visualisation techniques for the computer simulation of bushfires in two dimensions. MS Thesis University of New South Wales, Australian Defence Force Academy, 140 pp.

Gardner, R. H., Hargrove, W. W., Turner, M. G. and Romme, W. H. (1996). Climate change, disturbances and landsape dynamics. In *Global change and terrestrial ecosystems*, ed. B. H. Walker and W. L. Steffen. IGP Book series No. 2. Cambridge: Cambridge University Press.

Geldenhuys, C. J. (1994). Bergwind fires and the location patterns of forest patches in the Southern Cape landscape, South Africa. *Journal of Biogeography*, **21**, 49–62.

Green, D. G. (1983). Shapes of simulated fires in discrete fuels. *Ecological Modelling*, **20**, 21–32.

Green, D. G. (1989). Simulated effects of fire, dispersal and spatial pattern on competition within forest mosaics. *Vegetatio*, **82**, 139–53.

Green, D. G., Gill, A. M. and Noble, I. R. (1983). Fire shapes and the adequacy of fire-spread models. *Ecological Modelling*, **20**, 33–45.

Haines, D. A. (1982). Horizontal roll vortices and crown fires. *Journal of Applied Meteorology*, **21**, 751–63.

Hartford, R. A. and Rothermel, R. C. (1991). Moisture measurements in the Yellowstone Fires in 1988. USDA Forest Service Research Note INT-396.

Heinselman, M. L. (1973). Fire in the virgin forests of the Boundary Waters Canoe Area, Minnesota. *Quaternary Research*, **3**, 329–82.

Hornby, L. G. (1936). Fire control planning in the Northern Rocky Mountain region. Progress Report 1. USDA Forest Service Ogden UT, 179pp.

Huff, M. H., Ottmar, R. D., Alvarado, E., Vihananek, R. E., Lehmkul, J. F., Hessburg, P. F. and Everett, R. L. (1995). Historical and current forest landscapes in eastern Oregon and Washington. Part II. Linking vegetaion characteristics to potenital fire behavior and related smoke production. USDA Forest Service General Technical Report PNW-GTR-355.

Hunter, M. L. (1993). Natural fire regimes as spatial models for managing boreal forests. *Biological Conservation*, **65**, 115–20.

Johansen, R. W. (1987). Ignition patterns and prescribed fire behavior in southern pine forests. Georgia Forestry Commission Forest Research Paper 72.

Johnson, E. A., Miyanishi, K. and Weir, J. M. H. (1995). Old-growth, disturbance, and ecosystem management. *Canadian Journal of Botany*, **73**, 918–26.

Karafyllidis, I. and Thanailakis, A. (1997). A model for prediction forest fire spreading using cellular automata. *Ecological Modelling*, **99**, 87–97.

Keane, R. E., Garner, J. L., Schmidt, K. M., Long, D. L., Menakis, J. P. and Finney, M. A. (1998). Development of the input spatial data layers for the *FARSITE* fire growth model for the Selway–Bitterroot Wilderness complex, USA. USDA Forest Service General Technical Report INT-GTR-3.

Keane, R. E., Morgan, P. and Running, S. W. (1996a). FIRE-BGC – a mechanistic ecological process model for simulating fire succession on Northern Rocky Mountain coniferous forest landscapes. USDA Forest Service Research Paper INT-RP-434.

Keane, R. E., Ryan, K. C. and Running, S. W. (1996b). Simulating effects of fire on northern Rocky Mountain landscapes with the ecological process model FIRE-BGC. *Tree Physics*, **16**, 319–31.

Kessell, S. R. (1976). Gradient modeling: a new approach to fire modeling and wilderness resource managment. *Environment Management*, **1**, 39–48.

Knight, I. and Coleman, J. (1993). A fire perimeter expansion algorithm based on Huygens' wavelet propagation. *International Journal of Wildlife Fire*, **3**(2), 73–84.

Kourtz, P. and O'Regan, W. G. (1971). A model for a small forest fire to simulate burned and burning areas for use in a detection model. *Forestry Science*, **17**(2), 163–9.

Kourtz, P., Nozaki, S. and O'Regan, W. (1977). Forest fires in the computer – a model to predict the perimeter location of a forest fire. Fisheries and Environment Canada. Information Report FF-X-65.

Kushla, J. D. and Ripple, W. J. (1997). The role of terrain in a fire mosaic of a temperate coniferous forest. *Forest Ecology And Mamagement*, **95**, 97–107.

Lesica, P. (1996). Using fire history models to estimate proportions of old growth forest in Northwest Montana, USA. *Biological Conservation*, **77**, 33–9.

Li, C., Ter-Mikaelian, M. and Perera, A. (1997). Temporal fire disturbance patterns on a forest landscape. *Ecological Modeling*, **99**, 137–50.

Linn, R. R. and Harlow, F. H. (1998). FIRETEC: A transport description of wildfire behavior. In *Proceedings of the 2nd Symposium on fire and forest meteorology*, pp. 14–19, American Meteorological Society.

McAlpine, R. S. and Wakimoto, R. H. (1991). The acceleration of fire from point source to equilibrium spread. *Forestry Science*, **37**(5), 1314–37.

Martin, R. E. and Sapsis, D. B. (1987). A method for measuring flame sustainability of live fuels. In *Proceedings of the ninth conference on fire and forest meteorology*, pp. 71–74. Boston, MA: American Meteorological Society.

Methven, I. and Feuenkes, U. (1988). Fire games for park managers: exploring the effect of fire on landscape vegetation patterns. In Landscape ecology and management. *Proceedings of the first symposium of the Canadian Society For Landscape Ecology and Management*, ed. M. R. Ross, University of Guelph, May 1987, Montreal Canada: Polyscience Publications Inc.

Minnich, R. A. (1988). *The biogeography of fire in the San Bernardino Mountains of California: A historical study*. University of California: Publications in Geography 28. 121p.

Minnich, R. A. and Chou, Y. H. (1997). Wildland fire patch dynamics in the chaparral of southern California and northern Baja California. *International Journal of Wildlife Fire*, **7**(3), 221–48.

Mladenoff, D. J., Host, G. E., Boeder, J. and Crow, T. R. (1996). *LANDIS: A spatial model of forest landscape disturbance, succession, and management. GIS and environmental modeling: progress and research issues*, pp. 175–9. Fort Collins CO: GIS World Books.

Morrison, P. H. and Swanson, F. J. (1990). Fire history and pattern in a Cascade range landscape. USDA Forest Service General Technical Report PNW-GTR-254. 77p.

Omi, P. N. (1996). Landscape-level fuel manipulations in Greater Yellowstone: opportunities and challenges. In *Proceedings of the second biennial conference on the*

Greater Yellowstone Ecosystem: The ecological implications of fire in Greater Yellowstone. Sept., 19–21, 1993, ed. J. Greenlee. Fairfield, WA: International Association of Wildland Fire.

O'Regan, W. G., Kourtz, P., Nozaki, S. (1976). Bias in the contagion analog to fire spread. *Forestry Science*, **2**(1), 61–8.

Peet, G. B. (1967). The shape of mild fires in Jarrah forest. *Australian Forests*, **31**(2), 121–7.

Peterson, D. L. and Ryan, K. C. (1986). Modeling postfire conifer mortality for long-range planning. *Environmental Management*, **10**(6), 797–808.

Ratz, A. (1996). Long-term spatial patterns created by fire: A model oriented towards boreal forests. *International Journal of Wildlife Fire*, **5**(1), 25–34.

Richards, G. D. (1988). Numerical simulation of forest fires. *International Journal of Numerical Methods in Engineering*, **25**, 625–33.

Richards, G. D. (1990). An elliptical growth model of forest fire fronts and its numerical solution. *International Journal of Numerical Methods in Engineering*, **30**, 1163–79.

Richards, G. D. (1995). A general mathematical framework for modeling two-dimensional wildland fire spread. *International Journal of Wildlife Fire*, **5**(2), 63–72.

Richards, G. D. and Bryce, R. W. (1995). A computer algorithm for simulating the spread of wildland fire perimeters for heterogeneous fuel and meteorological conditions. *International Journal of Wildlife Fire*, **5**(2), 73–80.

Roberts, S. (1989). A line element algorithm for curve flow problems in the plane. CMA-R58-89. Centre for Mathematical Analysis, Australian National University (Cited from French 1992).

Romme, W. H. (1982). Fire and landscape diversity in subalpine forests of Yellowstone National park. *Ecology Monographs*, **52**, 199–221.

Rothermel, R. C. (1972). A mathematical model for predicting fire spread in wildland fuels. USDA Forest Service Research Paper INT-115.

Rothermel, R. C. (1983). How to predict the spread and intensity of forest and range fires. USDA Forest Service General Technical Report INT-143.

Rothermel, R. C. (1991). Predicting behavior and size of crown fires in the northern Rocky Mountains. USDA Forest Service Research Paper INT-438.

Rothermel, R. C. (1993). Mann Gulch Fire: A race that couldn't be won. USDA Forest Service General Technical Report INT-299.

Rothermel, R. C., Wilson, R. A., Morris, G. A. and Sackett, S. S. (1986). Modeling moisture content of fine dead wildland fuels input to the BEHAVE fire prediction system. USDA Forestry Service Research Paper INT-359.

Ryan, K. C. and Reinhardt, E. D. (1988). Predicting postfire mortality of seven western conifers. *Canadian Journal of Forestry Research*, **18**, 1291–7.

Sanderlin, J. C. and Sunderson, J. M. (1975). A simulation for wildland fire management planning support (FIREMAN): Volume II. Prototype models for FIREMAN (PART II): Campaign Fire Evaluation. Mission Research Corp. Contract No. 231–343, Spec. 222. 249 pp.

Sanderlin, J. C. and Van Gelder, R. J. (1977). A simulation of fire behavior and suppression effectiveness for operation support in wildland fire management. In *Proceedings of the first international convention on mathematical modeling*, pp. 619–630. St. Louis, MO.

Simard, A. J., Haines, D. A., Blank, R. W. and Frost, J. S. (1983). The Mack Lake Fire. USDA Forestry Service General Technical Report NC-83. 36p.

Suffling, R., Lihou, C. and Morand, Y. (1988). Control of landscape diversity by catastrophic distrubance: a theory and a case study of fire in a Canadian boreal forest. *Environmental Management*, **12**(1), 73–8.

Swetnam, T. W. (1993). Fire history and climate change in giant sequoia groves. *Science*, **262**, 885–9.

Swetnam, T. W. and Betancourt, J. L. (1990). Fire-southern oscillation relations in the southwestern United States. *Science*, **204**, 1017–20.

Timoney, K. P. and Wein, R. W. (1991). The aerial pattern of burned tree vegetation in the subarctic region of Northwestern Canada. *Arctic*, **44**(3), 223–30.

Turner, M. G. and Dale, V. H. (1991). Modeling landscape disturbance. In *Quantitative methods in landscape ecology*, ed. M. G. Turner and R. H. Gardner, pp. 325–51. New York: Springer-Verlag.

Turner, M. G. and Romme, W. H. (1994). Landscape dynamics in crown fire ecosystems. *Landscape Ecology*, **9**(1), 59–77.

Turner, M. G., Gardner, R. H., Dale, V. H. and O'Neill, R. V. (1989). Predicting the spread of disturbance across heterogeneous landscapes. *Oikos*, **55**, 121–9.

Van Wagner, C. E. (1969). A simple fire growth model. *Forestry Chronicle*, **45**, 103–4.

Van Wagner, C. E. (1973). Height of crown scorch in forest fires. *Canadian Journal of Forestry Research*, **3**, 373–8.

Van Wagner, C. E. (1977). Conditions for the start and spread of crownfire. *Canadian Journal of Forestry Research*, **7**, 23–4.

Van Wagner, C. E. (1978). Age-class distribution and the forest fire cycle. *Canadian Journal of Forestry Research*, **8**, 220–7.

Van Wagner, C. E. (1983). Fire behavior in northern conifer forests and shrublands. In *The role of fire in northern circumpolar ecosystems*, ed. R. W. Wein and D. A. MacLean, Chap. 4, pp. 65–80. New York: John Wiley.

Van Wagner, C. E. (1993). Prediction of crown fire behavior in two stands of jack pine. *Canadian Journal of Forestry Research*, **23**, 442–9.

Van Wagtendonk, J. W. (1996). Use of a deterministic fire growth model to test fuel treatments. Sierra Nevada Ecosystem Project: final report to Congress, vol II. *Assessments and scientific basis for management options*, pp. 1155–65. Davis: University of California, Centers for Water and Wildland Resources.

Vasconcelos, M. J., Guertin, P. and Zwolinski, M. (1990). FIREMAP: simulation of fire behavior, a GIS supported system. In *Proceedings of the effects of fire in management of southwestern natural resources conference*, pp. 217–21. Tucson AZ. USDA Forest Service General Technical Report RM-191.

Von Niessen, W. and Blumen, A. (1988). Dynamic simulation of forest fires. *Canadian Journal of Forest Research*, **18**, 805–12.

Wade, D. D. and Ward, D. E. (1973). An analysis of the Air Force Bomb Range Fire. USDA Forest Service Research Paper SE-105.

Wallace, G. (1993). A numerical fire simulation model. *International Journal of Wildlife and Fire*, **3**(2), 111–16.

Wilson, R. (1980). Reformulation of forest fire spread equations in SI units. USDA Forestry Service Research Note INT-292.

Xu, J. and Lathrop, R. G. (1994). Geographic information system based wildfire spread simulation. *Proceedings of the twelfth conference on fire and forest meteorology*, pp. 477–84.

9

Achieving sustainable forest structures on fire-prone landscapes while pursuing multiple goals

John Sessions, K. Norman Johnson, Jerry F. Franklin and John T. Gabriel

Introduction

A number of recent studies (Green *et al.*, 1993; Keane *et al.*, 1996) have demonstrated spatial simulations of wildfire across landscapes. Others have described considerations in the likely spread of disturbances across landscapes (Turner and Romme, 1994). Still others have addressed the spatial simulation of vegetation condition across simulated and real landscapes (Turner *et al.*, 1989). A number of efforts have incorporated non-spatial fire effects into forest planning approaches that seek management actions that achieve multiple goals (Reed and Enrico, 1986; Johnson *et al.*, 1986; Boychuk and Martell, 1996). Most of the spatial efforts have come from landscape ecology, while the non-spatial approaches were developed through management planning. An attempt has been made to merge these two approaches by combining the spatial simulation of forest development on a large landscape, including wildfire occurrence and effects, with the search for management actions that achieve multiple goals.

This work was done as part of the Sierra Nevada Ecosystem Project (SNEP). SNEP was commissioned by Congress to undertake a "scientific review of the remaining old growth in the national forests of the Sierra Nevada in California, and for a study of the entire Sierra Nevada ecosystem . . ." (SNEP, 1994). In interpreting this instruction, and letters from various Committee Chairmen, the Steering Committee guiding the work of SNEP charged the Science Team to develop a range of alternative management strategies to maintain the health and sustainability of Sierra Nevada ecosystems while providing resources to meet human needs (SNEP, 1994). Development of management strategies for late-successional forests and watersheds was especially emphasized in Congressional instructions.

In our analysis, the authors focused on five goals for Sierra Nevada ecosystems identified by SNEP scientists for improving the health and sustainability of these ecosystems (SNEP, 1996a, 1996b): (i) increasing the general extent and complexity of late-successional forests, (ii) reducing the potential for high-severity fire, (iii) restoring riparian areas and watersheds, (iv) reintroducing historical ecosystem processes, and (v) producing a sustainable supply of timber in a cost-effective manner. The Eldorado National Forest and intermingled lands in the mid-Sierras – a total of approximately 400 000 hectares – are used to illustrate how attempts were made to achieve these goals.

The methods and results reported here draw heavily from our work in the SNEP as reported in Cousar *et al.* (1996), Johnson *et al.* (1996), and Sessions *et al.* (1996). These three papers provide extensive detail on our work which we have summarized in this chapter.

Late-successional forests and fire: focus of the analysis

In response to the Congressional direction for a scientific review of the remaining late-successional and old-growth (LS/OG) forest on the national forests of the Sierra Nevada, Franklin and Fites-Kaufmann (1996) led a comprehensive assessment of the state of these forests (Fig. 9.1(*a*), (*b*)). Their analysis found a significant decline in the amount and complexity of LS/OG forest in the commercial forest types, especially mixed conifer and east-side pine. They found that key structural features of LS/OG forests – such as large-diameter trees, snags, and logs – were generally at low levels and that much of the remaining high quality LS/OG forest on national forests is unreserved and potentially available for harvest. On the positive side, they found that the forest cover in most areas was not heavily fragmented by clear-cutting and that stands generally have sufficient structural complexity to provide at least low levels of LS/OG forest function.

Franklin *et al.* (1996) then proposed a number of different conservation strategies for late-successional forests and evaluated their ability to: (i) provide sufficient, well-distributed high-quality LS/OG forest to sustain the organisms and functions associated with such ecosystems for the next century, and (ii) provide conditions which facilitate connectivity for organisms moving between old-growth forest areas.

These conservation strategies all involve increasing the general extent and complexity of late-successional forests in the Sierra Nevada. Some conservation strategies identify relatively large areas of late successional emphasis (ALSEs) where late-successional forests will be emphasized, while also increasing the late-successional attributes of the intervening forest (called the "matrix"). Variations on this strategy call for more or less prescribed fire and mechanical treatment (timber harvest) in the ALSEs to accelerate development of these characteristics. Other strategies call for a more dispersed late-successional system.

From the beginning of the SNEP assessment of late-successional forests, it was clear that the threat of severe fire from the build up in fuels and decrease in fire periodicity in some types would be major considerations in any strategy to rebuild the late-successional forests of the Sierra Nevada (Franklin and Fites-Kaufmann, 1996; Weatherspoon and Skinner, 1996). While opinions may vary somewhat

Fig. 9.1(a). Stand shot of Sierran Mixed-Conifer Forest in Yosemite National Park, California. Shows dominant sugar pines (Pinus lambertiana) and abundance of white fir (Abies concolor) associates. Also shows the extensive development of canopy fuels which now extend from ground to the top of the crown as the result of fire suppression during the last 100 years.

Fig. 9.1(b). *Aerial view of old-growth mixed-conifer stand in the Sierra Nevada (Tahoe National Forest, California). This area was ranked as "high quality" in terms of its structural complexity and other old-growth characteristics. Note the complex pattern of small (1/2 ha) stand structural units.*

about the current extent of the threat of severe fire, it is clear that there is a need to understand the survivability of late-successional forests, and the forest in general under different forest management strategies, including strategies explicitly aimed at reducing fuels and limiting the extent of fires that do occur (Weatherspoon and Skinner, 1996). Therefore, an analysis has been undertaken to assess the likelihood that various policies will achieve late-successional goals over time while simultaneously reducing the potential for high severity fire.

An integrated assessment of forests, streams, and watersheds

SNEP (1996) assessed the state of Sierra Nevada ecosystems from a variety of perspectives. Other issues identified in SNEP that intersect with the issues surrounding forest management relative to late-successional forests and fire are: (i) declines in aquatic biodiversity and existing and potential threats to riparian-associated species and ecosystems, (ii) existing and potential difficulties from watershed disturbance, (iii) declines in terrestrial biodiversity and existing and potential threats to terrestrial wildlife species and ecosystems, (iv) production of timber as an objective on some lands including the sizes and species that might be harvested and the associated costs and revenues, (v) the potential effect of budget constraints on fuel treatments and other activities on federal lands.

These issues were considered simultaneously with strategies for late-successional forests and wildfire because of their potential interaction. Mechanical treatment to improve LS/OG rank, decrease fuel loadings, and/or produce timber can impact riparian areas and watersheds. Aquatic goals for riparian zones can affect the amount of LS/OG forest and the freedom to treat the zones to reduce fire hazards. LS/OG goals for ALSEs and the matrix can influence fire hazard in those areas and the distribution of hectares among seral stages and among different wildlife habitats. Creation of defensible fuel profile zones (fuel breaks) can increase the survivability of the ALSEs and the forest in general and produce timber volume and value, but at the cost of potentially reducing LS/OG rank. Budget constraints can influence the ability to undertake activities of any sort.

Consideration of wildlife

Rather than focusing on particular wildlife species, the focus was on rebuilding late-successional forests, restoring riparian areas and watersheds, and reintroducing historical ecosystem processes, because restoring these habitats and processes were identified in the SNEP reports as the key steps in the recovery of Sierra Nevada wildlife. Further analysis may be needed to determine how different management strategies meet requirements of particular species.

Consideration of disturbance processes other than fire

In addition to wildfire, insects and pathogens have also been major factors in shaping the development of Sierra Nevada forests (SNEP, 1996b). The effects of insects and pathogens were represented only to the degree that they are reflected in the existing condition of the vegetation classes and in projected changes in these classes associated with the stand simulations. It is hoped to add the capability to represent periodic outbreaks of insects and pathogens in future versions of the methodology presented here.

Recognition of multiple spatial scales

To address the major goals of the analysis, we recognized four spatial scales (Fig. 9.2): (i) the Sierra Nevada range for understanding the condition of the various ecosystems, (ii) the Eldorado National Forest for estimating sustainable timber harvests, (iii) large landscape units called "LS/OG polygons" for assessing LS/OG condition and aggregate watershed disturbance, and (iv) combinations of riparian zone and forest condition within LS/OG polygons for projecting forest structure, yield, fire hazard, and contribution to watershed disturbance.

Approaches to forest-level optimization

Since the early days of the national forests, forest plans have guided the level of timber harvest and the scheduling of timber harvest activities. Relatively simple formulas were used to set the harvest levels based on controlling the volume harvested, the area cut, or both volume and area (Davis and Johnson, 1987).

During the last few decades in both public and private forest planning, optimization models have become dominant in the development of forest plans (Davis and Johnson, 1987). These models attempt to maximize or minimize some quantity, subject to reaching specified policy goals, which are represented as constraints, given the choices for management that are allowed for each part of the forest. Typical objectives have been to maximize timber harvest, minimize cost, or maximize present net value. Typical policy goals have been to maintain a non-declining yield of timber harvest over time, limit the rate of harvest in different portions of the forest, and attain some distribution of hectares among age-classes or seral stages.

As the policy goals have become more complex, the optimization models have often been reformulated as "goal programs." Then the policy goals that were previously represented as absolute constraints are transformed to allow for under- or over-achievement with an associated penalty. The overall objective is then to minimize the total penalty. This formulation is especially useful in achieving a distribution of forest hectares among seral stages over time.

A classification of model structures for forest optimization

Johnson and Scheurman (1977), Clutter *et al.* (1983), Johnson and Tedder (1983), and Davis and Johnson (1987) described the lineage of strategic planning systems based on forest-level optimization models where timber harvest is a major activity. Two important variables were: (i) the model formulation (Model I or Model II) and (ii) the solution technique (linear programming or binary search).

216   J. Sessions *et al.*

Fig. 9.2. The four spatial scales recognized in the analysis: (i) the Sierra Nevada Range in California, (ii) a national forest within the range, (iii) a landscape unit created from analysis of the late-successional/old growth (LS/OG) forests of the national forest, and (iv) the finer scale detail of the landscape unit (LS/OG polygon) composed of forest stands overlaid with riparian influence zones.

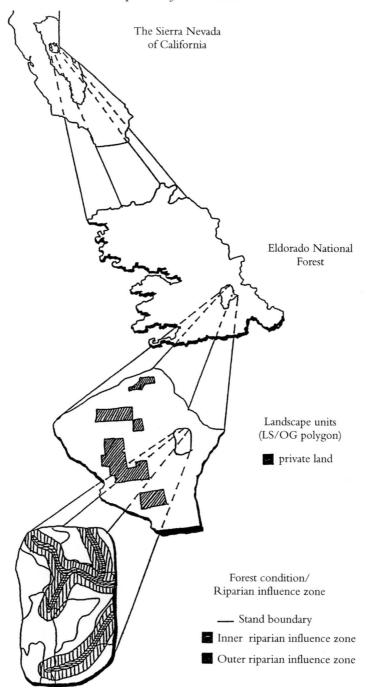

In both Model I and Model II, forest vegetation is classified according to variables like species type, density, and age. Each unique combination of these classification variables forms a "stratum". Overlaying the classification system on the forest divides the vegetation into these strata. All instances of each stratum over some geographic area, such as the entire forest or a watershed, are aggregated into inventory categories in the analysis and are often called "analysis areas".

Johnson and Scheurman (1977) coined the terms "Model I" and "Model II" to label fundamentally different ways to define decision variables for the analysis areas for scheduling timber harvest and investment, the distinction being the way in which the regenerated (future) stands are handled. In Model I, regenerated stands are coupled directly to, and identified by, the existing stands to which they are associated. In Model II, regenerated stands are detached from the existing stands and new decision variables are defined for them.

Thus, Model I defines decision variables that follow the life history of a hectare over all planning periods while Model II defines decision variables that follow the history of a hectare over the life of a stand growing on the hectare, from its birth (or start of the planning periods) through its death (or end of the planning periods). In Model I, a decision variable traces the activities on a hectare over the entire planning horizon; in Model II, a hectare may pass through several decision variables before reaching the planning horizon as stands are born, live, and die.

Any problem formulated with Model I can be formulated with Model II and vice versa. It should be noted, though, that the power of Model II comes from merging hectares of like characteristics from across the planning area as they are regeneration harvested. Through this process, fewer activities need be created to represent the problem compared to Model I, but at the cost of losing some spatial definition in the management of future stands. When merging hectares as they are regeneration harvested is acceptable, Model II is a powerful tool; when such merging is not acceptable, Model I is usually preferable.

Incorporating wildfire into forest optimization

The models presented so far do not explicitly consider risk of loss from fire. Any effect of fire was handled through adjusted yield functions; the notion that fire could cause the "premature" death of a stand was not recognized. Reed and Enrico (1986, 1989) broke from this tradition by creating a linear programming model of timber management in which the expected burned area was subtracted from each age class in each time period, and added along with the cutover area to the youngest age class in the following period. While describing their model as stochastic, they actually solved its equivalent "mean value problem", i.e., the random proportion burned was replaced with its expected value (Boychuk and Martell, 1996).

Johnson et al. (1986) built an extension of FORPLAN, the forest planning model used by the Forest Service, that would represent fire and other stochastic losses and changes deterministically in a manner similar to that proposed by Reed and Enrico. Johnson and Stuart (1986) showed the transfer of hectares among stand ages and classes for reasons other than harvest could actually be represented as a generalized version of Model II, although this formulation is sometimes called Model III (Boychuk and Martell, 1996).

This mean value formulation of Model II is now in wide use, especially by the Forest Service in California as it has been made available in FORPLAN.

While it is a major improvement toward incorporating fire risk compared with the traditional approaches to timber harvest scheduling, some problems with this approach have been noted. Pickens and Dress (1988) and Hof et al. (1988), in their studies of the effect of stochastic technological coefficients on forest-level timber management models, found that attempting to implement mean value problems in a stochastic system leads to a high probability for infeasible solutions.

Boychuk and Martell (1996) compared the results of a stochastic programming problem (SPP) and the corresponding mean value problem when fire risk is considered in forest planning analysis. Since they felt that replanning is inevitable, they compared only the first period solutions. Boychuk and Martell (1996) found that the mean value solution generally gave a good approximation to the stochastic programming problem, but consistently over-harvested under some conditions.

The mean value formulation does not enable the measurement of the variability of the solution. To assess the variability in forest plan outcomes, Boychuk and Martell (1996) formulated a stochastic programming version of the forest planning problem with risk of fire. As with the mean value formulations, they developed a model with linear objectives and linear constraints. They represented stochastic fire loss by a discrete two-point probability distribution that yielded the desired mean and coefficient of variation.

Selection of a model structure for forest optimization with wildfire

Recent activity in forest optimization including fire risk effects has emphasized mean value and stochastic programming formulations, which can be viewed as variations in Model II, as discussed above. While these approaches have increased our understanding of how fire affects management plans, the authors did not choose these approaches for a number of reasons.

The mean value approach was not chosen in our modeling for three reasons. (i) The mean-value approach discussed above, adjusts stratum condition for fire mortality. In reality, fires in the Sierra Nevada do not occur by strata; rather, they occur in a particular place and affect, to one degree or another, all strata they encounter. (ii) The mean-value approach depends on fires causing only a few different outcomes. When many different outcomes are possible, model size can explode. In our analysis, the effect of fire on stand condition was uniquely tailored in each stratum in each time period under each prescription. It was felt that much of this fine detail would be lost by collapsing fire results into a form amenable to a mean value approach. (iii) The mean-value approach does not explicitly consider variability in fire occurrence in terms of time, place, and size. Fires in the Sierra Nevada have shown considerable variability in the amount of fire per decade.

Also, the stochastic programming approach of Boychuk and Martell (1996), was not used even though it has proven interesting and valuable for research insights, as this approach is not practicable for applications with a large number of different starting conditions as in our analysis.

Given the difficulties seen with using mean value and stochastic programming in our analysis, a different approach has been taken. First, a Model I version of a strategic forest planning problem without fire is solved, given the goals of the analysis. Then fire is allocated onto the landscape using a number of stochastic variables including weather, fire size, and ignition probability. The fires and their effects are simulated for a number of periods as the fires sweep across LS/OG polygons, running through the strata found in the polygons, and leaving a differential mark on the landscape depending on vegetative condition (fuels, crown composition, and structure) in the different strata in each polygon burned. Finally, the strategic plan is reworked, salvaging where permitted, and the prescriptions previously chosen inside the LS/OG polygons that burn are adjusted to allow for the effect of wildfire and to make any needed changes in post-fire activities. Adjustments are made to better achieve the goals of the prescription, much as a manager would react to an unforeseen activity. For fires that create conditions which exceed watershed disturbance limits, future activities which would create additional disturbance are postponed until watershed conditions improve. A number of simulations of fires, with the associated reactions, are done to help understand the mean and range of potential fires and their effects.

The strategic plan that is developed for each alternative given the likely fire effects is not, however, completely reoptimized. Rather, these effects are used to help understand the likely influence of fire on the strategic plan.

Thus it is not claimed that our approach will formulate strategic plans that are "best", given our goals and the likely fire effects, as might be claimed for mean value or stochastic programming. It is possible, though, to have a much fuller understanding of the spatial effects of fire on large landscapes under different strategies, while recognizing the differential effect that fire can have on each area that burns.

Solution techniques for forest optimization problems

The formulations shown above for Model I and Model II have a linear objective function and linear constraints. As such, linear programming can be used to find the mathematically optimal solution, i.e., the solution that gives the maximum (or minimum) value for the objective function given the constraints (see Dykstra, 1985 or Davis and Johnson, 1987 for more discussion). Many harvest scheduling models, like FORPLAN, rely on linear programming as their solution technique.

Attempting to solve large problems with linear programming has led to a number of difficulties. First, linear programming software to solve these problems has often been costly or unavailable. Second, the size of the problem can easily exceed the capabilities of commercial software in terms of columns (choices for management of the analysis areas) or rows (number of area constraints and policy constraints). Due to the large problem size caused by the spatial detail recognized in the problem and our desire to be able to represent non-linear relationships, linear programming was not used here as has often been done in forest planning. Rather, a gradient search heuristic was used which seeks good (i.e., near-optimal) solutions based on the work of Reeves (1993).

The SAFE FORESTS model

To undertake this analysis, we constructed a model that we call the Simulation and Analysis of Fire Effects in the FORESTS of the Sierra Nevada (SAFE FORESTS). This section covers our classification of land, types of goals, measurement criteria, types of activities, mathematical formulation, and solution methodology. In the discussion, the Eldorado NF of the central Sierra Nevada Mountains of California is used to illustrate the model, its data requirements, and its outputs.

Classification of land: spatial units

For the purposes of this model, spatial units include polygons and lines (vectors). Spatial units have geographic coordinates and associated attributes recognized in the modeling.

Landscape units: LS/OG polygons

SNEP incorporated an assessment of late-successional and old growth (LS/OG) forest conditions on federal lands in the Sierra Nevada (Franklin and Fites-Kaufmann, 1996). The assessment used structural criteria that recognized that successional changes in forest ecosystems are at least as much structural as they are compositional phenomena. A common pattern is for forests to develop from a relatively simple structural condition early in succession (more so following clear-cutting than most natural disturbances) to conditions that incorporate both a broader array of individual structures and greater spatial heterogeneity; shifts in tree species composition may not even occur during succession, such as on sites that only support a single species.

The SNEP assessment was based entirely on structural definitions of LS/OG forest conditions for a variety of practical and scientific reasons, including the desire to recognize a gradient in LS/OG forest function in a region where many manipulated forest stands have retained some level of LS/OG function. The structural criteria used were typical features of high-quality LS/OG forests in the Sierra Nevada mixed conifer zone – a range of tree sizes, but with many large and decadent trees, large snags and down logs, areas of high tree canopy density, and a high degree of heterogeneity in structural conditions. The latter point is a particularly important one – high-quality LS/OG forests in the mixed conifer zone are commonly mosaics of contrasting structural units rather than extensive uniform stands of large, old trees of either high or low density. Furthermore, both the pioneer species (typically pines) and more shade-tolerant associates (commonly firs) are typically represented in such forests. Frequent fires of light-to-moderate intensity typically maintain these compositional and structural conditions in perpetuity, although the spatial arrangement of the mosaic of structural units may change.

Mapping LS/OG conditions in SNEP was done at the scale of large landscape units (LS/OG polygons) which were identified and characterized by a team of more than 100 resource specialists from the land management organizations. Polygons were areas judged to be relatively uniform in type and distribution of vegetation patches (see Franklin and Fites-Kaufmann, 1996 for details). A composite LS/OG structural ranking was assigned to each polygon using a scale that extended from 0 (no contribution to LS/OG forest function) to 5 (very high level of contribution to LS/OG forest function).

On the Eldorado National Forest, 180 LS/OG polygons were identified through this process averaging approximately 12 000 hectares in size (Fig. 9.3). These polygons fell into four major forest types (Fig. 9.4). Franklin and Fites-Kaufmann (1996; see also Franklin et al. 1997) also identified clusters of LS/OG polygons with above-average levels of LS/OG characteristics (called areas of late successional emphasis or ALSEs) as potential focus areas for strategies to maintain high-quality LS/OG forest conditions in the Sierra Nevada (Fig. 9.5).

For our analysis, a quantitative description of the forest was needed within the LS/OG polygons to simulate LS/OG rank, growth, mortality, and harvest over time. Therefore, Forest Service information was used to identify forest vegetation condition classes within each LS/OG polygon (Fig. 9.6), the spatial distribution of these classes relative to streams, aspect, and topography, and the existing quantitative structure of each class. Each vegetation class was then evaluated within each LS/OG polygon for LS/OG rank using criteria provided by Franklin and Fites-Kaufmann (1996) including number of large trees, canopy closure, intermediate canopy, and number of snags. For westside mixed conifer, as an example, a rank 4 must have 2.4–3.6 trees over one meter in diameter (breast height) per hectare while a rank 1 must have 0.2–0.6 of these trees per hectare. Overall LS/OG polygon rank for the simulations was estimated as an area-weighted average across vegetation classes. This approach gave initial LS/OG rankings for the polygons similar to, but not identical to, those developed by Franklin and Fites-Kaufmann (1996): it was found the modeled ranks more tightly distributed around the middle rank values (rank 2 and 3) than the ranks determined by mappers and largely lacking ranks 4 and 5.

Our simulations had a goal of reversing the significant decline in the amount and complexity of LS/OG forest in the commercial forest types found by Franklin and Fites-Kaufmann (1996). Achievement of higher LS/OG ranks could be assisted by stand growth, prescribed fire, or certain types of timber harvest. If the potential for high-severity fire is an issue, care must be taken to develop prescriptions that control understory structure and canopy density without losing key features of LS/

222 J. Sessions et al.

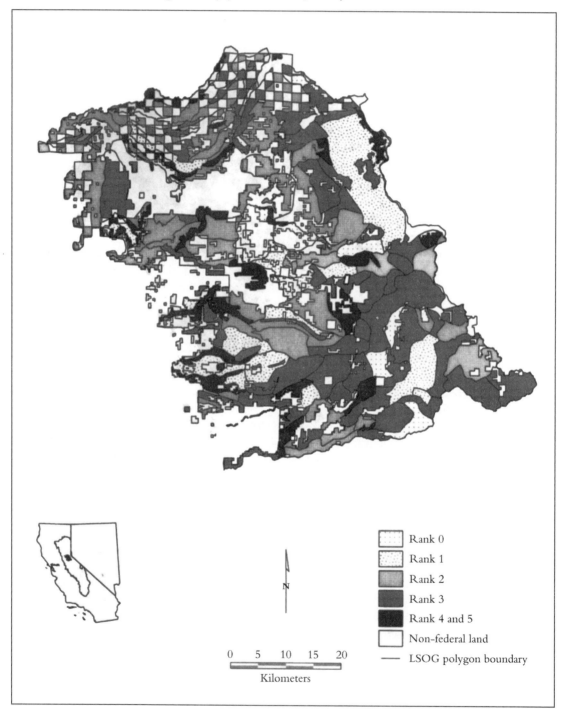

Fig. 9.3. Late successional/old growth (LS/OG) polygons for the Eldorado National Forest ranked by local experts as to their level of LS/OG structural complexity and contribution to late successional forest function (rank 0 = lowest level of complexity and contribution, rank 5 = highest level) (Sessions et al., 1996).

Fig. 9.4. Forest types on the Eldorado National Forest recognized in the analysis. Much of the analysis focused on the montane mixed conifer type where most of the commercial forest land is located and where issues of fire hazard and interactions with private landowners are strongest (Sessions et al., *1996).*

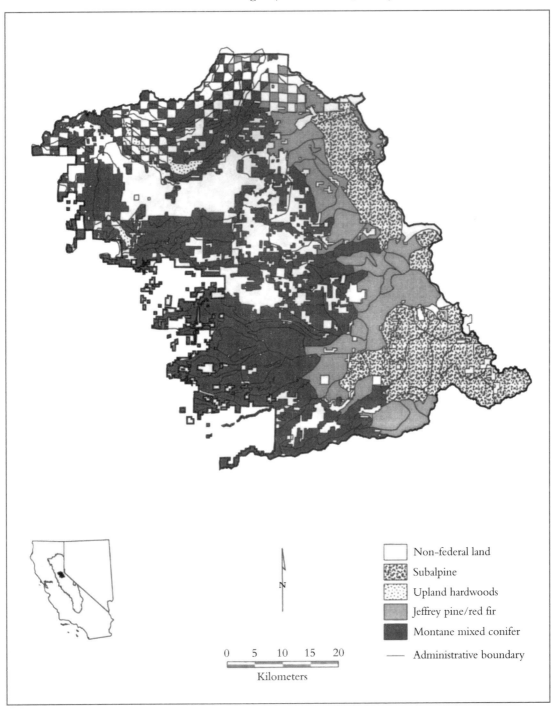

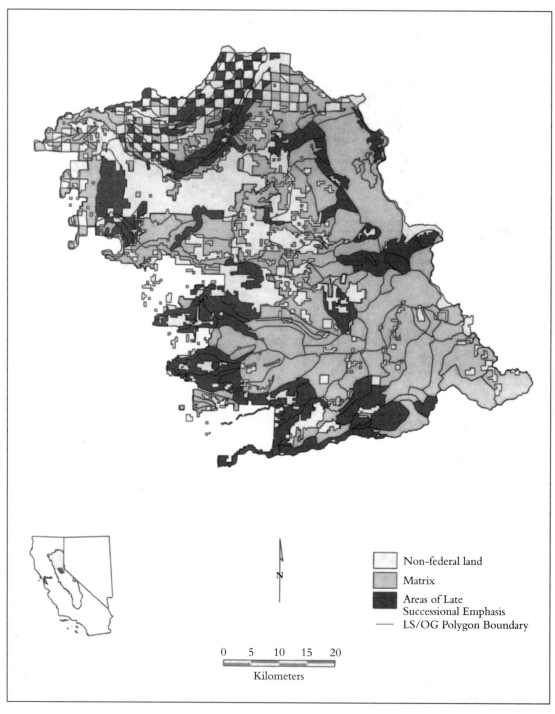

Fig. 9.5. *Areas of late successional emphasis (ALSE) on the Eldorado National Forest. These areas were chosen because of their relatively high level of late successional/old growth (LS/OG) characteristics and their location. They anchor the strategies considered in this paper to restore the LS/OG forests of the Sierra Nevada (Sessions et al., 1996).*

Fig. 9.6. The vegetative detail recognized in the analysis. Vegetative polygons were delineated by the US Forest Service using satellite imagery. Three classification variables were used: (i) forest type, (ii) tree size, and (iii) density. As an example, M3N refers to a stand of mixed conifer (M), small sawtimber (3), of medium density (N) (Sessions et al., 1996).

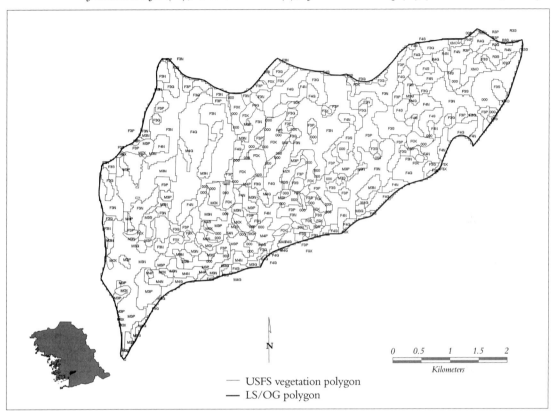

OG structures. The tension between the two goals of achieving key LS/OG features and reducing the potential for high-severity fire led us to use an LS/OG rank 4 as the highest target for late-successional forests instead of an LS/OG rank 5.

Fuel breaks (defensible fuel profile zones)

The Eldorado National Forest LS/OG polygons are overlain with a GIS coverage of potential fuel break polygons (defensible fuel profile zones) (Fig. 9.7). These zones would be 0.4 kilometers wide based on simulations by van Wagtendonk (1996) and were developed in cooperation with Forest Service personnel (Bahro, 1996; Weatherspoon and Skinner, 1996). They permit simulation of fire containment strategies. In alternatives that use fuel breaks, these strips of forest would be modified to achieve low levels of canopy cover and understory fuels with the goal of reducing flame length, fire intensity, spotting, and crown fire. The resulting stand structure would also provide a safe defensible area for fire suppression activities. They require periodic maintenance to remain effective. Fuel breaks are placed

Fig. 9.7. *A ridge-top network of defensible fuel profile zones (fuel breaks) on the Eldorado National Forest. The forests in these strips, approximately 0.4 kilometers wide, were modified in alternatives that recognized fuel breaks to achieve low levels of canopy cover and understory fuels, with the goal of reducing flame length, fire intensity, spotting, and crown fire (Sessions et al., 1996).*

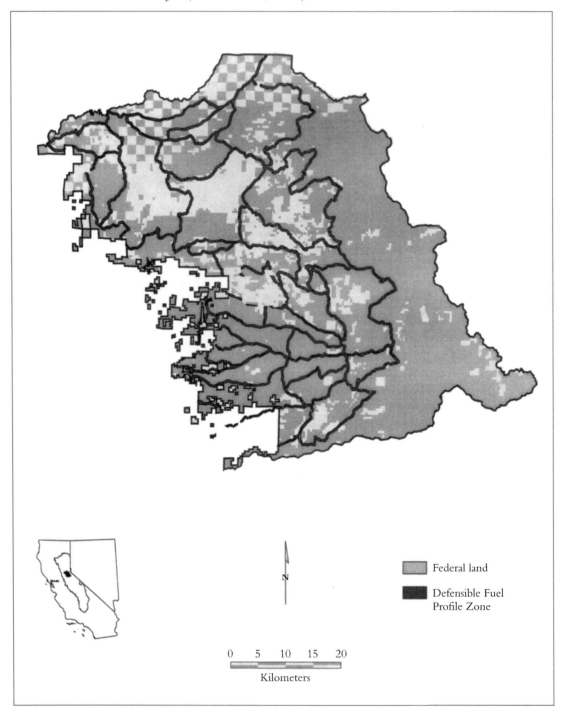

mainly on dominant ridge lines or strong intermediate ridges on the Eldorado NF. They have suitable access to facilitate safe fire suppression and also provide anchor points for large-scale prescribed burning programs.

Sub-division of LS/OG polygons for fire simulation

In our fire simulations, an entire LS/OG polygon is burned when a fire enters it, with the effects of the fire varying within the polygon depending on the condition of the different vegetative strata within it. The delineation of potential fuel breaks resulted in a sub-division of LS/OG polygons when a fuel break ran though it, increasing the number of LS/OG polygons (for simulation purposes) from 201 to 621 on the Eldorado NF (Fig. 9.8).

Each LS/OG simulation polygon is linked to an attribute table containing descriptive information such as LS/OG rank, size, dominant forest type, designation as an area of late successional emphasis (ALSE), selection as a fuel break polygon, and other characteristics.

Classification of land: non-spatial information

In addition to the spatial information, a set of non-spatial strata was recognized within each LS/OG polygon. The geographic locations of non-spatial data are not recognized in the SAFE FORESTS model. For example, it is known that 60% of the area of LS/OG polygon 99 is north-facing. However, where this north-facing land is located within the polygon is not tracked precisely. The different strata layers include land-use zones, ownership, USDA Forest Service vegetation classification, slope, aspect, and roadless areas.

Land-use zones

Each simulation polygon was further sub-divided into as many as four land use zones, three of which relate to potential influence on aquatic environments. First, there is a category called "reserved areas" which is composed of wilderness areas and areas considered too unstable for road building or logging in non-wilderness areas. Next, three aquatic influence zones are recognized that exhaustively divide the landscape outside of reserved areas. They are based on the aquatic and riparian system developed by D. Erman and reported in Kondolf *et al.* (1996).

Erman and his colleagues suggested a number of overlapping zones that are defined in relation to their influence on the adjacent aquatic ecosystem. The community influence zone is the area recognized as clearly riparian, with its distinctive flora and fauna and with many organisms that use both terrestrial and aquatic habitats on a regular basis. The energy influence zone includes all area near the stream that is likely to contribute energy and structure to the aquatic ecosystem, usually defined as the height of the tallest tree that can be grown on the site. It generally includes the community influence zone. The land use influence zone is the area adjacent to the energy influence zone in which human activity is likely

Fig. 9.8. Polygons used in the simulation of actions and effects under different strategies on the Eldorado National Forest. They were formed by the intersection of the LS/OG polygons and the fuel breaks (Sessions et al., *1996).*

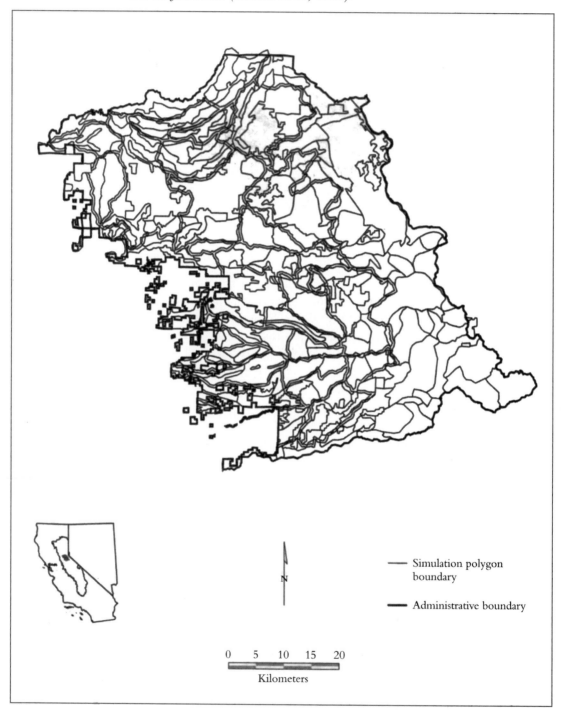

to influence the aquatic ecosystem by increasing nutrient and sediment inputs and other factors.

These three zones were mapped in a GIS exercise as follows: (i) a zone of approximately 46 meters (height of one site-potential tree) on each side of all streams to represent the community influence zone *and* the energy influence zone called the community/energy zone, (ii) a zone of variable width that begins just outside the community/energy zone, called the land use influence zone. This zone is calculated on the basis of modeled stream width and adjacent slope steepness using methodology from Kondolf et al. (1996), and (iii) the remainder of the watershed called the uplands. On the Eldorado NF (Fig. 9.9), the three zones divide the landscape approximately as follows: (i) community/energy zone (13% of the area), (ii) land use influence zone (34%), and (iii) uplands (53%).

Strata

For each land-use zone within each LS/OG polygon, four strata were recognized: forest condition (50–60 choices), owners (Eldorado National Forest/other), slopes (less than 40%, greater than 40%), and aspect (northeast/southwest).

In defining forest condition, we used the USDA Forest Service Region 5 forest classification of species type, size class, and percent of crown closure (USDA Forest Service 1994) to develop our vegetative classes. An aerial inventory of the Eldorado NF using this forest classification creates approximately 40 000 polygons on federal land reflecting the fine scale mosaic of the Sierra Nevada (Fig. 9.6). Forest inventory analysis (FIA) plot information was used to estimate existing forest condition for each vegetative class in terms of a tree list by species and diameter class and other information. These conditions were used as the starting point for developing silvicultural prescriptions to reach forest management goals appropriate for the different management strategies (see Cousar et al., 1996).

Although forest types were recognized on both federal and non-federal lands, forest management choices were modeled only on federal lands. Non-federal lands, mostly private land, were carried through the analysis for the purpose of reflecting an assumed contribution to watershed disturbance.

Slope classifications were recognized to reflect different fire behavior and logging methods. Aspect was recognized to reflect different burning conditions due to fuel moisture and temperature and different goals for late-successional forests.

Using GIS, areas that reflect unique intersections of these four land use and five stratum variables were identified. The model does not spatially track these variables within the simulation polygons beyond their presence in a land use zone. Rather, all occurrences of each unique intersection are grouped into a stratum. During simulations, an entire stratum within a land-use zone within each LS/OG polygon was assumed to receive the same treatment over a 10-year period. This permitted mapping activities to a location within an LS/OG polygon.

Several hundred strata can occur in each land-use zone within each LS/OG polygon. Generally, though, each land-use zone within each LS/OG included only a subset of the possible strata. On the Eldorado, the 621 simulation polygons

Fig. 9.9. A classification of riparian influence zones on the Eldorado NF. Three zones were recognized: (i) the community/energy zone of approximately 46 meters (one tree height) on each side of all streams, (ii) the land use influence zone of variable width based on stream width and slope steepness, and (iii) the uplands. Limits on watershed disturbance were keyed to these zones with higher levels of disturbance permitted in zones farther from the streams (Sessions et al., 1996).

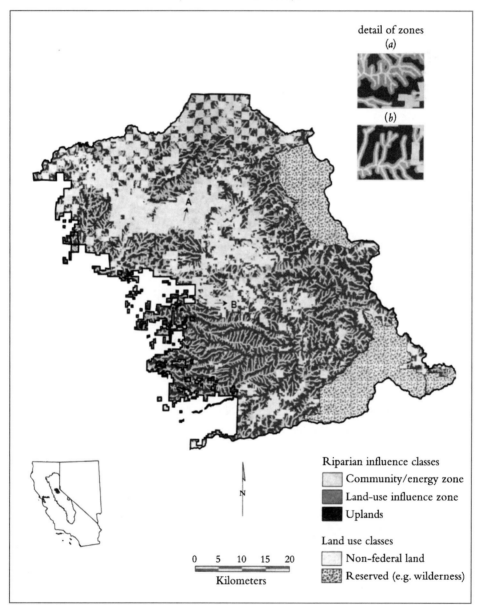

contained approximately 83 000 strata considering all land-use zones, with more than 80% of the strata representing the federal land component. The strata, in turn, were composed of an average of three patches spread across a land-use zone within a simulation polygon. Thus, average patch size recognized in our analysis of the Eldorado was about 1 hectare.

Fires and their effects: ignition, fuel, and fire size

Large fires account for most of the area burned in forest fires. In a study of several climatic regions of the western United States, Strauss et al. (1989) concluded that the proportion of area burned by the 1% of the largest fires ranged from 80%–96% of the area burned. In our simulations only large fires were modeled. Our definition of large fires on the Eldorado NF was a fire of at least 1000 hectares (Bahro, 1996).

Ignition probabilities for large fires were estimated for each simulation polygon based on ignition history, ratio of large fires to ignitions, and three point estimates of fire size (mean fire) in nine vegetation strata (Sapsis et al., 1996). Vegetative strata were based on three variables: (i) life form, (ii) weather zone, and (iii) population density class. These ignition probabilities are used in the SAFE FORESTS model as relative ignition probabilities and weighted to calibrate the model against historical fire data. After the model decides to burn a wildfire (based on weather probabilities) the relative ignition probabilities direct where the fire will occur.

Custom fuel models were developed for all the vegetative types (Sapsis et al., 1996). Fuels are the energy source for fire. Fuel models were assigned to simulation polygons to estimate potential fire behavior under conditions conducive to large fire occurrence. These estimates of fire behavior, based on fuel model, topography, and weather conditions, drive the effects on forest resources expected to result from large fire occurrence.

Potential damage to forests from extreme-weather wildfires, in our analysis, can come from two sources: (i) surface fire and (ii) crown fire. The potential for damage from fire under extreme weather conditions was estimated for each forest vegetation class based on likely flame length. Flame length and scorch height were calculated using the BEHAVE model to correct custom fuel models for elevation, slope, and aspect (Sapsis et al., 1996; Cousar et al., 1996). Potential damage to the live trees was then estimated as a function of species, diameter, height to the live crown, canopy closure, and topography, with the potential modified over time as the characteristics of the stand change. Stem mortality was based on fire effects probabilities for non-crown fires based on the USDA First Order Fire Effects Model (see Cousar et al., 1996 for more discussion). Crown fires could occur under certain combinations of flame lengths and crown closures. They could also occur when ladder fuels were present at heights that would carry fire to the live crown of the overstory based on relationships in Alexander (1988).

In SAFE FORESTS, attempts were first made to achieve specified goals without recognition of wildfire. Given the planned actions that best meet the goals for an

Fig. 9.10. A three-step approach to wildfire simulation. For each year, a choice was made between a drought (extreme weather) year and a non-drought year, the simulation polygons were analyzed for fire ignition, and wildfire acreage was generated up to the limit on acreage chosen for that year. At each step, a probability distribution, created from an analysis of historical records, was sampled for the needed information.

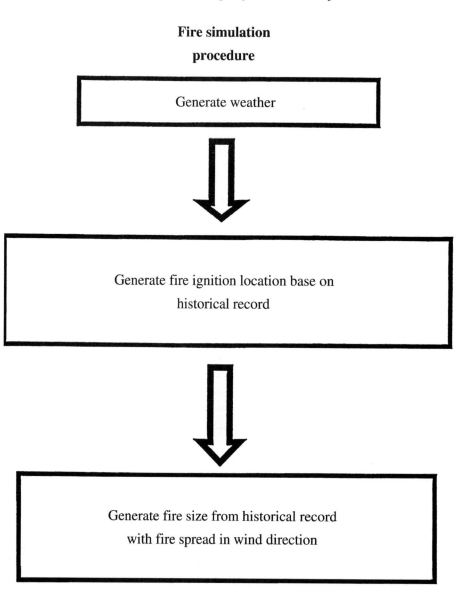

analysis, a series of wildfires was then generated under extreme weather conditions for each period from a probabilistic simulator that considers weather, fire size, and ignition probability to select the amount, individual size and location of fires that will occur in a period (Fig. 9.10). Average decadal amount of wildfire and the range in this amount were obtained from national forest records to help construct the probability distribution of fire size (Barho, 1996). It was assumed that this fire area occurred under extreme weather conditions, i.e., conditions that often enable fire to escape initial suppression attempts.

Fire spread under extreme weather is difficult to predict. In our model, a fire spreads from polygon to polygon, in the direction of the likely wind direction in extreme weather, to fulfill the selected fire size. This should be viewed as a first approximation; much work remains to be done here.

Fire has differing effects on the forest within the polygons that burn depending on the forest condition of the vegetation classes recognized in the polygon. At the end of each period, decisions are made about whether to salvage and/or postpone green tree harvest on the federal forest that burned depending upon the goals for the area and watershed conditions.

Goals and their attainment

Five goals explicitly guided the analysis: (i) increasing the general extent and complexity of late-successional forests, (ii) reducing the potential for high-severity fire, (iii) restoring riparian areas and watersheds, (iv) reintroducing historical ecosystem processes, and (v) producing a sustainable supply of timber in a cost-effective manner.

In each analysis, these goals were specified in hierarchical fashion such that achievement of the higher-order goal could not be compromised by attempting to achieve a lower-order goal. As an example, assume that we are interested in the upper riparian zone in the matrix. In some analyses, the primary goal was control of watershed disturbance, the secondary goal was achievement of some late-successional rank, and the tertiary goal was achievement of an even-flow timber harvest. In all cases, the problem was structured such that we tried to achieve as much of the secondary goal as possible given that achievement of the primary goal was not compromised and that we tried to achieve as much of the tertiary goal as possible given that achievement of the secondary goal was not compromised. Ironically, this optimization problem is stated such that an attempt is made to "maximize" (or "minimize") attainment of the lowest-order goal after first attaining specified values for the higher-order goals.

Increasing the complexity and extent of LS/OG forests

Forest structure goals were measured through LS/OG rank with these ranks primarily depending on number of large trees and canopy closure. Achievement of LS/OG ranks could be assisted by stand growth, prescribed fire, or certain types of timber harvest. Much of the prescription development undertaken was aimed at finding ways to accelerate the restoration of late-successional structures (increase LS/OG ranks) in the different types and forest conditions in the Sierra Nevada. Without attention to understory structure and canopy density, however, achievement of high levels of late-successional structural complexity can be associated with high likelihoods of severe fire (Agee, 1996; Arno et al., 1996). Therefore attempts were made to develop prescriptions that would enhance LS/OG structures while having moderate to low levels of basal area likely to be consumed by wildfire. This generally involved surface fuel treatment (hazard reduction) in conjunction with selective harvest of small diameter trees. These small trees, in addition to being highly sensitive to fire-induced mortality, are also a significant hazard because they link fire to the canopy (Alexander, 1988; Weatherspoon and Skinner, 1996). Additional details of prescription development can be found in Cousar et al. (1996).

Reducing the potential for high-severity fire

Potential damage from extreme-weather wildfires was measured by percent of basal area that would be killed if a wildfire would occur. In classifying fire regimes based on fire severity, Agee (1993) states that "stands in low severity fire regimes have 20% or less of the basal area removed by fire, while stands in high severity fire regimes have 70% or more of their basal area removed" and that stands in moderate-severity regimes often experience a mixture of high and low severity. For the purposes of this study, we considered a basal-area loss of greater than 60% over a large landscape as "high-severity".

In our analysis, fire behavior could be modified by human intervention in three ways: (i) commercial timber harvest, (ii) fuel reduction treatments including mechanical treatment and prescribed burning, and (iii) creation of fuel breaks (defensible fuel profile zones) – 0.4 kilometer strips on which the overstory canopy closure would be maintained at low levels, as would understory fuels (Weatherspoon and Skinner, 1996). In our analysis, the first two methods affect fire severity; the third mainly affects fire size. Timber harvest, though, can cause excessive watershed disturbance or retard achievement of late-successional goals. Thus, treatments involving harvest were often limited by pursuit of other goals.

Restoring riparian areas and watersheds

Watershed disturbance was measured by the percentage of equivalent roaded acres (ERA) in the watershed by riparian influence zone. ERA is an index of watershed disturbance used extensively by the national forests of California. Each proposed activity is given an ERA coefficient to measure its disturbance potential as is the existing condition of the landscape. (For a discussion of the ERA method, see Menning et al. (1996) and Sessions et al. (1996).)

Watershed conditions were aggregated and controlled by LS/OG polygon. LS/OG polygons are of the appropriate scale (1000 to 2500 hectares) recommended by Chatoian (1995) for use with the ERA method. The existing condition of the watershed was taken into account in establishing the initial ERA levels. Future activities then contributed ERA amounts based on their projected degree of disturbance. The analysis reported below allowed timber harvest activity only when the aggregate ERA for an LS/OG polygon would not exceed the average ERA "threshold of concern" on the Eldorado National Forest.

Much private land occurs within and adjacent to the Eldorado National Forest, almost all of it as forest. We assumed that the lands were well-roaded, with frequent re-entries and mechanical site preparation. It was further assumed that the federal forests would react to actions on private lands within the watershed (here LS/OG polygon) by limiting their actions so that overall watershed limits were not violated. Given the cumulative effect considerations required of federal forest managers, we feel this is a reasonable assumption. Under our assumptions about activity on non-federal land, very little federal action that contributes ERA coefficients, such as timber harvest, could occur where non-federal forests covered more than half of an LS/OG polygon.

Reintroducing historical ecosystem processes

Rebuilding LS/OG forests, reducing the potential for high-severity fire (where it historically was believed to be uncommon or absent), and limiting accelerated disturbance in watersheds should help reintroduce historical ecosystem processes. Employing prescribed fire and silvicultural methods that simulate, to some degree, the historical effect of wildfires on Sierra Nevada forests should also help to achieve this goal.

Providing sustainable, cost-effective timber harvests

The highest sustainable timber harvest for 50 years, compatible with watershed disturbance limits and late-successional and fire hazard targets, could be specified as a goal. A wide variety of harvest intensities and timings were available to help find the highest sustainable level given other goals. The highest sustainable level was calculated before wildfires occur. Stands that burn severely before their scheduled harvest were deducted from the estimated sustainable level.

Salvage after wildfire occurred if it was consistent with overall goals. Thus, the timber harvest volume available for any period was the sum of two components: (i) the "green" timber harvest associated with the estimated sustainable timber harvest for 50 years, with the harvest level for a period reduced by the amount scheduled for harvest in the period that burned severely and (ii) salvage timber harvest associated with reaction to wildfire. Thus, the overall expected harvest for a period could vary somewhat depending on the extent of severe fires.

Activities and their effects

In SAFE FORESTS, two general types of human intervention (activities) can be used to meet stated goals: (i) timber harvest, and (ii) prescribed fire. Depending upon topography, mechanical fuel reduction can be combined with timber harvest. Undisturbed growth is also considered a possible "activity". The development of prescriptions that can be used to meet the goals of an analysis is described in Cousar *et al.* (1996).

Each activity is represented in the SAFE FORESTS model by its decadal contribution to forest structure (LS/OG rank), its contribution to watershed disturbance (ERA), its flame length if a wildfire occurs under extreme weather, and its contribution to timber production (board feet harvested). Activities are strung together for five 10-year periods, in keeping with a Model I formulation, to form what we call a "prescription". As an example, two prescriptions for the mixed conifer strata with existing vegetative condition M3G on a gentle slope with a southwest aspect that had previous salvage harvesting were: (i) let the forest grow without intervention "NNNNN", and (ii) active vegetative management to maintain the forest structure at LS/OG rank 3, reduce the potential for severe wildfire, and provide timber harvest volume. This prescription consists of commercial timber entries each 20 years (H) coupled with prescribed burning in intervening periods (P) or "HPHPH".

Silvicultural methods employed

A wide variety of harvest timings and intensities were developed for each vegetation class. Harvest was oriented toward achieving the goals for stand development over time using a stand optimizer overlaid on the USDA FVS (Wessin variant) individual tree simulator (Cousar et al., 1996). Generally, these harvests concentrated on reducing shade-tolerant trees in the understories.

Removal of selected species and sizes in each cutting enabled achievement of the late-successional goals to be addressed, which call for the continuous presence of large trees across the landscape, while also addressing fire hazard concerns. In practice, this approach would require both individual tree selection and group selection. Work done for SNEP (Helms and Tappeiner, 1996) points out a variety of silvicultural systems, including various selection methods, could be used, in conjunction with prescribed fire, to sustain the forests of the Sierra Nevada and reduce fire hazard there.

Stages in the SAFE FOREST model

A four-stage forest-level model was built to analyze the implications of different policies and objectives for managing the federal lands of the Sierra Nevada (Fig. 9.11):

(i) The first stage uses a Model I formulation for the strata within each ALSE polygon, in which each prescription represents all actions, inputs, and effects over the planning periods, to maximize the achievement of LS/OG rank within the polygon subject to goals on watershed disturbance.

(ii) The second stage uses a Model I formulation for the remainder of the forest (all non-ALSE LS/OG polygons) which sets goals for LS/OG rank and watershed disturbance for each LS/OG polygon and goals for timber harvest level (even-flow) over time for the entire forest. The timber harvest level includes the results from the first stage analysis. Subject to achieving these goals, the second stage maximizes timber harvest from the forest given the permitted activities.

Fig. 9.11. The four stages of the SAFE FORESTS model. In the first two stages, a search is conducted for the set of activities over the planning periods that best meets the goals of the simulation. This search is first done for the areas of late successional emphasis and then for the rest of the forest. In the last two stages, fires are simulated for each planning period, based on a randomly selected weather stream and randomly selected ignitions and wildfire sizes, and the schedule of activities are adjusted where fire occurs to meet the goals for the remaining periods.

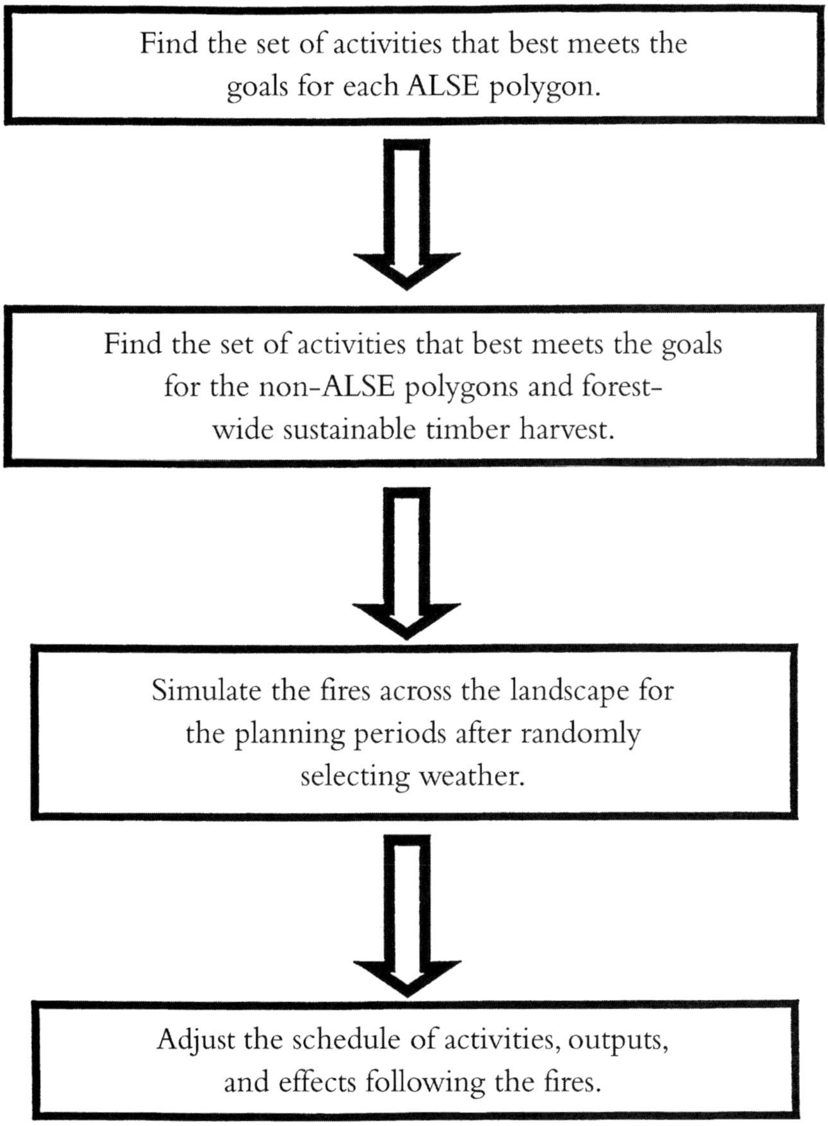

Find the set of activities that best meets the goals for each ALSE polygon.

Find the set of activities that best meets the goals for the non-ALSE polygons and forest-wide sustainable timber harvest.

Simulate the fires across the landscape for the planning periods after randomly selecting weather.

Adjust the schedule of activities, outputs, and effects following the fires.

(iii) The third stage simulates the stochastic application of fire to the forest for five decades. It is assumed that the planned schedule of activities, forest growth, and effects from stages one and two will occur. We then simulate the size, distribution, and intensity of wildfires for the forest for a series of randomly selected weather streams, with a weather stream identifying the distribution of weather between normal and extreme in each period.

(iv) Finally, we incorporate the implications of these fires for the actions on the forests, outputs, and effects. The effects of fire depend on the stand condition when the fire occurs. Salvage occurs when permitted. Then, the prescribed activities from the first two stages are adjusted to take the "next best" solution given that a fire has occurred. That is, in each burned stratum, a substitute prescription is chosen which maintains the same pre-fire activities as the original prescription, but considers the fire effects over the decades remaining in the planning horizon. Both in-stratum and cumulative LS/OG polygon effects are considered in the post-fire prescription choice.

Modifying the prescriptions after a fire involves substituting a Model I vector (column) of activities that maintains the pre-fire activities while changing the post-fire activities to meet the goals of the prescription given that a fire has occurred. Following a fire, and evaluation of fire effects on forest structure, SAFE FORESTS draws from among a large number of pre-generated stratum-level prescriptions that represent pathways to stratum-level forest structure goals given that either a human-caused or natural disturbance occurs. The prescription chosen must maintain the emphasis (structure goal for the stratum) and recognize the watershed goals for the LS/OG polygon. See Cousar et al. (1996) for additional details on the goal-oriented strata level prescription development used here.

Solution methodology

A heuristic is used to assign activities to each stratum of each land-use zone of each polygon to reach goals for the administrative unit. Only activities on federal land are simulated.

The solution procedure has four stages:

(i) Assignment of activities to strata in ALSE polygons in the absence of fire (Fig. 9.12).
(ii) Assignment of activities to strata in the non-ALSE polygons in the absence of fire (Fig. 9.13).
(iii) Simulation of fire (Fig. 9.14).
(iv) Simulated management response to fire (Fig. 9.15).

The heuristic has similarities to binary search in that strata are ordered in terms of priority for treatment. It improves on binary search, and has similarities to the approach used by Hogason and Rose (1984), in that it allows for consideration of

Fig. 9.12. First stage of the SAFE FORESTS model. Activities are assigned to strata in Areas of Late Successional Emphasis (ALSE) polygons to meet the goals of the analysis. Strata are combinations of forest condition, owner, slope, and aspect. Goals considered include achieving late successional/old growth (LS/OG) rank and limiting the equivalent roaded acre (ERA) measure of watershed disturbance. (Sessions et al., 1996).

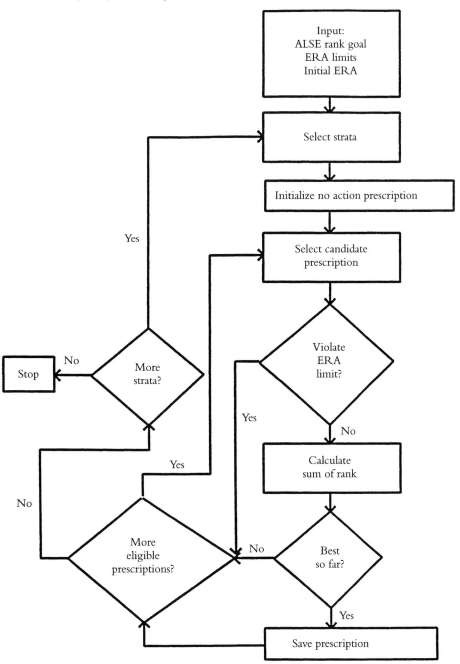

Fig. 9.13. Second stage of the SAFE FORESTS model. Activities are assigned to the forest outside of areas of late successional emphasis to meet the goals of the analysis. Strata are combinations of forest condition, owner, slope, and aspect. Goals considered include achieving late successional/old growth (LS/OG) rank, limiting the equivalent roaded acre (ERA) measure of watershed disturbance, and finding the even-flow harvest level for the forest. (Sessions et al., 1996).

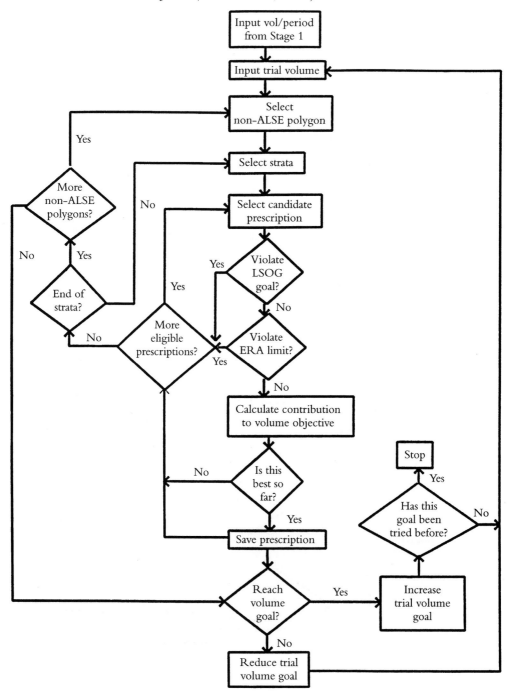

Fig. 9.14. Third stage of the SAFE FORESTS model. Wildfires are simulated using historical information on weather, ignitions, and fire size, all represented as probability distributions that are sampled during each simulation. When fuel breaks are part of the scenario, their effectiveness is also represented as a probability distribution (Sessions et al., *1996).*

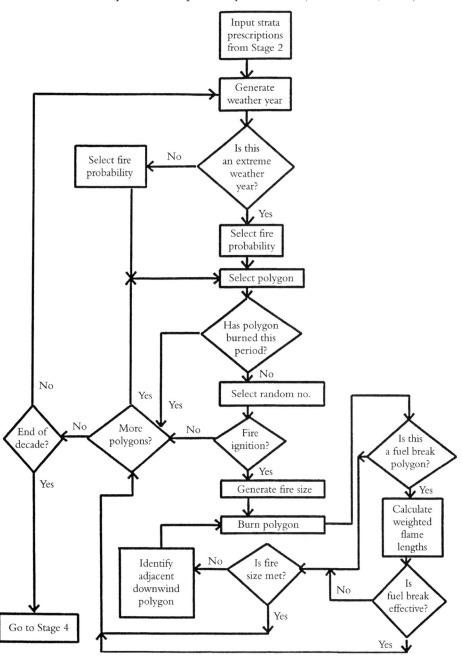

Fig. 9.15. Fourth stage of the SAFE FOREST model. Once a simulation polygon burns, adjustments are considered in the management of the strata within it to allow salvage of timber and the selection of new activities for the remainder of the planning periods, considering the goals of the simulation and the current state of the strata. When a wildfire occurs in simulation polygon, the analysis assumes that the fire covers the entire polygon, with the effect of the fire being a function of fuel conditions in the strata within the polygon (Sessions et al., 1996).

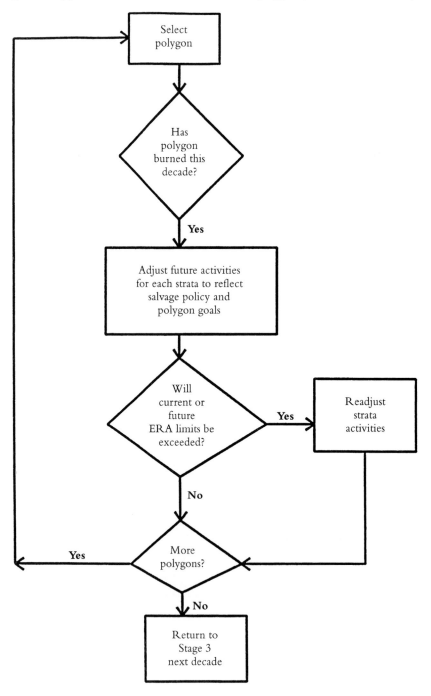

multiple prescriptions for each stratum instead of only one for each stratum as done in binary search.

First and second stages

The first stage applies to each ALSE polygon. All strata are given an initial prescription of NO ACTION. LS/OG rank is maximized subject to ERA limits. Prescriptions considered are limited to those that will satisfy certain tests for rank. Each of three zones that relate to distance from stream has a single limit on ERA; thus each stratum has only one ERA limit in each period. All periods are considered simultaneously to control on ERA. Given the order of consideration for the strata, prescriptions are identified that maximize rank, that is, reach the desired rank earlier than any other candidate prescription while not exceeding the cumulative ERA limit for the land zone of the ALSE polygon in any period. If more than one prescription can achieve the desired rank at the earliest time, the prescription that has the lowest average ERA is chosen.

The second stage applies to the non-ALSE polygons, in terms of selection of activities, and to the entire forest in terms of the harvest flow goal. All non-ALSE polygons are given an initial prescription of NO ACTION. As with the ALSE polygons, each of three land zones that relate to distance from stream has a single limit on ERA and all periods are considered simultaneously to control on ERA.

Given the order of consideration for the strata, prescriptions are selected which meet LS/OG goals, any other limits on permitted activities and make the largest contribution to providing harvest in those periods that have lower than average harvest while not exceeding the ERA in any period. In the event of a tie, the prescription that has the lowest average contribution to ERA for the strata is chosen. Aggregate harvest is then updated and the calculations move to the next strata.

This heuristic uses binary search with an embedded gradient subsearch. The analysis begins with the existing forest structure and known condition of each watershed. The strata are ordered within each LS/OG polygon, land class, vegetation strata, aspect, and slope class. To guide the gradient subsearch toward a sustainable harvest flow, a forest wide objective function of minimizing decadal deviations from a trial decadal harvest volume is specified. The sequence of steps is as follows:

Specify trial decadal timber harvest volume:

For LS/OG polygon = ALSE to non-ALSE
  For aspect = south to north
    For slope = flat to steep
      For landclass = uplands to near stream
        For veg. type = first to last
          For activity = first to last

            If polygon is ALSE polygon then

identify contribution this activity would make to ALSE goal for strata while not exceeding watershed goal for land class

If polygon is not ALSE

identify the maximum contribution this activity would make to reducing deviations to trial sustainable timber production goal while meeting LS/OG goal and not exceeding watershed goal for land class

Next activity

Update solution by allocating best activity to entire strata (0/1)

    Next veg. type
   Next landclass
  Next slope
 Next aspect
Next LS/OG polygon

If trial volume not sustainable, then reduce trial volume and repeat procedure. If volume sustainable, increase trial volume and repeat procedure.

These procedures do not guarantee to find the solution that yields the highest value of the objective function given the constraints, as does linear programming. These solutions have not been systematically compared to those that would result from using linear programming as the solution technique. In early use of the model, however, a second pass was taken through the strata to see whether the solution could be improved with little effect. A positive feature of the model is that the solution is closely tied to the ground through the large number of polygons, spatially defined riparian land classes within polygons, and slope and aspect classes. Experience with management plans developed with more coarse land stratifications indicated as much as a 30% deviation between the linear programming solution and the actual harvest plan.

The priorities for consideration of different strata recognize a number of objectives. In terms of topographic aspect and slope, the priority generally is south flat, north flat, south steep, north steep. These priorities reflect both cost of treatment and risk of fire. The priority for treatment of the major species groups are mixed conifer, ponderosa pine, red fir, and hardwood. Alpine receives only the NO ACTION prescription.

Third and fourth stages – fire simulation

A sequential approach has been taken to simulate the occurrence, extent and effects of fire. Following assignment of activities, fire ignitions occur based upon historical probabilities within polygons and spread into contiguous areas.

Fire effects are then estimated using the vegetation structure and composition in each stratum at the time of the fire along with the topographic variables of slope and aspect. The simulations are then repeated a number of times to obtain an estimate of the variability of the outcomes.

The solution from part one is then subjected to a set of weather streams, fire ignitions, and fire spread. The fire effects are calculated and management reaction is simulated.

The fire simulation procedure is:

For decade = first to last
  For year = 1 to 10
    For subpolygon = first to last

     If this is an extreme weather year

        test random number against probability of a large fire starting under extreme weather conditions in this polygon

     If this is a normal year

        test random number against probability of a large fire starting under extreme weather conditions in this polygon

     If fire starts then

        If polygon has already burned in this period do not reburn

        identify fire size from a list of historical fire sizes for administrative unit using frequency distribution of historical fire sizes

        grow fire by burning entire polygon fire starts in

        update veg status for burned strata by selecting new activity that maintains management emphasis for that veg type given that a fire has occurred.

        If ERA for LS/OG polygon is not exceeded, then salvage

        If ERA for LS/OG polygon is exceeded, postpone green harvest for other strata while maintaining same management emphasis

        If fire has not reached target fire size, spread fire to adjacent polygon according to prevailing winds for that area (administrative unit or subunit).

        If adjacent polygon is a fuel break then calculate average weighted flame length for strata in burned polygon and average weighted flame for strata in fuel break

            test random number against probability of fire crossing fuel break given approaching flame length, fuel break flame length, and the potential for spotting, based on the work of Van Wagtendonk (1996),

            If random number less than probability of crossing fuel break then burn fuel break and fire stops

            Update polygon veg. type

If fire has not reached target fire size, spread fire to adjacent polygon according to prevailing winds for that area (administrative unit or subunit).

   Next year
  Next decade
See Bahro (1996) for specific coefficients used in modeling fire behavior.

A sample analysis

One strategy has been chosen to illustrate our approach of the many that we simulated (Johnson et al., 1996). It is representative of the results that we found. The strategy attempts to achieve an LS/OG rank of 4 in the ALSEs and an LS/OG rank of 3 in the matrix, while pursuing an aggressive strategy to reduce the likelihood of high-severity fire. It employs prescribed fire across the landscape, allows timber harvest in the matrix, and implements the system of fuel breaks (DFPZs) in the first period. Watershed disturbance is limited to the threshold of concern.

Our discussion emphasizes the pine and mixed-conifer forests on the Eldorado National Forest (see Fig. 9.4). These forest types contain most of the commercial forest and have experienced a large decline in late-successional complexity, increase in fuel loadings, and reduction in fire frequency. A strategy was used that allowed a combination of prescribed fire, timber harvest, and undisturbed growth. One pass of the wildfire simulator was used that had an average level of fire over the planning horizon, to create the results reported here. The variability in these results from repeated runs of the fire simulator is discussed below.

Average LS/OG rank in the pine and mixed-conifer forests increased over time (Figs. 9.16 and 9.17: see color section), but fully achieving an LS/OG goal of 4 in the ALSEs will take longer than 50 years. Attempts to decrease the potential for high-severity fire limited the rate of increase somewhat.

Much of the pine and mixed-conifer forest is currently susceptible to high-severity fire (Figs. 9.18 and 9.19: see color section). Without active management, the extent of these forests susceptible to high-severity fire would increase over time. With fuel treatments (prescribed fire or a combination of prescribed fire and timber harvest), however, the potential for these forests to experience high-severity fire greatly decreased (Figs. 9.18 and 9.19).

Keeping watershed disturbance within commonly suggested limits for federal forests significantly affected the role of timber harvest in achieving the goals of the analysis – these limits cut in half the timber harvest that would otherwise occur. Historic and expected actions on non-federal land were especially important in determining the amount of action on federal land consistent with limits on watershed disturbance.

Timber harvest was concentrated in the smaller diameter classes, at least for the next few decades, although the creation of DFPZs would result in the harvest of some large trees. A total of about 30–40 million board feet per year would be harvested, approximately one-third to one-half of historical amounts.

Overall timber harvest activity was measured in terms of the number of times the stands in the pine and mixed conifer forests were entered for harvest over 50 years. Under this strategy, approximately 20% of the forest outside of Wilderness was entered for harvest at least once. All hectares receive prescribed fire on a one- or two-period cycle except in the period of harvest.

Effects of repeated simulations

The results reported so far have been based on one set of wildfires through time such as those shown in Fig. 9.20. To assess the variability in wildfires that might occur, ten simulations of wildfire were performed. Each of these simulations reflects a different weather stream and selection of fire size and area of occurrence based on the probabilistic fire simulator. The acreage burned on the Eldorado NF, under the strategy described above, varied from 7.5–25% of the forest acreage with an average of 15.6%. In terms of the range of fire effects on LS/OG rank, the highest variability was found in the lowest (0,1) and highest ranks (4,5) while the middle ranks (2,3) showed considerable stability.

Using ten wildfire simulations enabled a map to be developed showing probability of wildfire under extreme weather conditions during a fifty-year period (Fig. 9.21). Looking in detail at the pine and mixed-conifer forests, the simulations suggest that more than two-thirds of these forests have at least a 10% chance of fire (one burn) under extreme weather conditions in the next 50 years and almost 20% of them have at least a 30% chance of fire.

Effect of the DFPZs

Ten simulations with DFPZs and ten without them were done, with all other elements of the strategy as described above. With DFPZs, as mentioned above, the acreage burned over 50 years varied from 7.5 to 25% for an average of 15%. Without DFPZs, the acreage burned varied from 10 to 30% with an average of 21%. Thus, adding DFPZs to the strategy reduced average acreage burned on the Eldorado National Forest by 26%. This percentage difference varied from simulation to simulation depending on where the fire started, relative to the location of the nearest DFPZs that it would encounter. Over all simulators, the acreage reduction from introducing DFPZs varied from 8 to 49%.

The potential for achieving multiple goals for Sierra Nevada forests

Five goals explicitly guided the analysis: (i) increasing the general extent and complexity of late-successional forests, (ii) reducing the potential for high-severity fire, (iii) restore riparian areas and watersheds, (iv) reintroducing historical ecosystem processes, and (v) producing a sustainable supply of timber in a cost-effective

Fig. 9.20. Two periods of a wildfire simulation showing the location and size of wildfires under extreme weather. The amount of severe weather in a period and the total amount and location of wildfire are all represented as probability distributions based on historical records. The effect of wildfire on forest condition is a function of fuel conditions in each stratum within the polygon (Sessions et al., 1996).

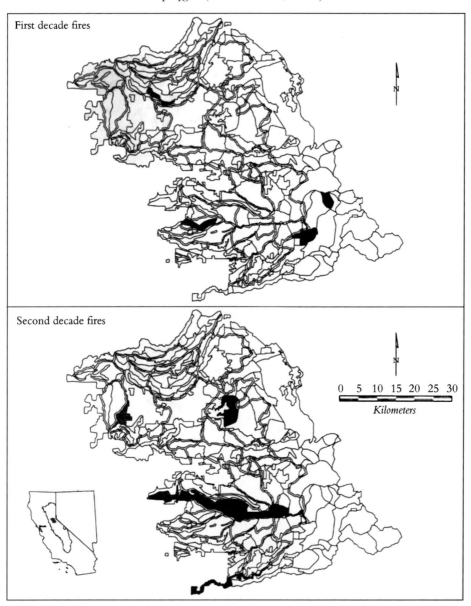

Fig. 9.21. Number of burns in the simulation polygons over five 10-year periods in ten simulations. These results can be interpreted as the probability of a simulation polygon burning under extreme weather over 50 years. As an example, three to four burns in the polygon translates into 30–40% chance of a wildfire in the polygon over 50 years (Johnson et al., 1996). Reprinted from the Journal of Forestry, **96**(1), *44–5 published by the Society of American Foresters, 5400 Grosvenor Lanes, Bethedsa, MD 20814–2198. Not for further reproduction.*

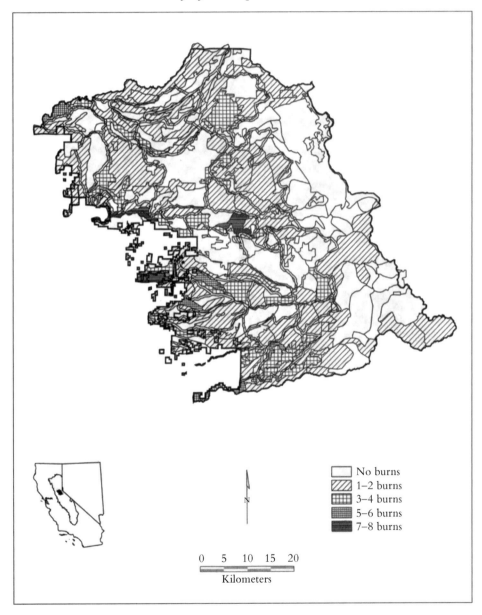

manner. To a significant degree, it is felt that it is possible to achieve these goals, based on our analysis. Some of the key points are:

(i) It appears possible to simultaneously increase the general extent and complexity of late-successional forests and reduce the potential for high-severity fire. This possibility has been the subject of much debate – this analysis should help point the way to achieve both goals.

(ii) A sustained supply of timber could be produced, but at reduced levels compared to recent history. Timber harvest consumed only a small proportion of overall growth, with much of the remainder going toward rebuilding late-successional forests and some being consumed by prescribed fire and wildfire. Once late-successional forests are rebuilt, the timber harvest level consistent with maintaining these forests could increase.

(iii) Timber harvest generally paid for itself, although the net revenue fluctuated widely depending on the size and species harvested. On the other hand, sizable investments would be needed for prescribed fire. The assumed relationship between timber harvest and watershed disturbance caused significant reliance on prescribed fire to reach ecosystem goals. It appears a significant increase in the use of prescribed fire over planned levels would be needed to achieve these goals.

(iv) A combination of prescribed fire and timber harvest would reduce the likelihood of high-severity fire more rapidly than prescribed fire alone. Obviously, we would need a combination of practices to reduce fire hazard as neither practice fully meets all needs: the current buildup of fuels often requires some initial timber harvest to safely apply prescribed fire; smoke management will limit the use of prescribed fire; the major investment needed by prescribed fire will be hard to sustain; and timber harvest offers a variety of benefits, but also has limitations relative to watershed effects and representation of historical ecosystem processes.

(v) Building DFPZs reduced the size of wildfires as a first step toward limiting the extent of high-severity fires and increasing the safety of fighting wildfires. With an aggressive fuel-reduction program, however, little long-term difference in fire effects on forest structure was found between alternatives that used DFPZs and those that did not: in the long-run under an aggressive fuel reduction program, few wildfires reach the crowns so little damage to LS/OG structures occurs.

(vi) The cumulative effects of public and private activities could greatly limit the federal timber harvest, especially in watersheds with mixed ownerships. The authors assumed that the federal forests would react to expected actions on private lands within the watersheds by reducing their harvest actions so as not to violate watershed limits. Federal activities were not shifted to very low impact methods, but that seems a key to federal harvests going forward in mixed ownerships.

(vii) Management of intermingled and adjacent non-federal lands will impact the success of any strategy for federal lands. Management of non-federal lands can affect the performance of DFPZs, the functioning of late-successional forests on federal land, and, as noted above, the level of watershed disturbance in watersheds shared with federal lands. Ecosystem management will be difficult without co-operation across ownerships.

(viii) The methodology employed here (Figs. 9.16, 9.19) should aid collaborative learning about alternative strategies for managing Sierra Nevada forests. Recognition of the episodic and probabilistic nature of wildfire increases the credibility and realism of the analyses and should help the public understand the relative risks of alternative strategies for these forests.

Discussion

An analysis of forest structure and composition has been presented over time under different management objectives for the federal forests of the Sierra Nevada emphasizing the interaction of forests, fire, watersheds, and people. In this work, considerations of fire behavior and effects are integrated into a dynamic, spatial landscape analysis for a relatively large area (approximately 1 000 000 acres). Our contribution is to develop and apply methodology to guide the management of a large landscape in the Sierra Nevada toward multiple goals, reflecting a suite of values, while simultaneously recognizing significant spatial detail in the landscape simulation and portraying the patchy and episodic nature of wildfire.

A number of previous efforts have conducted spatial simulations of wildfire across a forested landscape. Other efforts have incorporated non-spatial fire effects into forest planning which seeks management actions that achieve multiple goals. Attempts have been made to merge these two approaches by combining the spatial simulation of forest development on a large landscape, including wildfire occurrence and effects, with the search for management actions that achieve multiple goals. It is believed that this work advances the incorporation of natural disturbance into strategic forest planning, and, as a result, should prove useful in assessing forest management strategies for the fire-prone landscapes of the West.

The potential of this work, though, is limited by our ability to develop summary measures of ecosystem integrity. Take the measurement of "health and sustainability" for Sierra Nevada ecosystems. SNEP made only modest progress in developing measures of "health and sustainability of Sierra Nevada ecosystems", both because of the limited time available and the difficulty of the task. Two of our key measures – LS/OG rank and ERA level – lack the scientific studies to help evaluate the appropriateness of these measures. Despite the long-term study of the forests of the Sierra Nevada, no commonly accepted measures of late-successional forests were found there, either in terms of stand structure or patch size. SNEP scientists had to create both measures; much discussion and evaluation of their adequacy will and should occur. Despite the long-term importance of measuring and controlling cumulative effects from federal lands, few measures of cumulative watershed effects exist that are accepted by scientists and useable by managers. Measures like ERA rest on a combination of understanding of ecosystem processes, logic, and a very modest amount of evidence. It is believed that the difficulties in developing commonly accepted measures of ecosystem health and sustainability will continue to limit our ability to understand the implications of alternative policies for managing large ecosystems like the Sierra Nevada.

Model limitations and future work

The SAFE FORESTS model should be considered a work in progress. Improvements could be made in almost all areas. The spatial incorporation of wildfire has been a central part of this model. Improved ways of distributing fire on the landscape should undoubtedly be made. More explicit spatial recognition of the strata within polygons may be required.

There has been no linkage between suppression effort and fire size. Although cumulative effects of wildfire on watershed disturbance and forest stocks of both live and dead wood are tracked, the financial costs of wildfire suppression are not addressed, so trade-offs between presuppression efforts, suppression efforts, and forest condition cannot be made.

It was assumed that fuel treatments altered the effect of fire on forest structure, but not fire size. Since it would be expected that fuel reduction would make fires easier to suppress, even under extreme weather, this assumption should be refined.

Questions can always be raised in spatial simulations about whether the landscape pattern produced over time reflects likely reality or merely the model architecture. As an example, LS/OG polygons were important landscape units for both fires and treatments in our analysis. To a significant degree, these units were created through past treatments and fires, so it is felt there is some objective reality to recognizing them for analysis. Still, future work should examine how modifying the approach to delineating landscape units might influence the results.

An attempt has been made to recognize weather as a major driver in natural disturbance. Although normal and extreme weather years are generated during simulations, the ignition frequency probabilities have been assumed constant with time, and fuel conditions associated with weather do not specifically recognize the cumulative effects of multi-year drought on tree stress, insect cycles, and the dynamics of forest mortality. There has been no accounting of forest smoke from both presuppression and suppression activities to gauge the effectiveness of management activities.

The development of the SAFE FORESTS model has concentrated upon forest structure, fire hazard, watershed condition, and timber output as measures of ecosystem health and sustainability. Explicit relationships between these measures of ecosystem condition and performance and wildlife remain to be incorporated.

Additional development will be needed if spatial relationships between types of wildlife habitat are desired.

In summary, many useful improvements could be made. This study should be considered an initial effort into largely uncharted terrain.

References

Agee, J. (1993). *Fire ecology of Pacific Northwest forests*. Island Press. 493 p.

Agee, J. (1996). The influence of forest structure on fire behavior. In *Proceedings of the 17th Forest Vegetation Management Conference*, pp. 52–68. Jan. 16–18, 1996. Redding CA.

Alexander, M. (1988). Help with making crown fire hazard assessments. In *Proceedings "Protecting people and homes from wildfire in the Interior West"*. Compilers W. Fisher and S. Arno, pp. 147–56. USDA Forest Service General Technical Report INT-251.

Arno, S., Harrington, M., Fiedler, C. and Carlson, C. (1996). Using silviculture and prescribed fire to reduce fire hazard and improve health in ponderosa forests. In *Proceedings of the 17th Forest Vegetation Management Conference*, pp. 114–17. Jan. 16–18, 1996. Redding, CA.

Bahro, B. (1995). Unpublished tables on file with Eldorado National Forest, Placerville, CA.

Bahro, B. (1996). A summary of Eldorado National Forest and Plumas National Forest firebehavior characteristics for the Sierra Nevada Ecosystem Project fire modeling effort. Unpublished report on file with SNEP.

Berg, N. H., Roby, K. B. and McGurk, B. J. (1996). Cumulative watershed effects: Applicability of available methodologies to the Sierra Nevada. In *Sierra Nevada Ecosystem Project: Final report to Congress*, vol. III, pp. 39-78. University of California, Centers for Water and Wildland Resources, Davis, CA.

Boychuk, D. and Martell, D. (1996). A multi-stage, stochastic programming model for sustainable forest-level timber supply under risk of fire. *Forestry Science*, **42**(1), 10–26.

Chatoian, J. (1995). Personal communication from a meeting of watershed specialists convened by the Sierra Nevada Ecosystem Project and chaired by John Chatoian on May 17, 1995 at the PSW Station, Albany, CA.

Clutter, J. L., Forston, J. C., Pienaar, L. V., Brister, G. H. and Bailey, R. L. (1983). *Timber management: a quantitative approach*. New York: John Wiley. 333 pp.

Cousar, P., Sessions, J. and Johnson, K. N. (1996). Methodology for estimating stand projection under different goals. In *Sierra Nevada Ecosystem Project: Final report to Congress*. Addendum. University of California, Centers for Water and Wildland Resources, Davis, CA.

Davis, L. S. and Johnson, K. N. (1987). Forest management, 3rd edn. McGraw-Hill. 790 pp.

Dykstra, D. (1985). *Mathematical programming for natural resource management*. McGraw-Hill. 320 p.

Franklin, J. F. and Fites-Kaufmann, J. A. (1996). Analysis of late-successional forests. In *Sierra Nevada Ecosystem Project: final report to Congress*, vol. II, chap. 21, pp. 627–62. University of California, Centers for Water and Wildland Resources, Davis, CA.

Franklin, J., Graber, D., Johnson, K. N., Fites-Kaufmann, J., Menning, K., Parsons, D., Sessions, J., Spies, T., Tappeiner, J. and Thornburgh, D. (1997). Alternative approaches to conservation of late successional forests in the Sierra Nevada and their evaluation. In *Sierra Nevada Ecosystem Project: Final report to Congress*. Addendum. University of California, Centers for Water and Wildland Resources, Davis, CA.

Green, K., Finney, M., Cambell, J., Weinstein, D. and Landrum, V. (1993). Using GIS to predict fire behavior. *Journal of Forestry*, **93**(5), 21–5.

Helms, J. A. and Tappeiner, J. C. (1996). Silviculture in the Sierra. In *Sierra Nevada Ecosystem Project: Final report to Congress*, vol. II, chap. 15, pp. 439–76. University of California, Centers for Water and Wildland Resources, Davis, CA.

Hof, J.G., Robinson, K. S. and Betters, D. R. (1988). Optimization with expected values of random yield coefficients in renewable resource linear programs. *Forestry Science*, **34**(3), 634–46.

Hoganson, H. H. and Rose, D. W. (1984). A simulation approach for optimal timber management scheduling. *Forest Science*, **30**, 220–38.

Johnson, K. N. and Scheurman, H. L. (1977). *Techniques for prescribing optimal timber harvest and investment under different objectives – discussion and synthesis*. Forestry Science Monograph, 18.

Johnson, K. N. and Tedder, P. L. (1983). Linear programming vs. binary search in periodic harvest level calculation. *Forestry Science*, **29**, 569–82.

Johnson, K. N. and Stuart, T. (1986). *FORPLAN (Version II) – a mathematical programming guide*. Land Management Planning Systems Section, USDA Forest Service, Wash., DC. 80.

Johnson, K. N., Sessions, J. and Franklin, J. F. (1996). Initial results from simulation of alternative management strategies for two national forests of the Sierra Nevada. In *Sierra Nevada Ecosystem Project: Final report to Congress*. Addendum. University of California, Centers for Water and Wildland Resources, Davis, CA.

Johnson, K. N., Stuart, T. and Crim, S. (1986). *FORPLAN (Version II) – an overview*. Land Management Planning System Section, USDA Forest Service, Washington, DC. 73pp.

Keane, R. E., Morgan, P. and Running, S. (1996). *FIRE-BGC – A mechanistic ecological process model for simulating fire succession on coniferous forest landscapes of the northern Rocky Mountains*. USDA FS INT-RP-484. 122pp.

Kondolf, G. M., Kattelmann, R., Embury, M. and Erman, D. C. (1996). Status of riparian habitat. In *Sierra Nevada Ecosystem Project: Final report to Congress*, vol. II, chap. 36, pp. 1009–32. University of California, Centers for Water and Wildland Resources, Davis, CA.

Menning, K., Erman, D. C., Johnson, K. N. and Sessions, J. (1996). Modeling aquatic and riparian systems, assessing cumulative watershed effects, and limiting watershed disturbance. In *Sierra Nevada Ecosystem Project: Final report to Congress*. Addendum. University of California, Centers for Water and Wildland Resources, Davis, CA.

Pickens, J. B. and Dress, P. E. (1988). Use of stochastic production coefficients in linear programming models: objective function distribution, feasibility, and dual activities. *Forestry Science*, **34**(3), 574–91.

Reed, W. J. and Enrico, D. (1986). Optimal harvest scheduling at the forest level in the presence of the risk of fire. *Canadian Journal of Forestry Reseearch*, **16**, 266–78.

Reed, W. J. and Enrico, D. (1989). A new look at whole forest modeling. *National Research Modeling*, **3**(3), 399–427.

Reeves, C. (1993). *Modern heuristic techniques for combinatorial problems*. New York: John Wiley.

Sapsis, D., Bahro, B., Gabriel, J., Jones, R. and Greenwood, G. (1996). An assessment of current risks, fuels, and potential fire behavior in the Sierra Nevada. In *Sierra Nevada ecosystem project: Final report to Congress*, vol. III, pp. 759–86. University of California, Centers for Water and Wildland Resources, Davis, CA.

Sessions, J., Johnson, K. N., Sapsis, D., Bahro, B. and Gabriel, J. T. (1996). Methodology for simulating forest growth, fire effects, timber harvest, and watershed disturbance under different management regimes. In *Sierra Nevada Ecosystem Project: Final report to Congress*. Addendum. University of California, Centers for Water and Wildland Resources, Davis.

Sierra Nevada Ecosystem Project (SNEP). (1994). *Progress report*. University of California, Centers for Water and Wildland Resources, Davis, CA.

Sierra Nevada Ecosystem Project (SNEP). (1996a). Summary. In *Sierra Nevada Ecosystem Project: Final report to Congress*. University of California, Centers for Water and Wildland Resources, Davis, CA.

Sierra Nevada Ecosystem Project (SNEP). (1996b). Assessment summaries and management strategies. In *Sierra Nevada Ecosystem Project: Final report to Congress*, vol. I. University of California, Centers for Water and Wildland Resources, Davis, CA.

Strauss, D., Bednar, L. and Mees, R. (1989). Do one percent of forest fires cause ninety-nine percent of the damage. *Forestry Science*, **35**(2), 319–28.

Turner, M. and Romme, W. (1994). Landscape dynamics in crown fire ecosystems. *Landscape Ecology*, **9**(1), 59–77.

Turner, M., Gardner, R., Dale, V. and O'Neil, O. (1989). Predicting the spread of disturbance across heterogeneous landscapes. *Oikos*, **55**, 121–9.

USDA Forest Service. (1994). *R-5 timber management plan inventory handbook (FSH 2409.21b)*. R-5 Regional Office, San Francisco.

USDA Forest Service. (1995). *Managing California spotted owl habitat in the Sierra Nevada National Forests of California: an ecosystem approach*. Draft Environmental Impact Statement.

van Wagtendonk, J. W. (1996). Use of a deterministic fire growth model to test fuel treatments. In *Sierra Nevada Ecosystem Project: Final report to Congress*, vol. II, chap. 43, pp. 1155–66. University of California, Centers for Water and Wildland Resources, Davis, CA.

Weatherspoon, C. P. and Skinner, C. N. (1996). Landscape-level strategies for forest fuel management. In *Sierra Nevada Ecosystem Project: Final report to Congress*, vol. II, chap. 56, pp. 1471–92. University of California, Centers for Water and Wildland Resources, Davis, CA.

10

Modeling the driving factors and ecological consequences of deforestation in the Brazilian Amazon

Virginia H. Dale and Scott M. Pearson

Introduction

Land use is having an increasing impact on forest ecosystems worldwide. The destruction and fragmentation of native habitats are recognized as being among the greatest threats to biological diversity and ecosystem processes (McNeeley et al., 1990; Soulé, 1991; National Research Council, 1992). Forested habitats are being destroyed during the extraction of forest products and to open additional lands for agricultural and suburban development. Whereas some minimum area of native habitat in a landscape is necessary for maintaining species richness and population viability, the spatial pattern of habitat is also important. Habitat fragmentation is recognized as a major threat to biodiversity (Whitcomb et al., 1981; Skole and Tucker, 1993; Kaiser, 1997). Habitat fragmentation results when an area with one continuous forest cover is changed to a mosaic of different land-cover types. This change can occur by natural processes (e.g., fires, windthrows), but human activities (e.g., urbanization, agricultural expansion) are primarily responsible for the present rapid changes in the abundance and spatial distribution of forest cover. Human activities create patches that have vegetation cover and shapes unlike natural patches. For example, human-created patches differ from natural patches by having more regular edges (Krummel et al., 1987). This pattern is clearly evident when one compares square agricultural fields with heterogeneous shapes of patches created by small fires.

Anthropogenic changes in forest cover and land use affect processes in forest ecosystems. Disturbances, such as fire and windthrow, maintain spatial heterogeneity (Pickett and White, 1985; Turner, 1987). The frequency, intensity, and spatial extent of disturbances result from interactions between biotic and abiotic elements. For example, fires are related to the amount of fuel wood present, the flammability of the vegetation, and the frequency of lightning strikes. Windthrows are generally related to above- and below-ground structure (e.g., root depth), strength of prevailing winds, and condition of the vegetation (e.g., trees weakened by insect outbreaks are more susceptible to wind). Human activities can affect the occurrence of these natural disturbances. For example, windthrows are more common on the edges of clearcuts (Franklin and Forman, 1987), and the frequency and severity of forest fires are altered by silviculture and fire suppression activities (Christensen et al., 1989). Deforestation changes the storage of carbon and the ability of vegetation to sequester atmospheric carbon (Post et al., 1990).

The social drivers of land-use change are complex, but essential, to understanding and predicting these pervasive landscape-level changes. The major anthropogenic drivers include population growth and economic affluence, technology, national economy, political structure, and attitudes and values (Turner et al., 1993). The relative importance of these factors varies with the situation and the spatial scale of analysis. Human population growth can be considered an ultimate cause for most land-use changes; however, local demographics as well as variability in per capita resource consumption can modify the effects of population. The three major proximate causes of land-use change are economic exploitation of natural resources (e.g., logging, mining, hydroelectric power), population expansion (e.g., urbanization and colonization), and expansion of agriculture (e.g., permanent agriculture, shifting cultivation). All of these proximate causes have been responsible for some of the land-use changes in regions of the world as disparate as the southern Appalachia and western Amazonia.

The Brazilian state of Rondônia, comprising an area of 243 000 km^2 in the western Amazon Basin, has experienced extensive and rapid deforestation. Skole and Tucker (1993) estimated that the total deforested area rose from 6281 km^2 in 1978 to approximately 24 000 km^2 in 1988. This deforestation results from an immigrant human population that clears forested land to establish farms. Government-supported road construction has improved travel in the region and has been responsible for the large influx of people into the region (Fearnside, 1987). The Brazilian government has actively promoted settlement in the area. Government-sponsored colonization programs were started in 1968 to: (a) establish people on the land, (b) increase the standard of living, (c) promote economic growth, and (d) use the land's resources (Leite and Furley, 1985). A consequence of this immigration is deforestation along the roads and also in adjacent areas as the roads provide access to land for farming. The broad-scale loss of tropical forest can have local effects including the disruption of hydrological regimes (Shukla et al., 1990), degradation of soils (Hecht, 1990), and loss of biological diversity. Forest loss could also have global effects by releasing greenhouse gases into the atmosphere thereby affecting climate change (Post et al., 1990; Dale et al., 1991). Therefore, the loss of these forests has both local and global significance.

Modeling techniques for deforestation provide tools for understanding the causes and consequences of rapid land-cover changes. The patterns of change are often complex and reflect an interaction between ecological, economic, historical, and social factors. Models of land-cover change have been useful for disentangling the complex suite of social factors that influence the rate and spatial pattern of deforestation and in estimating the ecological impacts of changes in forest cover (Wilkie and Finn, 1988; Southworth *et al.*, 1991; Baker, 1992; Lee *et al.*, 1992; Dale *et al.*, 1993, 1994a; Gilruth *et al.*, 1995; Turner *et al.*, 1996; Wear *et al.*, 1996; Gaston *et al.*, 1998). Valid models can be used to explore the consequences of altering the driving mechanisms of landscape change and can provide input for other analyses designed to interpret the ecological and social implications of these changes.

Modeling spatial patterns of deforestation

Modeling land-use changes such as deforestation requires combining spatially explicit ecological information with socioeconomic factors. Most previous models have simulated land-cover change as a surrogate for land use. These models are usually landscape-transition models, as classified by Dale and Rauscher (1994). Some models use socioeconomic information to modify ecological processes. For example, Baker (1992) used a disturbance-succession model to simulate changes in the age and spatial patterning of forests. The socioeconomic effects in that model were manifest as differences in fire regimes that resulted from varying levels of fire suppression effort. Other models have sought to integrate ecological and socioeconomic factors more closely. For example, land-cover changes in a southern Appalachian watershed were simulated using a probabilistic model (Turner *et al.*, 1996; Wear *et al.*, 1996). In that model, the probability of change between land-cover types for a given map cell was influenced by the ecological (e.g., elevation, slope) and economic (e.g., distance to road) parameters associated with that cell. Models that show feedbacks between ecological and socioeconomic modules are rare, but several authors have outlined approaches that include feedbacks between ecological function and society (e.g., Lee *et al.*, 1992; Riebsame *et al.*, 1994).

In some cases, it may be difficult to separate the relationship between environment and society into different modules. Models of shifting cultivation in Africa have incorporated these two components in a single model of agricultural land use (Wilkie and Finn, 1988; Gilruth *et al.*, 1995). Wilkie and Finn (1988) modeled a society in Zaire in which decisions to clear a site, maintain cultivation there, or leave it fallow depended on land ownership, travel time from village to site, and site productivity. Productivity declined after several seasons of cultivation but could be restored by leaving the site fallow and allowing it to revert to forest. In addition, the length of fallow period required to restore fertility varied between sites because of ecological factors. Thus, the agricultural system was a complex product of both ecological and social influences. Gilruth *et al.* (1995) modeled a similar system in Guinea, West Africa. Both of these models explored the sustainability of these agricultural systems under the influence of social change in the form of human population growth.

The model used in this study simulates deforestation in which the decisions to clear forest incorporate both the ecological features of sites and the socioeconomic characteristics of the local farm community (i.e., human population, transportation network, management alternatives). Because the modeling design cuts across various disciplines (sociology, economics, ecology, and geography), focus will be on the interactions between these disciplinary approaches rather than on the details of each subcomponent. Therefore, all the submodels are as simple as was deemed appropriate. The definition of appropriateness is determined by the issue at hand. For example, if the concern is with loss of endangered species then the model may not incorporate detailed soil processes unless they affect the focal species. The particular issue also determines the temporal and spatial scale of the model. In this study, the model is being used to determine influences on carbon storage and animal species persistence. Thus, concern is mostly with changes in the pattern of forests on the landscape.

The land-use model

To understand the effects of alternative forms of land management, a simulation model called DELTA (dynamic ecological land tenure analysis) has been developed that integrates a socioeconomic model of colonization and an ecological model of forest clearing and subsequent carbon change. The modeling system estimates patterns and rates of deforestation under different immigration policies, land tenure practices, and road development scenarios (Southworth *et al.* 1991; Dale *et al.*, 1993, 1994*a* describe the model in detail). The code is written in Fortran and runs on a personal computer. A user-friendly interface written in the C compiler language allows changes to initial conditions or parameter values and provides graphical output.

DELTA can be classified as a stochastic, dynamic simulation model. The dynamics of lot use and tenant farmer movement are simulated within DELTA at the scale of farm lots and tenants. By tracking the history of individual lots and tenants, the resulting aggregate patterns of land-use change are more likely to reflect the human settlement process, offering the possibility for prescriptive application of the modeling outputs to real-life management. By introducing stochastic elements into many of the lot selection and land-use decisions, DELTA allows realistic simulation of ecological as well as socioeconomic impacts by averaging over multiple computer runs for a single set of parameter inputs.

DELTA consists of four linked sub-components that together simulate causes and effects of changes in forest cover. The four components are: (a) a settlement diffusion model, (b) a model of ecological impacts, (c) changes in transportation infrastructure (which come from blueprints of the road network), and (d) land-use changes.

The settlement diffusion sub-model allocates and tracks tenant farming families among lots. Selection of a particular lot is based upon lot size, three indices of agricultural suitability based upon soil quality and physical aspect, distance to the nearest market along paved and feeder roads, and length of an occupant's current tenure. The simulation allows tenants to move between lots, multi-tenant lots to occur, and lots to be coalesced into large pastures.

The measure of ecological impact implemented within DELTA is a relationship between carbon in the trees and land use. The estimate of carbon released from

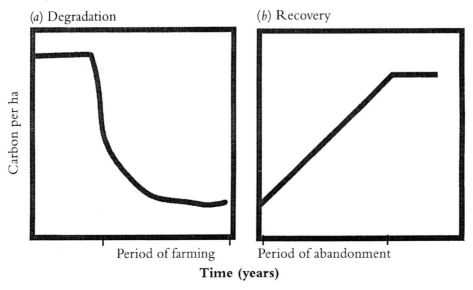

Fig. 10.1. Continuous carbon release curve for (a) *degradation and* (b) *recovery.*

terrestrial sources to the atmosphere is based on a piecewise continuous curve of the carbon per hectare (ha) in soil and vegetation under pristine conditions, farming or pasture use, and abandonment (Fig. 10.1) (based on the approach of Houghton *et al.*, 1983). For the Rondônia example, the pristine forest is estimated to have 293 Mg/ha biomass (Brown *et al.*, in press), much more than the previous estimate of 170 Mg/ha of biomass (Brown and Lugo, 1992). This value is multiplied by the number of hectares in the lot to determine initial biomass values for the lot. Forest clearing tends to occur first closest to the road where farmers have easy access to the land. With farming or pasture use, the initial level of carbon (x) undergoes a negative exponential decline. A carbon loss rate per hectare of $e^{-0.19x}$ is obtained by assuming that long-term farms or pasture would be in place 15 years at which time 10 Mg/ha of carbon remain on the lot (following the relationship for productivity given in Serrão and Toledo (1990) and using the values presented in John (1973), and Lang and Knight (1979)). Once a lot is abandoned by a tenant, if it is not immediately reoccupied, its carbon content is then assumed to increase linearly as vegetation regenerates itself slowly (Fig. 10.1(*b*)). This recovery rate will vary with the type and intensity of land-clearing practice. If a lot is reoccupied, the carbon content will proceed along an exponential decline beginning at the level of carbon that is in place when the farmer settles on the lot.

Changes in transportation infrastructure are currently inputs to the model. Given suitable data, the changes in such infrastructural elements as the road networks and city locations can also be computed based on realized net revenues represented as crop prices minus crop transportation and production costs.

Fig. 10.2. The inputs for, and outputs from, the DELTA model. Spatial information is included in the geographic information system (GIS).

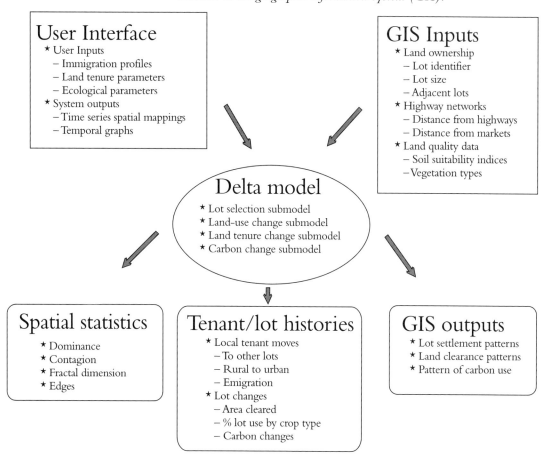

Land-use changes in DELTA relate to the proportion of a lot that is used for farming of annuals or perennials, animal grazing, fallow, or undeveloped forests. They also portray the proportion of the lot that is cleared in any one year. The potential for coalescing or fragmentation of lots into different ownerships is also a component in this model. The values for this aspect of the model are based upon available data for Central Rondônia, but a worst-case and a best-case situation are also simulated as is discussed in the application section below.

An overview of the DELTA model is shown in Fig. 10.2. The user supplies information specific to the situation, such as the profiles of the immigrant farmers moving into the region, the land tenure parameters, and ecological conditions. The user also provides the necessary information for the computer interface between the Fortran model code and the information being input and output so that the time series and temporal graphs are obtained at the correct intervals. The DELTA model uses spatial data, so there is a geographic information systems (GIS) interface provided. Currently, the mapped information is manipulated and stored in ARC\INFO and translated as ASCII code to the DELTA model which is written in Fortran. The DELTA model then simulates which lots are selected by the farmers and effects on those lots in land use, land tenure, and carbon storage. The outputs from the model can be viewed as either spatial statistics, histories of the tenants and lots, or GIS maps. Thus, the user has some choices as to how the model results are viewed and interpreted.

There are some major differences between the DELTA model and other land use models of this type. The DELTA model deals only with a very short time frame of 10 to 50 years. In fact, we find that many of the land changes have stabilized by 20 or 30 years in the model. One reason that these short time periods are important is because the model focuses on social and economic interactions that tend to occur on fairly short time intervals. The model does not consider natural disturbances, which are the concern of many other types of ecological models. As a result of the focus on social/economic interactions, the ecological relationships, such as carbon flux, are extremely simple. This simplicity is appropriate for modeling tropical systems where there is relatively little information on species' growth rates, response to environmental conditions, or successional dynamics.

The strength of the DELTA modeling approach is that it has been developed to address land management issues and thus is a useful planning instrument. The model allows one to consider the ecological causes and implications of human interactions. The model is presented in a format that is easily understood and used by land managers. The maps that are produced and the changes over short periods of time (e.g., decades) are amenable to land-planning decisions. The model uses data that are generally available, including remote sensing data, and the projections have been compared to LANDSAT imagery (Frohn *et al.*, 1997). Thus, although it differs from many ecological models that focus on natural interactions and examine changes over very long terms, it is found that the short-term nature and emphasis on readily available data make the DELTA model a worthwhile planning tool.

Model application

The model was applied in the Ouro Preto colonization area, which is located along the BR-364 highway in central Rondônia, Brazil. Ouro Preto was created in 1970 and was conceived as an ideal colonization project. It contained 4011 km^2 of approximately 100-ha lots on some of the territory's most fertile soils. As word spread concerning the good agricultural conditions, more and more colonists moved to the region. Although initial plans were to have only 500 families in Ouro Preto, by 1974 an estimated 4000 migrant families had obtained lots (Mueller, 1980), and by 1987 the number of settled families was 5098 (Becker, 1987). Soon other colonization projects were established to meet the ever increasing demand for land. Local soil conditions, hydrology, and fertility were not considered in designing the lot layout. The lots were rigidly laid out along roads in a grid 4 km apart. Although much of the land in the original colonization area was fertile, this was not universal; yet it was assumed by planners that the colonists could be equally productive on all soil types. The model allows farmer productivity to differ according to initial soil conditions and past land uses.

The model itself was designed to be generally applicable to situations where lots are laid out along transportation routes; however, the input data made it specifically applicable to central Rondônia. Using ARC/INFO, a spatially explicit database was constructed, consisting of vegetation, transportation networks for different time periods, pasture suitability, and agricultural suitability of Rondônia and lot boundaries of Ouro Preto. These data set the initial conditions for each of the lots tracked in the Fortran model. The vegetation and suitability data were digitized from the RadamBrasil 1:1 000 000 maps obtained from a radar image study of the Amazon (RadamBrasil, 1978a, b). Of course, it would have been more desirable to use *in situ* data for each lot, but these were not available. However, the soil suitability data do reveal the relatively good conditions for growing crops in central Rondônia as compared to the rest of the Brazilian Amazon. The transportation networks were digitized from road maps produced by the Department of Roads in Rondônia (Departmento De Estradas de Rodagem, Rondônia, 1988) and provided statewide coverage for 1979, 1982, and 1988. The lot boundaries in the study sites were digitized from blueprint maps by the National Council for Colonization and Agrarian Reform in Brazil.

Model scenarios

The model was applied to a set of 294 lots within the colonization area of the Ouro Preto Integrated Colonization Project. The model contrasts three scenarios meant to represent (A) typical farmers in Rondônia, (B) farmers who use innovative, sustainable farming techniques and grow perennial tree crops but have no income from milk or cattle, and (C) activities in the worst situations experienced in Brazil along the TransAmazon Highway. The typical scheme of colonists (A) is that they burn the tropical forest, plant annual and perennial crops, followed by pasture, and finally abandon their lots (Leite and Furley, 1985; Millikan, 1988; Coy, 1987). The sustainable scenario (B) simulates an innovative farm management system based on observations of a few farmers in the area (Dale *et al.*, 1993, 1994a). Sustainable cultivation practices occur where farmers plant a diversity of crops and allow some of their farm land to grow into perennial tree crops from which products can be harvested (e.g., rubber, cocoa). The farmers did not have pasture or cattle but averaged more than twice the income in 1990 from perennial crops as the other farmers in the region. A worst-case scenario (C) simulates farmers clearing 20 ha of forested land each year for the first 14 years of ownership and planting only annual crops or pasture. None of the cleared land is planted for agroforestry purposes or put in fallow. This case is meant to be an extreme event but mimics the

Fig. 10.8. *Comparison of model projections with remote sensing data for 294 lots in Rondônia, Brazil (from Frohn et al., 1997). (White is forests and gray is cleared area.)*

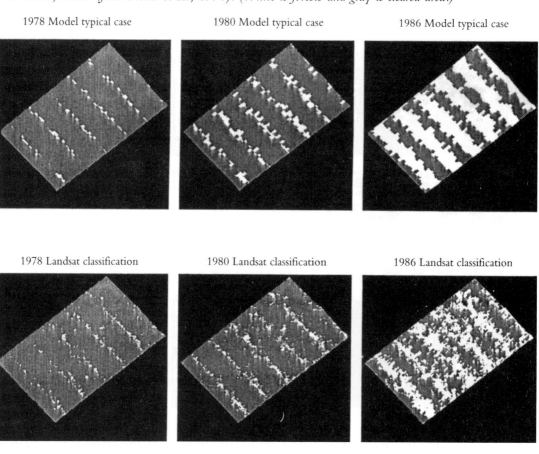

Fig. 10.9. *Forested areas (black) and non-forested areas (white) in three 2025 km² study areas that have unique levels of agricultural development in the state of Rondônia, Brazil.*

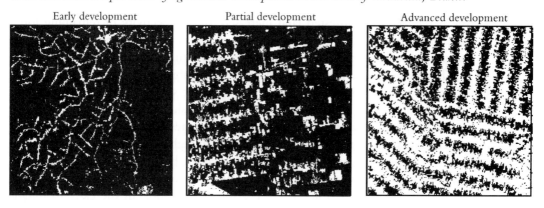

Fig. 10.10. Projected habitat loss for species with (a) *both moderate area requirements and moderate gap-crossing ability and* (b) *large area requirements and low gap-crossing ability (from Dale et al., 1994a).*

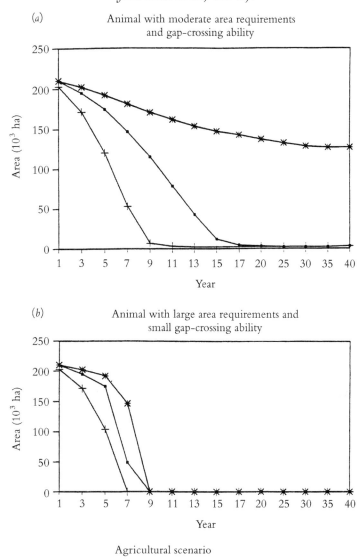

stages of agricultural development in Rondônia, large patches of habitat remain because forest loss is restricted to roadsides (Fig. 10.7) and few species were affected by fragmentation. However, as deforestation proceeded in the typical and worst-case scenarios only species with small area requirements could persist in the remaining small forest patches. Thus, the spatial arrangement of forest loss is important, and the interpretation of these landscape changes depends on the species and ecological process in question.

Conclusions

Models that describe and predict patterns of deforestation could play an important role in making wise decisions about future development. Through modeling, the implications of national or regional policies can be evaluated and discussed by knowledgeable persons worldwide. Modeling land-use changes in the Brazilian Amazon highlights the similarities between a worst-case scenario and the typical management practices. Technological advances in remote sensing and ecological and economic analyses, not available during the early exploitation of North America and Europe, can be applied to tropical development. For example, satellite imagery permits land-cover changes to be viewed in real time. Today the dramatic rate of deforestation in the Amazon Basin is well known (e.g., Skole and Tucker, 1993), whereas the large-scale clearing of forest in the northern hemisphere went largely unnoticed by the developed world in past centuries. Technological advances in computer science allow ecologists and land managers to explore the implications of specific land management strategies over time and space with simulation models (e.g., Dale *et al.*, 1993, 1994*b*). By detailing the implications of particular management schemes, these simulations can provide insights to land managers on the potential outcomes of particular practices.

Acknowledgments

Bob Frohn and Woods Hole Research Laboratory provided a land-cover map of Rondônia. M. G. Turner, R. V. O'Neill, C. T. Hunsaker, J. T. Croom, W. F. Laurance, and T. S. Plaut provided useful suggestions for improving this paper. R. H. Gardner provided software used in the quantification of spatial patterns. This research received support from the US Man and the Biosphere Program and from the Office of Health and Environmental Research, US Department of Energy under contract No. DE-AC05-84OR21400 with Lockheed Martin Energy Research, Inc., and from a grant from the National Science Foundation DEB 9416803. This is publication number 4786 of the Environmental Sciences Division of Oak Ridge National Laboratory.

The submitted manuscript has been authored by a contractor of the US Government under contract No. DE-AC05-96OR22464. Accordingly, the US Government retains a non-exclusive, royalty-free licence to publish or reproduce the published form of this contribution, or allow others to do so, for US Government purposes.

References

Addicott, J. F., Aho, J. M., Antolin, M. R., Padilla, D. K., Richardson, J. S. and Soluk, D. A. (1987). Ecological neighborhoods: Scaling environmental patterns. *Oikos*, **49**, 340–6.

Baker, W. L. (1992). Effects of settlement and fire suppression on landscape structure. *Ecology*, **73**, 1879–87.

Becker, B. K. (1987). Estrategica do Estado e Povoamento Espontaneo na Expansao da Fronteira Agricola em Rondônia: Interacao e Conflito. Tübingen Geographische Studien. In *Homen e Natureza na Amazonia Simposio International e Interdisciplinar Blaubeuren 1986*, ed. G. Kohlepp and A. Schraeder, vol. 95, pp. 237–351. Aebeitsgemeinschaft Deutsche Lateinamerika Forschung (ADLAF).

Brown, S. and Lugo, A. (1992). Aboveground biomass estimates for tropical moist forests of the Brazilian Amazon. *Interciencia*, **17**, 8–18.

Brown, I. F., Martinelli, L. A., Thomas, W. W., Moreira, M. Z., Ferreira, C. A. C. and Victoria, R. A. (1998). Studies of a southwestern Amazonian forest: Uncertainy in estimates of biomass. *Forest Ecology and Management*. In press.

Christensen, N. L., Agee, J. K., Brussard, P. F., Hughes, J., Knight, D. H., Mishall, G. W., Peek, J. M., Pyne, S. J., Swanson, F. J., Thomas, J. W., Whells, S., Williams, S. E. and Wright, H. A. (1989). Interpreting the Yellowstone fires of 1988. *BioScience*, **39**, 678–85.

Coy, M. (1987). Rondônia: Frente Pioneira e Programa POLONOROESTE. O Processo de Diferenciação Sócio-Econômica na Periferia e os Limites do Planejamento Público. Tübingen Geographische Studien. In *Homen e Natureza na Amazonia Simposio International e Interdisciplinar Blaubeuren 1986*, ed. G. Kohlepp and A. Schraeder, vol. 95, pp. 53–270. Aebeitsgemeinschaft Deutsche Lateinamerika Forschung (ADLAF).

Dale, V. H. and Rauscher, H. M. (1994). Assessing the impacts of climate change on forests: the state of biological modeling. *Climatic Change*, **28**, 65–90.

Dale, V. H., Houghton, R. A. and Hall, C. A. S. (1991). Estimating the effects of land use change on global atmospheric CO_2 concentrations. *Canadian Journal of Forest Research*, **21**, 87–90.

Dale, V. H., O'Neill, R. V., Pedlowski, M. and Southworth, F. (1993). Causes and Effects of Land-Use Change in Central Rondônia, Brazil. *Photogrammetric Engineering and Remote Sensing*, **56**, 997–1005.

Dale, V. H., O'Neill, R. V., Southworth, F. and Pedlowski, M. (1994a). Modeling effects of land management in the Brazilian amazonian settlement of Rondônia. *Conservation Biology*, **8**, 196–206.

Dale, V. H., Pearson, S. M., Offerman, H. L. and O'Neill, R. V. (1994b). Relating patterns of land-use change to faunal biodiversity in the central Amazon. *Conservation Biology*, **8**, 1027–36.

Departmento De Estradas De Rodagem, Rondônia. (1988). Estado De Rondônia. Scale 1:1 000 000. Porto Vehlo, Brazil. Map.

Fearnside, P. M. (1987). Causes of deforestation in the Brazilian Amazon. In *Geophysiology of Amazonia: vegetation and climate interactions*, ed. R. F. Dickinson, pp. 214–21. New York: John Wiley.

Fearnside, P. M., Leal, N. and Fernandes, F. M. (1993). Rainforest burning and the global carbon budget: biomass, combustion efficiency, and charcoal formation in the Brazilian Amazon. *Journal of Geophysical Research*, **98**, 16733–43.

Franklin, J. R. and Forman, R. T. T. (1987). Creating landscape patterns by forest cutting: Ecological consequences and principles. *Landscape Ecology*, **1**, 5–18.

Frohn, R. C., McGwire, K. C., Dale, V. H. and Estes, J. E. (1997). Using satellite remote sensing analysis to evaluate a socio-economic and ecological model of deforestation in Rondônia, Brazil. *International Journal of Remote Sensing*, **17**, 3233–55.

Gardner, R. H., Milne, B. T., Turner, M. G. and O'Neill, R. V. (1987). Neutral models for the analysis of broad-scale landscape pattern. *Landscape Ecology*, **1**, 19–28.

Gaston, G., Brown, S., Lorenzini, M. and Singh, K. D. (1998). State and change in carbon pools in the forests of tropical Africa. *Global Change Biology*, **4**, 97–114.

Gilruth, P. T., Marsh, S. E. and Itami, R. (1995). A dynamic spatial model of shifting cultivation in the highlands of Guinea, West Africa. *Ecological Modeling*, **79**, 179–97.

Hansen, A. J. and Urban, D. L. (1992). Avian response to landscape pattern: the role of species' life histories. *Landscape Ecology*, **7**, 163–80.

Hecht, S. B. (1990). Indigenous soil management in the Latin American tropics: Neglected knowledge of native peoples. In *Agroecology and Small Farm Development*, ed. M. A. Altieri and S. B. Hecht, pp. 151–8. Boca Raton, FL: CRC Press.

Houghton, R. A., Hobbie, J. E., Melillo, J. M., Moore, B., Peterson, B. J., Shavers, G. R. and Woodwell, G. M. (1983). Changes in the carbon content of terrestrial biota and soils between 1860 and 1980: Net release of CO_2 to the atmosphere. *Ecological Monographs*, **53**, 235–62.

John, D. M. (1973). Accumulation and decay of litter and net production of forests in tropical West Africa. *Oikos*, **24**, 430–5.

Kaiser, J. (1997). When a habitat is not a home. *Science*, **276**, 1636–8.

Krummel, J. R., Gardner, R. H., Sugihara, G., O'Neill, R. V. and Coleman, P. R. (1987). Landscape patterns in a disturbed environment. *Oikos*, **48**, 321–4.

Lang, G. E. and Knight, D. H. (1979). Decay rates for tropical trees in Panama. *Biotropica*, **11**, 316–17.

Laurance, W. F., Laurance, S. G., Ferreira, L. V., Rankin-de Merona, J. M., Gascon, C. and Lovejoy, T. E. (1997). Biomass collapse in Amazonian forest fragments. *Science*, **278**, 1117–18.

Lee, R. G., Flamm, R. O., Turner, M. G., Bledsoe, C., Changler, P., DeFerrari, C., Gottfried, R., Naiman, R. J., Schumaker, N. and Wear, D. (1992). Integrating sustainable development and environmental vitality. In *New perspectives in watershed management*, ed. R. J. Naiman, pp. 499–521. New York: Springer-Verlag.

Leite, L. L. and Furley, P. A. (1985). Land development in the Brazilian Amazon with particular reference to Rondônia and the Ouro Preto colonization project. In *Change in the Amazon basin Volume II. The frontier after a decade of colonization*, ed. R. Hemming, pp. 119–40, Manchester, UK: Manchester University Press.

Li, H., Franklin, J. F., Swanson, F. J. and Spies, T. A. (1993). Developing alternative forest cutting patterns: A simulation approach. *Landscape Ecology*, **8**, 63–75.

McNeely, J. A., Miller, K. R., Reid, W. V., Mittermeier, R. A. and Werner, T. B. (1990). *Conserving the world's biological biodiversity*. The International Union for Conservation of Nature and Natural Resources, World Resources Institute, Conservation International, World Wildlfie Fund–US and the World Bank: Gland, Switzerland and Washington, DC.

Millikan, B. H. (1988). The dialectics of devastation: tropical deforestation, land degradation, and society in Rondônia, Brazil. MA Thesis, University of California, Berkeley.

Moran, E. F. (1981). *Developing the Amazon*. Bloomington, Indiana: Indiana University Press.

Mueller, C. C. (1980). Recent frontier expansion in Brazil: the case of Rondônia. In *Land, people and planning and contemporary Amazonia*, ed. F. Barbira-Scazzochio, Cambridge: Cambridge University Press.

National Research Council. (1992). *Conserving biodiversity: a research agenda for development agencies*. Washington, DC: National Academy Press.

Offerman, H. L., Dale, V. H., Pearson, S. M., Bierregaard, R. O. and O'Neill, R. V. (1995). Effect of forest fragmentation on neotropical fauna: current research and data availability. *Environmental Reviews*, **3**, 191–211.

O'Neill, R. V., Milne, B. T., Turner, M. G. and Gardner, R. H. (1988). Resource utilization scales and landscape pattern. *Landscape Ecology*, **2**, 63–9.

Pearson, S. M. (1993). The spatial extent and relative influence of landscape-level factors on wintering bird populations. *Landscape Ecology*, **8**, 3–18.

Pearson, S. M. (1994). Ecological perspective: understanding the impacts of forest fragmentation. In *Remote-sensing and GIS in ecosystem management*, ed. V. A. Sample, pp. 178–91. Covelo, California: Island Press.

Pearson, S. M., Turner, M. G., Gardner, R. H. and O'Neill, R. V. (1996). An organism-based perspective of habitat fragmentation. In *Biodiversity in managed landscapes: theory and practice*, ed. R. C. Szaro, pp. 77–95. Oxford: Oxford University Press.

Pickett, S. T. A. and White, P. S. (1985). *The ecology of natural disturbance and patch dynamics*. New York: Academic Press, Inc.

Post, W. M., Peng, T. H., Emanuel, W., King, A. W., Dale, V. H. and DeAngelis, D. L. (1990). The global carbon cycle. *American Scientist*, **78**, 310–26.

RadamBrasil. (1978). Brasil, Departamento Nacional da Producão, Mineral, Projeto RADAMBRASIL, Folha SC.20 Porto Velho; geologia, geomorfologia, pedologia, vegetacão e uso potencial da terra. Rio de Janeiro.

RadamBrasil. (1979). Brasil, Departamento Nacional da Producão Mineral, Projeto RADAMBRASIL, Folha SC.20 Guapore; geologia, geomorfologia, pedologia, vegetacão e uso potencial da terra. Rio de Janeiro.

Riebsame, W. E., Parton, W. J., Galvin, K. A., Burke, I. C., Bohren, L., Young, R. and Knop, E. (1994). Integrated modeling of land use and cover change. *BioScience*, **44**, 350–6.

Serrão, E. A. and Toledo, J. M. (1990). The search for sustainability in Amazonian pastures. *Alternatives to deforestation: steps toward sustainable use of the Amazon rain forest*, ed. A. B. Anderson, pp. 195–214, New York: Columbia University Press.

Shukla, J., Nobre, C. and Sellers, P. (1990). Amazon deforestation and climate change. *Science*, **247**, 1322–5.

Skole, D. and Tucker, C. (1993). Tropical deforestation and habitat fragmentation in the Amazon: Satellite data from 1978 to 1988. *Science*, **260**, 1905–10.

Soulé, M. E. (1991). Conservation: tactics for a constant crisis. *Science*, **253**, 744–50.

Southworth, F., Dale, V. H. and O'Neill, R. V. (1991). Contrasting patterns of land use in Rondônia, Brazil: simulating the effects on carbon release. *International Social Sciences Journal*, **130**, 681–98.

Turner, B. L. II, Moss, R. H. and Skole, D. L. (1993). Relating land use and global land cover change: A proposal for an IGBP-HDP core project. International Geosphere–Biosphere Programme, Stockholm.

Turner, M. G., ed. (1987). *Landscape heterogeneity and disturbance.* New York: Springer-Verlag.

Turner, M. G., Wear, D. N. and Flamm, R. O. (1996). Land ownership and land-cover change in the Southern Appalachian Highlands and the Olympic Peninsula. *Ecological Applications,* **6**, 1150–72.

Wear, D. N., Turner, M. G. and Flamm, R. O. (1996). Ecosystem management with multiple owners: Landscape dynamics in a southern Appalachian watershed. *Ecological Applications,* **6**, 1173–88.

Wiens, J. A. (1989). Spatial scaling in ecology. *Functional Ecology,* **3**, 385–97.

Whitcomb, R. F., Robbins, C. S., Lynch, J. F., Whitcomb, B. L., Klimkiewicz, M. K. and Bystrak, D. (1981). Effects of forest fragmentation on the avifauna of the eastern decious forest. In *Forest island dynamics in man-dominated landscapes,* ed. R. L. Burgess and D. M. Sharpe, pp. 125–205. New York: Springer-Verlag.

Wilkie, D. S., and Finn, J. T. (1988). A spatial model of land use and forest regeneration in the Ituri Forest of northeastern Zaire. *Ecological Modelling,* **41**, 307–23.

11

Spatial simulation of the effects of human and natural disturbance regimes on landscape structure

William L. Baker

Introduction

A primary source of temporal variation in landscape patterning is the shifting mosaic of patches produced by natural and human disturbances. Viewing the landscape from a high hill or from an airplane, one can see landscape patchiness from several sources, including the underlying topography and its consequent array of micro-environments, the vegetation itself, and superimposed upon these, a mosaic of disturbance patches (Forman, 1995). At an instant-in-time, of course, the shifting mosaic of disturbance patches is seemingly fixed, and this "present snapshot" view has had an enduring influence on our approach to land management, as well as our understanding of ecosystem functioning.

The present snapshot of landscape pattern, of course, changes, and through rephotography and historical analysis (e.g., Hastings and Turner, 1965), it is now understood that cumulative changes in landscapes brought about by a series of disturbance events over long time periods can be substantial (e.g., Heinselman, 1996). To obtain this view that transcends the limits of our human-scaled perception of the rates and patterns of landscape change, the scale that is relevant to the ecosystems themselves must be approached. One approach, emphasized throughout this book, is to develop and use quantitative models of landscape change, and the particular subject of this chapter is the long-term changes on the landscape scale (the scale of kilometers) related to human and natural disturbances.

Concept of a disturbance regime at the landscape scale

A disturbance regime can be defined as the set of disturbances produced over some time period in a land area of interest. To characterize the regime the attributes of this set of disturbances need to be obtained from a sample of the regime at the appropriate temporal and spatial scale (Baker, 1992a). The most relevant temporal scale of the disturbance regime is the turnover time, or the time required to disturb an area equal to the landscape of interest. Turnover time varies with the type of ecosystem, but is probably between a half century and two millennia for most of the earth's ecosystems. In temperate zone forests subject to fires, turnover times are typically about 60 to nearly 500 years (Baker, 1995). In temperate zone landscapes where wind is the primary disturbance agent, turnover times may be more than 1200 years (Canham and Loucks, 1984). In tropical forests where treefalls and gap dynamics predominate, turnover times may be about a century (Lieberman *et al.*, 1985). The mosaic is unlikely to be temporally stable at any particular spatial extent (Baker, 1989). However, an adequate sample of the disturbance regime is best obtained from a land area that is several times the maximum disturbance size, as this ensures that the spectrum of variation in disturbance attributes will be best represented (Baker, 1989).

The major attributes of the disturbance regime that have been simulated are disturbance size and interval, and intensity. By observing or reconstructing these attributes of individual disturbances over time and space, distributions of attributes that summarize variation in the disturbance regime can be derived for use in modeling the regime (Baker *et al.*, 1991; Baker, 1992*a*). Other important attributes include shape, placement in the landscape, amount of edge, and orientation (Baker, 1992*a*), yet these are little studied. Increasing attention to these spatial ecological attributes of human and natural disturbances is needed.

Natural and human disturbances as spatial processes in landscapes

Natural disturbances have a spatial and topographic logic that has seldom been fully captured in models. Hurricane effects on forests, for example, are in part a function of topographic exposure, as well as the larger-scale arrangement of coastlines and mountain ranges (Boose *et al.*, 1994). Fires are preferentially initiated in certain topographic locations, spread more rapidly upslope than on flat surfaces, and are influenced by terrain-modified airflow patterns (Barrows, 1951; Baughman, 1981). The probability of natural disturbance is also influenced by the characteristics of the vegetation on top of the topographic surface. Lightning ignition rates are higher, for example, in ponderosa pine forests, than in adjacent and otherwise similar mixed fir forests (Meisner *et al.*, 1994).

Human disturbances also are spatially arrayed in a logical way. Human disturbances that affect forests at the landscape scale include certain types of timber harvesting, conversion of forests to agriculture, urbanization and rural subdivision construction, road construction, and livestock grazing (Forman, 1995). The topographic and physical constraints on road construction, as well as the obvious need for transportation efficiency, underlie spatial patterns of forest fragmentation in the Western United States (Reed *et al.*, 1996*a, b*). Deforestation in the Amazon basin is similarly arrayed along major arterial roads (Stone *et al.*, 1991). Agricultural settlement and deforestation of the Midwestern United States were geometrically constrained by the rectangular land survey, and also by topographic limits to effective cultivation (Sharpe *et al.*, 1987). Urbanization and rural subdivision development processes occur as spatial spread, but with spotting (Medley *et al.*, 1995). While the general patterns of these processes have been identified, spatial analysis of the patterns of land use change can lead to a much better foundation for spatial modeling (Mertens and Lambin, 1997).

Since both natural and human disturbance processes are spatial, a variety of methods for quantifying changing landscapes in a spatial way has arisen, particularly with the advent of geographical information systems (GIS). Major trends in landscape structure to be expected along a gradient of human modification have been hypothesized (Godron and Forman, 1983). Many of these expected trends, such as a decline in patch size, increasing homogeneity in patch size, and decline in connectivity have been found in case studies (e.g., Medley *et al.*, 1995; Forman, 1995; Reed *et al.*, 1996*a,b*). Software is now readily available for quantifying changes in landscape structure (e.g., Baker and Cai, 1992; McGarigal and Marks, 1995).

Long-term models of natural and human disturbances

Models of natural disturbances

Several models have been developed that consider generic disturbances in abstract landscapes, but the best-developed models of real disturbances are fire models. Generic disturbance models have usually been used for theoretical analysis (Turner *et al.*, 1989*a*, 1993; Baker, 1995; Boychuk and Perera, 1995, 1997; Boychuk *et al.*, 1997). Spatial fire models are more developed than models of other types of disturbances, but many spatial fire models are designed to only model the spread of one fire, rather than an ongoing regime of fires. The most complex spatial models designed for individual fires include topographic effects, fuels data, wind-driven spread, atmospheric conditions, and other factors that may modify fire behavior (Finney, this volume). However, a variety of other fire spread models has been developed over the last two decades (e.g., Kessell, 1979; Burgan and Rothermel, 1984; Ohtsuki and Keys, 1986; Vasconcelos and Guertin, 1992; Clarke *et al.*, 1994; Clark *et al.*, 1996; Gardner *et al.*, 1996).

Models that focus on fire regimes, or more general models that include more than one fire and longer time frames, are fewer. The author's model, DISPATCH, has been used in long-term fire regime simulations, although it contains a minimum of fire-specific realism (Baker, 1992*b*, 1993, 1994). A related model has also been developed by Ratz (1995). More realistic long-term simulation of fires in landscapes has been done by Antonovski *et al.* (1992), Gardner *et al.* (1996), Keane *et al.* (1996), and Li *et al.* (1996, 1997). Keane *et al.*'s FIRE-BGC contains a fire initiation algorithm and then uses *FARSITE* (Finney, this volume) to spread the fire, providing potentially more realistic fire behavior than in other long-term models. Spatial fire initiation and spread algorithms are also included in more general landscape models (e.g., He and Mladenoff, 1999; Mladenoff and He, this volume; Roberts, 1996*a*, 1996*b*, this volume).

The only other major spatial disturbances that have been a focus of modeling are insect outbreaks. Spatial models of gypsy moth outbreaks in the northeastern United States (Zhou and Liebhold, 1995) and spruce budworm in eastern Canada (Clark *et al.*, 1977) have been developed.

Human disturbances (timber harvesting, deforestation)

Although forest managers have been thinking in spatial terms for some time, the need for spatial thinking on landscape scales emerged as a central issue only recently. In the United States in the late-1980s to early 1990s, "new forestry" (Franklin, 1989) gave way to the US Forest Service's "New Perspectives" program (Salwasser, 1991), and then to "ecosystem management" (Overbay, 1992) as directions in managing public forest lands. All these approaches placed greater emphasis on planning and management on the landscape scale.

Formal tools and spatial models for timber harvesting and management on the landscape scale have also been developed only recently. Earlier forest planning tools, such as FORPLAN, that used optimization techniques to allocate harvesting, have been expanded to include explicit spatial arrangements as goals (Hof and Joyce, 1992). These tools can be used to identify feasible, if not optimal, solutions to planning timber harvest with spatial constraints (Bettinger *et al.*, 1997). Other tools, such as the SNAP program (Sessions *et al.*, 1997) can also be used to explore feasible solutions to spatial constraint problems. However, these approaches are difficult when multiple constraints must be considered or where long-term impacts must be explored, as may be the case where a variety of species have different responses at the landscape scale, and these responses play out over decades as the landscape changes.

An alternative approach is to simulate a particular pattern of forest harvesting, and then do post-hoc analyses of the implications for various resources or measures of ecological value over the length of the model run. Franklin and Forman's (1987) seminal work on the landscape response to alternative harvesting approaches in an abstract landscape led to more sophisticated computer models of timber harvesting (Li *et al.*, 1993; Wallin *et al.* 1994; Gustafson and Crow, 1994) and applications to real landscapes (Hansen *et al.*, 1993; Wallin *et al.*, 1996; Gustafson, 1996; Gustafson and Crow, 1996; Crow and Gustafson, 1997; Gustafson and Crow, this volume; Baskent, 1997).

These timber harvesting models do not include a socioeconomic component, except that different spatial harvesting strategies are included. Simulation of deforestation associated with shifting cultivation in Africa includes effects of land-tenure, population growth, travel distance, labor costs, and land suitability (Wilkie and Finn, 1988; Gilruth *et al.*, 1995). Similarly, simulation of deforestation due to clearing for agriculture in Brazil includes the effects of number of settlers and factors affecting their decisions about lot occupancy, additional clearing, and movement to new lots (Dale *et al.*, 1993*a*, *b*, 1994; Dale and Pearson, this volume). In part, these kinds of socioeconomic factors are excluded from timber harvesting models because the spatial pattern of harvesting is controlled by government agencies, but the decisions of these agencies may also be affected by the economic efficiency of different harvesting strategies and the spatial behavior of competing land uses. So, there is a potential for useful inclusion of socioeconomic components in spatial models of timber harvesting in the future.

Constraints to long-term simulation of natural and human disturbance regimes

Computer limits are disappearing

Computer technology is changing very rapidly, and, while many models a decade ago were limited to small rasters (e.g., 100 pixels × 100 pixels), now it is common to see models that work with larger arrays (e.g., 2000 pixels × 2000 pixels). This increase in area that can be simulated is very significant, because with the scale of resolution of Landsat TM imagery (30 m pixels), it is now feasible to run complex spatial models on land areas that are the size of typical US National Forests or the Western US National Parks. However, very complex spatial models may still require special approaches, such as parallel programming, to run effectively on large arrays at the scale of National Parks (Wu *et al.*, 1996). Since many US government agencies now have systematic programs for acquisition of digital data at these or finer pixel resolutions, there is now the possibility of simulations of major natural and human disturbance processes at scales relevant to government planning units (e.g., Gustafson and Crow, this volume and this chapter, p. 288). However, for this to be feasible, additional progress in spatial modeling of disturbance processes is needed.

Disturbance complexity, spatial knowledge, and algorithms remain limiting

Significant present limitations of spatial disturbance models are: (i) problems in simulating the processes that influence disturbance spread, (ii) an insufficient development of models for individual disturbance sub-processes, (iii) lack of spatial data and spatial understanding, and (iv) problems in the development of spatial algorithms. For long-term models of disturbance regimes, these problems are compounded, as simpler algorithms and approaches are needed to make long-term simulations feasible.

While spatial models of individual fires are the most developed spatial disturbance models, even these models cannot reproduce actual fire spread very well. This may be the case for most disturbances (e.g., fires, wind storms) that spread from a point over time under the influence of spatially and temporally fluctuating winds. Although there is progress in modeling winds in real terrain (Walker and Leone, 1996), linkage of spatial wind models and spatial fire models has not yet occurred, and it is unlikely that adequate spatial atmospheric data will be generally available in the near future. Thus, even at the scale of models of individual fires, only an approximation of actual spread is feasible at the present time (Finney, this volume).

In the case of long-term models of disturbance regimes, these models will be constrained for some time to predictions of general, aggregate attributes of landscape change (e.g., mean patch size, total number of patches), rather than the actual locations and final boundaries of disturbances.

Some of the essential component processes of disturbances in landscapes have yet to be modeled satisfactorily. For example, the factors that lead to successful fire ignition from an individual lightning strike are only known approximately (Fuquay et al., 1967), and there have been comparatively few efforts to model the relationship between lightning and the terrain-surface or vegetation (e.g., Meisner et al., 1994). Since global climate change may lead to changes in lightning frequency (Price and Rind, 1994), the linkage between lightning and the terrain surface needs to be modeled adequately before the effects of climate change on fires can be modeled spatially. In the case of human-initiated disturbances, such as the placement of timber harvesting units or the locations people choose to place swidden-agricultural units, there is also a need for better models, although there has been some progress in modeling these phenomena (e.g., Dale et al., 1993a). In general, there are less adequate models of the processes that initiate disturbances and that lead to them stopping than of disturbance spread, particularly for fires, but the modeling gaps vary with the type of disturbance.

Some of the gaps in our ability to model disturbances spatially come from knowledge gaps and data gaps, rather than lack of model development. Relatively few natural disturbance processes have been the subject of systematic spatial analyses (but see Foster and Boose, 1992 for hurricanes; Fowler and Asleson, 1984 for fires). Similarly, spatial analysis of human disturbance processes are few, but have considerable potential (e.g., Turner et al., 1996; Mertens and Lambin, 1997). Unfortunately, disturbance data are not always collected in a way that will facilitate these kinds of analyses. For example, our local National Forest maintains locations of fires that burned over the last few decades in a fire atlas, but there are no maps of the actual boundaries of these fires, so it is not possible to analyze where the fires stopped, and why. When our local silviculturalist was queried about how clearcuts are placed in the landscape, he indicated that they were placed following established standards and guidelines, but also using common sense. However, a spatial analysis would be required to determine these common sense rules sufficiently to place them in computer code, and a failure to do this correctly may lead to validation problems (see *Initial validation* below).

Spatial disturbance algorithms are difficult to construct, and may require a diversity of approaches, including multi-scale generalizations that go beyond typical cellular automata algorithms. The grid-cell format poses limits to modeling disturbance spread as a cell-to-cell process, as the typical row and column format of square pixels with eight neighbors does not reflect the reality of directional disturbance spread very well. Improved algorithms for fire spread that use additional neighbors and shape-constrained spread may provide one solution (Xu and Lathrop, 1994). Use of vector-format, rather than raster-format, may facilitate modeling some disturbance phenomena, such as spreading fire fronts (Finney, this volume). However, in some instances a disturbance spreads in relation to external cues or information not present in the grid-cell itself or its immediate neighbors. An example arose in my work with modeling clear-cut logging (*Initial validation* below), where it became apparent that clearcuts are placed in part into the existing patch structure. To effectively model phenomena in which information from several spatial scales affects the behavior at an individual grid-cell, or part of a vector, may require spread models that go beyond current cellular automata algorithms in which local dynamics are the primary influence. A similar need for multi-scale effects led to a broadening of cellular automata theory for modeling urban dynamics (Xie, 1996). However, whenever additional complexity must be added to spread algorithms, this may constrain their utility for long-term models where disturbances must be spread hundreds or thousands of times during a simulation run.

Validation of spatial disturbance models

Verification and validation of spatial disturbance models may or may not be needed, depending upon model purposes and goals (Rykiel, 1996). Theoretical models of natural disturbance regimes (e.g., Baker, 1995) need to be verified, in the sense of checking the mathematics and logic of the model, which should be both ecologically sound and have sound model logic (Loehle, 1983; Rykiel, 1996). Models designed to predict require validation, by comparison with real data, for at least the aggregate variables or indices that are to be predicted and possibly also for the actual spatial fit (Turner *et al.*, 1989*b*; Costanza, 1989).

The brief review of spatial disturbance modeling above suggests that spatial models of disturbances are seldom presently able to predict actual disturbance location and boundaries very accurately. Instead, most models, particularly long-term models, will only be able to be predictive of aggregate properties of landscape pattern or other variables for the near future. Validation using aggregate variables is thus most appropriate for such models, yet even this has seldom been attempted. Useful reviews of approaches to model validation are in Sargent (1988) and Rykiel (1996), and for spatial models, in Turner *et al.* (1989*b*).

The DISPATCH model

The DISPATCH model (DISturbance PATCH) is a spatial landscape-scale model designed to simulate the effects of changing disturbance regimes on landscape structure. It began primarily as a theoretical model to be used to explore the effects of climatic change on the structure of generic landscapes subject to natural disturbances (Baker *et al.*, 1991). As climate changes, the attributes of a disturbance regime, as described on p. 277, can be expected to change. DISPATCH can be used to analyse how long it takes for landscape spatial structure to adjust to a new disturbance regime and to explore in a quantitative way how changes occur.

Model development to the present

In the initial version of DISPATCH, the model contained five major components (Fig. 11.1). First, a climatic regime consists of a series of synoptic climatic types and the probability of their occurrence. Second, there is a disturbance regime corresponding to each synoptic climatic regime. The disturbance regime consists simply of a negative exponential disturbance size distribution. Third, the landscape is represented by a set of map layers that together determine the probability of disturbance initiation and spread. Typical map layers include vegetation type, time since last disturbance (age), and a digital elevation model, but additional maps can be included. The maps are maintained as raster layers in the GRASS GIS (USA–CERL, 1994). Fourth, these maps are combined, using a user-specified equation, to produce a single map that contains relative probabilities of disturbance initiation and spread. Fifth, the model contains a set of GIS programs to quantitatively analyze the spatial structure of the landscape as the model runs. These programs, originally called the GLE programs, now are called the r.le programs (Baker and Cai, 1992). The r.le programs, which are designed to analyze GRASS raster maps, enable the calculation of about 60 measures of landscape structure.

The DISPATCH model consists of: (i) main code written in a simulation language, SIMSCRIPT II.5 (CACI, 1987), (ii) data, maps, and programs contained in the GRASS geographical information system (USA-CERL, 1994), (iii) the r.le programs which work within GRASS, and (iv) external statistical and graphical software for analyzing the results of the model runs. The initial model ran with a weekly time step, and required about 6 hours on an early Sun Workstation to simulate a 250-year time period for a 200 × 200 pixel landscape.

Additions to the model since Baker et al. (1991) include: (i) probabilistic initiation and spread algorithms, (ii) additional options for disturbance regime parameters and distributions, including empirical size and timing distributions, and (iii) modifications that allow simulation of generic disturbance regimes in the absence of climatic change and for simulating clear-cut logging, described on pp. 288–303. Much of the subsequent work using DISPATCH uses a modified version in which the climatic driver is absent, and the disturbance regime consists of size and interval distributions specified by the user (Fig. 11.2).

Findings from using DISPATCH

The modified DISPATCH model has been used to analyze the effects of indiscriminant burning during the settlement period in the nineteenth century followed by fire suppression in the twentieth century (Baker, 1992b, 1993), as well as the potential restoration of the fire regime following the period of fire suppression (Baker, 1994), all in the Boundary Waters Canoe Area (BWCA) in northern Minnesota. This work used the fire history findings of Heinselman (1973). The settlement era, with its smaller and more frequent fires relative to the pre-EuroAmerican period, led to a lower mean landscape age, a more diverse landscape (both in

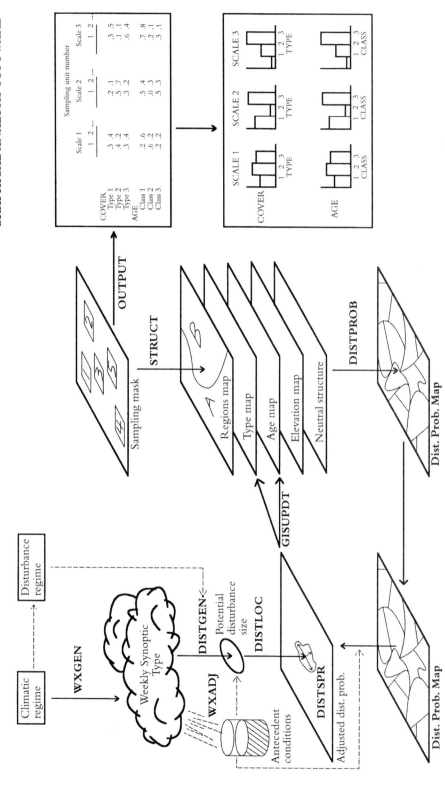

Fig. 11.1. Parts of the DISPATCH simulation model. The main components are simulation code, written using the SIMSCRIPT II.5 simulation language, the GRASS geographical information system, and external statistical analysis and graphical display software. The individual routines written in SIMSCRIPT are printed in bold capital letters. Details of the model are explained in Baker et al. (1991). Reprinted with permission from Baker et al. (1991).

Fig. 11.2. The three main components of the modified DISPATCH disturbance model are the disturbance regime, consisting of size and interval distributions, the disturbance probability map generated typically from an age map, and the r.le programs. Reprinted with permission from Baker (1994).

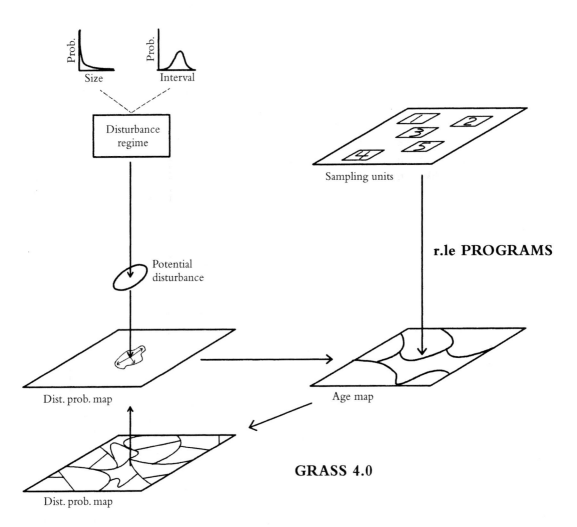

richness and Shannon diversity), and a few other changes in landscape structure (Baker, 1992b). The fire suppression period, with its much smaller and less frequent fires relative to the settlement period, led to a gradual rise in mean landscape age, a continued increase in both richness and Shannon diversity of the landscape, and a decline in mean patch size (Baker, 1992b). Restoration of the landscape structure present in the BWCA during the pre-EuroAmerican era can be achieved by simply reinstating the natural fire regime, but will require the length of the suppression period or longer to readjust (Baker, 1994).

As the scale of observation becomes finer, there is a lag between individual parts of the landscape in the time at which a settlement or suppression effect, or any other effect of management, actually occurs, because the response at finer-and-finer scales increasingly tracks the occurrence of individual fires (Baker, 1993). Lags in response between quarters of the BWCA may be as much as 450 years for measures, such as Shannon diversity, that are slow to adjust. An important finding from this work is that when a treatment is applied to a disturbance regime in a large land area, the response of the landscape is spatially heterogeneous, because disturbances are distributed as events in space and time, rather than appearing simultaneously over an entire landscape. The spatially heterogeneous response means that the snapshot landscape often contains a mosaic of patches that have been affected by the management action (e.g., suppression) for differing lengths of time, combined with patches that are not affected at all by the action. Landscapes are thus seldom likely to be fully adjusted to their disturbance regimes unless the disturbance regime has remained unchanged by external forces for a long period of time.

The rate and manner in which landscapes adjust to changes in their disturbance regimes was explored in a simulation of a generic, temperate-zone forested landscape subject to generic, patch-forming, catastrophic disturbances (Baker, 1995). Previous work in the BWCA showed that number-based measures of landscape structure, such as mean age or patch density, adjust more slowly, when the disturbance regime is changed, than do measures related to patch attributes (e.g., size or shape), so landscape structure does not respond simultaneously and uniformly (Baker, 1992b, 1993, 1994). The time it takes for the landscape to adjust fully to a changed disturbance regime is generally half to two rotations of the new disturbance regime (Baker, 1995). So, if fire suppression is started tomorrow in a landscape previously subject to natural fires, it will likely be a century to several centuries (based on typical fire rotations mentioned earlier in the chapter) before the landscape structure fully reflects the new fire regime. Alterations that shorten rotations, such as global warming, produce a more rapid response than do alterations that lengthen rotations, such as cooling or fire suppression. The rate of response may be more rapid in landscapes with a comparatively low density of patches at the time the alteration occurs. Suppressing disturbances, global cooling, small, prescribed fires, and fragmentation due to clear-cut logging all lead to landscapes with comparatively more patches that are smaller and closer together, as well as a more diverse landscape (Baker, 1995).

Future prospects using the DISPATCH approach

The applications of DISPATCH up to the present time have been primarily theoretical and in generic or abstract landscapes, but there is an increasing need for understanding the response of real landscapes to land management activities and global change. Missing from previous work has been the influence of topography on the disturbance regime. Many land management activities are constrained or modified by topographic setting. However, in order for DISPATCH to include topographic effects, appropriate data on the spatial ecology of disturbances are needed, and these data are very difficult to obtain. Also missing from DISPATCH applications to date has been the use of real digital databases, which are now increasingly available in GIS form for at least the Federal lands in the United States. The following application represents a step toward using real digital databases, real land management prescriptions, and topographic constraints in modeling the response of the landscape to human disturbances.

Simulating long-term effects of timber harvesting in a subalpine forest

Every 10 years the US Forest Service revises its Forest Plans, which guide management of individual National Forests, and re-evaluates the impacts of land uses. On the Medicine Bow National Forest (MBNF) in southeastern Wyoming some of the major land uses are timber harvesting, domestic livestock grazing, and recreation (USDA Forest Service, 1985). The MBNF is a significant timber-producing Forest in the Rocky Mountain Region, yet there are increasing pressures from environmental groups and some members of the public to de-emphasize timber harvesting, in part because of concerns that timber harvesting is negatively affecting biological diversity.

The Forest commissioned an analysis of the impacts of timber harvesting on landscape structure as part of background work on the status of biological diversity (Baker, 1994). This study, and other landscape research on the Forest, suggest that many parts of the Forest are significantly fragmented by timber harvesting and roads, with the primary symptoms being a loss of old, interior forest and an increase in the amount of edge habitat adjoining roads and clearcuts (Reed et al., 1996a, b).

National Forests do not, in general, have adequate planning tools to enable the evaluation of the long-term impacts of land management activities, such as timber harvesting, on the landscape, even though this kind of assessment is required by the National Environmental Policy Act (NEPA). Planning tools, such as FORPLAN, are inadequate for this purpose, largely because they are not focused on the consequences at the landscape scale, and do not model the landscape-scale processes represented by human disturbances. Forests need to be able to project alternative land management plans into the future, generally using a 50-year or longer planning horizon, and to be able to see the impacts of these plans on the landscape.

Using the ideas that are developing about how to model timber harvesting on the landscape scale (Li et al., 1993; Wallin et al., 1994, 1996; Gustafson, 1996; Gustafson and Crow, 1996), the author modified DISPATCH for modeling clearcut logging in the Medicine Bow National Forest. Here the derivation of the model and its verification and validation are explained, and some preliminary results from using the mode are presented. A full explication of a simulation experiment with the model presented elsewhere (Tinker and Baker, in press).

Modifying DISPATCH to simulate timber harvesting with spatial constraints

A number of modifications to DISPATCH were required to simulate clear-cut logging as it is practiced by the Forest. The modified model is called LANDLOG. The major modifications in LANDLOG include the ability to input empirical size and interval distributions, the use of a "mask" to constrain the simulation to areas that are designated "suitable for timber harvesting" and to apply other spatial constraints, buffering of cutting units to assure spatial placement, the use of timber sale areas, and a spread algorithm to place the clearcuts into the landscape.

To effectively analyze proposed alternative harvesting plans in a way that can be simulated, a Forest needs to be able to specify size and interval distributions for harvesting units. When LANDLOG is running, the model selects disturbance sizes from the size distribution at intervals specified by the interval distribution using a random number generator, so over a long period of time the simulated disturbance size and interval distributions approach those specified by the user. There is no *a priori* theoretical form for these distributions, so the ability to input empirical distributions was added to LANDLOG. Moreover, in most timber harvesting regimes there are many disturbances in one year, but the number of disturbances varies from year to year. LANDLOG allows the user to input a mean number of disturbances per year and the standard deviation from the mean, which is then used to specify a normal distribution sampled by the model.

National Forests have several spatial constraints that limit the area on which timber harvesting can take place. Some of the "suitable area" is fixed, and does not vary from year-to-year, but some constraints do vary. Because LANDLOG uses GRASS maps as part of the model, it is possible to make use of the GRASS "mask" to model fixed and changing spatial constraints. A mask in GRASS is simply a binary map containing 0s where the GIS is to omit using data in all other maps and 1s where information in other maps is to be used in GIS operations. Fixed constraints, such as topographic and soils restrictions, riparian buffers, non-Federal in-holdings, and areas permanently withdrawn from timber harvesting (e.g., Wilderness) are simply put in the 0 category in the mask. The GIS, of course, can be used to identify and combine these areas into a single mask.

The primary constraint that varies each year, in the simulation reported here, is that subsequent clearcuts are not placed next to clearcuts that are less than about 20 years old. This constraint is simply the traditional silvicultural practice on the Forest. This constraint can be simulated by simply using the GRASS r.buffer command at the end of each year of harvesting to create a buffer around the cumulative cut area that is less than 21 years old, including new cuts placed during the year. Once the buffer area for the year is identified, it is given the value "0" and added to the mask that contains the fixed constraints. More difficult to simulate is the requirement that cuts taking place during a single year also be placed apart. Buffering each cut, then adding the buffer to the mask prior to placing the next cut would be too time-consuming in the model, so a simpler algorithm that approximates this was devised. During a single year, the model keeps a list of the initial placement (row, column) for each added cut, and each subsequent cut is placed so that it is at least a minimum distance, specified by the user, from all previous cuts during the year.

Many National Forests sell timber to be cut over a period of a few years, with the cuts clustered in parts of the Forest, and sold as a single "timber sale area". LANDLOG allows the user to specify the size (rows and columns) of rectangular timber sale areas to be used to allocate harvesting. LANDLOG then randomly locates a timber sale area within the area suitable for timber harvesting, then places cuts within the timber sale area until the area is filled to capacity, before locating a new timber sale area.

The algorithm in DISPATCH that spreads the disturbance from an initial point had to be modified to better approximate the patterns produced by clearcutting. The algorithm in the earlier version of DISPATCH was probabilistic, based on the disturbability of adjoining cells, which was determined by an equation (e.g., $0.5 \times age + 0.78 \times slope$) specified by the user. The Forest, however, chooses areas to harvest simply by whether they meet a minimum age requirement, which for the lodgepole pine (*Pinus contorta*) forests simulated here, is 90–125 years. To speed up the model and replicate this, the algorithm was simplified to spread randomly into any forest > 90 years old, rather than to spread probabilistically with greater spread into older forests. However, random spread would lead to very irregular cutting units, and clearcuts typically are relatively compact and placed in part along boundaries represented by existing patchiness from natural disturbances and older cuts. The algorithm that was devised thus begins with roughly rectangular cutting units, but modifies this shape to follow existing boundaries, and interjects a little randomness into the shape, so that not all units are the same shape (Fig. 11.3).

Openings created by harvesting are set to age 0 and then simply progress through the age-classes as time passes. As the opening ages it continues to form an edge with adjoining forest as long as there is a difference in age class (Table 11.2). Gustafson and Crow (1996) simply contrasted openings (< 20 years old) and closed forest, ignoring age-class differences. Neither approach is necessarily more correct.

The Medicine Bow National Forest verification and validation

DISPATCH is a logical model rather than a predictive tool, so it was never validated in the sense of comparing predictions with actual data, but LANDLOG does require validation. The logic of DISPATCH has been verified using a variety of the methods (complete list in Baker 1992*b*) suggested by Sargent (1988), and the model has been repeatedly tested through use over the last decade. Thus, the central parts of LANDLOG derived from DISPATCH have been well verified.

Fig. 11.3. Comparison of actual (gray) and simulated (black) clearcuts in a selected part of the study area. Individual pixels are 50 m on a side.

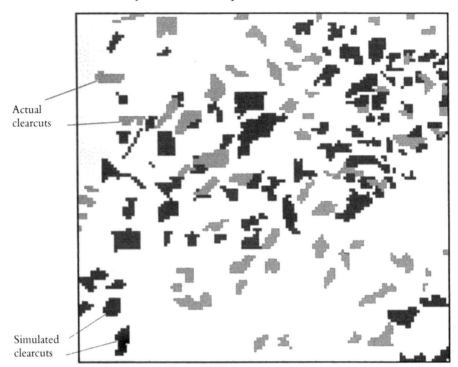

The time period and study area

The modifications that make LANDLOG a predictive model require further verification and validation. This is accomplished here by comparing real and simulated outcomes of clear-cut logging over 9 years in part of the Medicine Bow National Forest (MBNF). The period from 1985–1993 was used, as 1985 is the first year of implementation of the most recent forest plan (USDA Forest Service, 1985), and 1993 is the last year for which complete harvesting records were available at the time of this work. The harvesting strategy of the MBNF has been relatively consistent over this period.

The study area for this validation, and for the application reported on pp. 299–303, is the Snowy Range part of the MBNF (Fig. 11.4: see color section). This area represents the major part of the MBNF (the other is the Sierra Madre mountains to the west) on which clear-cut harvesting has been concentrated since the end of World War II. The actual area is represented in the GRASS GIS by maps whose overall dimensions are 44.7 km west-to-east × 60.45 km north-to-south. Maps are rasters of 894 columns × 1209 rows with a 50-m pixel resolution.

Data sources and maps

Data for the validation work and application were obtained from the MBNF. The required maps were obtained from the MBNF as digital ARC/INFO coverages

(ESRI, 1994), which were then converted into GRASS raster maps. The required maps included cover type (e.g., lodgepole pine, spruce-fir), forest age, suitable timber harvesting area, forest boundary, and the boundaries and dates of harvesting of all clearcuts in the study area. At the time of this work an adequate digital map of roads in the MBNF was not yet available, so the effects of roads, although substantial (Reed *et al.*, 1996*b*), could not be simulated.

A map of the Forest in 1985 was not available, so this map was recreated from the map of the Forest available in early 1994, the most recent complete map available at the time of this work. The age of the unharvested forest in 1985 was recreated by simply subtracting 9 years from the age of the forest in 1994, but it is more difficult to recreate the forest that existed prior to the harvests that took place between 1985 and 1993. These harvested areas needed to be re-assigned an age and cover type and merged with adjoining patches where appropriate. Since there were 788 individual clearcuts between 1985 and 1993, it was not feasible to manually examine each cut, so the GRASS r.neighbors and r.mapcalc programs were used to "grow", using a successive buffering operation, the unharvested adjoining forests back into the areas harvested. So, if a clear-cut was surrounded by a 220-year-old lodgepole pine forest, the cut area would be reassigned to this category and merged into that patch. This operation was generally successful with two exceptions. First, about 30% of the area of forest patches that was harvested from 1985 to 1993 was surrounded by forests too young to harvest; these harvested patches get reassigned an age that is too young. Second, some patches that were of harvestable age in 1985 were surrounded by land unsuitable for harvesting (e.g., riparian areas, meadows), so the grow operation could not reconstruct the ages of these patches.

The harvesting regime and the suitable timber harvesting area

To determine the essential parameters of the harvesting regime, the map of clear-cuts from 1985–1993 was analyzed using the GRASS GIS and the r.le programs (Baker and Cai, 1992). Using this map calculations were made of: (i) the mean and standard deviation of the number of cuts per year, (ii) the size distribution of the cuts, using 25 size classes, (iii) the mean edge-to-edge distance between the cuts, and (iv) the average north–south and east–west dimensions of timber sale areas. These data were then used as parameters for the validation work. An example of the parameters file for the validation runs is in Table 11.1.

While the total area in the study area is 292 561 ha, only 65 617 ha is covered by lodgepole pine forest that is considered suitable for timber harvesting. The area considered suitable for timber harvesting is determined by the MBNF based on a combination of factors related to soil type and soil stability, elevation, vegetation type, slope, and aspect (Carol Tolbert, GIS specialist, MBNF, pers. comm., 1996). An ARC/INFO coverage of the suitable area was obtained from the MBNF. Because timber harvesting almost never occurs in riparian areas, riparian vegetation was added to the area considered unsuitable for timber harvesting.

At the time of this work the MBNF had a policy that existing roadless areas

Table 11.1. *Typical parameters file for a LANDLOG validation model run*

Parameter value	Explanation
1 9	Starting and stopping year for run
1309 894	Rows and columns in area
1 2 3 4 5 6 1	Seeds for random number streams
v1	Name for model run
validate.final	Initial map
1	1=do display model output
0.0 0.0	This selects an empirical interval distr.
0.0 0.0 0.0	This selects an empirical size distr.
87.6 34.4	Mean and s.d. for number of cuts per year
9 75 21	Min. distance (pixels) between multiple disturbances in 1 year; distance (m) to buffer around young patches; age below which buffering is necessary.
47 47	Rows and columns in timber sale areas
"tmp"	Equation for "disturbability"
10 0 0 0	How often to save various maps
0 0.0 0.0	How often to use r.le programs
"999"	r.le.patch analysis; 999 = do not do this
"999"	r.le.pixel analysis; 999 = do not do this
"999"	r.le.dist analysis: 999 = do not do this
0.00 0 1 1.00 1	The disturbance interval distribution; probability is followed by interval (years)
0.000 0 0.041 2 0.109 6 0.162 10 0.133 14 0.131 18 0.093 22 0.080 26 0.057 30 0.037 34 0.039 38 0.019 42 0.018 46 0.020 50 0.011 54 0.013 58 0.010 62 0.008 66 0.008 70 0.004 74 0.000 78 0.003 82 0.000 86 0.001 90 0.001 94 0.002 150	The disturbance size distribution; probability is followed by size (pixels)

would not be entered for harvesting until completion of the Forest Plan revision. A map of existing roadless areas in digital form did not exist, so roadless areas were digitized using the map in the 1984 Forest Plan (USDA Forest Service, 1985) with the aid of other draft maps from the MBNF. This roadless area map was then

Table 11.2. *Landscape indices used to monitor changes in landscape structure*

Mean patch age
Standard deviation of patch age
Mean pixel age
Standard deviation of pixel age
Cover (decimal fraction) of forest by age class: 1–20, 21–40, 41–60, 61–80, 81–100, 101–150, 151–200, 201–250, 251–300, 301–350, 351–400, 400+ years
Total number of patches
Number of patches by age-class (age-classes as above)
Mean patch size (ha)
Standard deviation of patch size
Number of patches by size class: 0.25–2.49, 2.50–4.99, 5.00–7.49, 7.50–9.99, 10.00–12.49, 12.50–24.99, 25.00–49.99, 50.00–124.99, 125.00+ ha
Mean interior size (ha), assuming a 50 m depth of edge influence
Standard deviation of interior size
Mean interior size (ha) by age-class (age-classes as above)
Mean edge size (ha), assuming a 50 m depth of edge influence
Standard deviation of edge size
Shannon diversity of patch age-classes (age-classes as above)
Total edge length (ha)
Mean nearest neighbor distance between old forest (> 200 years old) patches
Standard deviation of nearest neighbor distances–old forest
Number of nearest neighbor distances–old forest–by distance class: to 0.09, 0.10–0.24, 0.25–0.49, 0.50–0.74, 0.75–1.24, 1.25–2.49, 2.50–4.9, > 5.0 km

---

categorized as "0" and added to the suitable map discussed above, so that roadless areas would not be available for harvesting in the model.

Monitoring landscape structure

DISPATCH and LANDLOG both focus on quantitative changes in landscape structure indices as the model runs, saving a copy of the landscape map at 10-year intervals for separate analysis using the r.le programs (Baker and Cai, 1992). A broad spectrum of indices was chosen for the validation work, to see whether the model would adequately predict changes in these measures (Table 11.2).

Most of the indices are self-explanatory, but a few require explanation. First, the model runs with the age of the forest, in years, as the attribute of each pixel, but it would be inappropriate to analyze the landscape structure of a raw age map, as two patches that differ by only a year in age would appear as an edge. Therefore, the raw age map was reclassed into a map with the attribute being the mid-point of each of 11 age-classes (Table 11.2), prior to analysis with the r.le programs. In the author's experience 50-year differences in the age of older forests appear as boundaries visible on high-altitude aerial photography, and can also be located on the ground, so this much age difference does represent a real edge in the forest.

Twenty-year differences in younger forests are also apparent on the ground. Second, it is assumed that an "edge" environment extends from the border of clearcuts into adjoining uncut forest a distance of 50 m. This is approximately the mean distance over which several micro-environmental and vegetational attributes are affected in MBNF forests (Vaillancourt, 1995).

Validation criteria

The validation trial consists of ten replicates starting with the same initial 1985 map, but ten different random number streams for the main functions that select the number of disturbances per year, disturbance size, disturbance placement, and disturbance spread, so results are variable. Initially, the model is considered to be validated for a particular measure of landscape structure if the actual value in 1993 is within one standard deviation of the mean simulated value from the ten replicates. After analyzing the results, however, additional criteria were added, as explained below. Since the goal of the simulation work is to model the aggregate changes in landscape structure resulting from clear-cut logging, rather than the exact placement of clearcuts, validation is restricted to aggregate indices of landscape structure (Turner *et al.*, 1989*b*).

Initial validation results and modifications

The initial validation trials revealed problems with the 1985 reconstruction and with the model functions that required modifications or interpretation. First, the most difficult problem with the validation was reconstructing the 1985 map, as discussed on p. 292. The MBNF does not track the age of harvested trees, so the age at the time of harvest is unknown and must be reconstructed. However, in many cases harvesting removes the entire patch, leaving no trace of the patch age in surrounding patches; thus, reconstruction based on surrounding patch ages may fail. The results will be interpreted in view of this problem, since no solution was found. This seems to be a general problem with validating harvesting simulation models. The only real solution may be simply to encourage forests to maintain better databases on the areas harvested.

Second, initial trials revealed that the spread algorithm may leave single-pixel or two-pixel patches that are isolated (Fig. 11.5). The Forest, however, appears to often harvest entire patches or contiguous areas of patches, rather than leaving small, isolated remnants. The result is that the initial simulated map in 1993 contained many more small patches than the actual 1993 map. Silvicultural approaches to placing harvest units clearly use several levels of thinking, including the existing patch pattern, spatial constraints, road locations, and topography. Harvest units do often overlap more than one patch, so simply harvesting entire existing patches would not replicate present silvicultural practices. A better harvesting algorithm is needed, but would be challenging to develop. The interim solution used here was to filter out isolated single-pixel patches, using the GRASS r.mapcalc program and a custom filter, before completing r.le analyses. This appears to bring the number of small patches produced by the model very close to the actual number produced.

Fig. 11.5. Illustration of the spread algorithm problem that produces isolated single-pixel patches or patches with only a few pixels. Shades represent age-classes, and lighter gray patches are new clearcuts produced by the model.

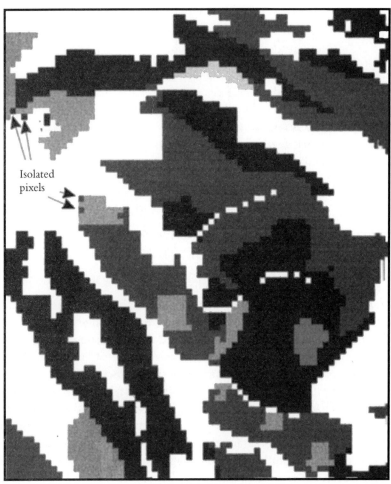

Validation results

The actual disturbance regime produced by the model during the ten replicates is very similar to the harvesting regime that actually occurred between 1985 and 1993. The simulated number of clearcuts (Fig. 11.6(*a*)), total clear-cut area (Fig. 11.6(*b*)), and mean clear-cut size (Fig. 11.6(*c*)) are all very close to the actual values over the 9-year simulation period. Simulated clear-cut shapes also appear similar to actual clear-cut shapes (Fig. 11.3), indicating that the spread algorithm produces realistic clear-cut patches.

The model does reasonably well at producing maps in 1993 that are similar in landscape structure to the actual 1993 map (Fig. 11.7). Of 62 aggregate indices (e.g., mean patch age) and age-class indices (e.g., cover of the 41–60 year age-class),

Fig. 11.6. Comparison of the actual and simulated clearcut logging regime for the 9-year simulation period. Error bars are one standard deviation, from n = 10 replicates: (a) *number of clearcuts,* (b) *total clearcut area, and* (c) *mean clearcut size.*

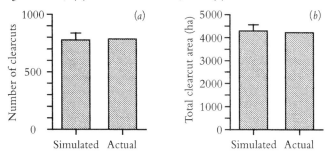

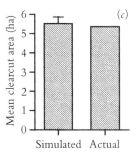

30 have values in 1993 that are within the one standard deviation error bars for the mean simulated 1993 value, and the model is certainly valid for these indices (Table 11.3). For another ten aggregate or age-class indices, the simulated 1985–1993 change in values is within 20% of the change in values that actually occurred between 1985 and 1993 (Table 11.3). With one exception (standard deviation of pixel age), these ten indices have values that all overestimate the actual change by 10–20%, suggesting the model has a consistent pattern. The most likely cause is the over-abundance of young, small patches that result from the spread algorithm's tendency to leave isolated patches containing a few pixels (see pp. 295–6). Evidence that this is the problem is the overabundance of patches in the 2.5–4.9 ha size class (Fig. 11.7(h p. 295–6)), and the slight overestimation of the number of patches in the 1–20 year age-class (Figure 11.7(d)). Since this problem could not be remedied without the development of a more complex spread algorithm, the model results will simply need to be interpreted with this result in mind; but the model is certainly usable for these indices, as the error is comparatively small.

Some differences between the simulated and actual 1993 index values arise simply because the 1985 reconstruction is only an approximation (Table 11.3). For example, the cover of patches in the 21–40 year age-class in the reconstructed 1985 map is probably too high largely because experimental strip cuts made during the 1950s and 1960s, especially in the northern part of the Forest, left isolated

298   W. L. Baker

Fig. 11.7. Landscape indices calculated for reconstructed map at beginning of 1985, actual map at end of 1993, and simulated map at end of 1993. Error bars for the 1993 simulated results are standard deviations for n = 10 *trials. The measures are: (a) cover (decimal fraction) of the suitable area by age class, (b) mean and standard deviation of patch age and pixel age, (c) total number of patches, (d) number of patches by age class, (e) mean and standard deviation of patch size, interior (with 50 m depth of edge buffer) size, and edge size (area of 50 m depth of edge buffer), (f) mean interior size (with 50 m depth of edge buffer) by age class, (g) number of distances by distance class for old forest patches (> 200 years old), measured as the distance (edge-to-edge) to the nearest neighboring old forest patch, (h) number of patches by size class, (i) Shannon diversity of patch age classes, (j) total length of patch edges, (k) mean and standard deviation of nearest neighbor distances for old forest patches (> 200 years old), measured as the distance (edge-to-edge) to the nearest neighboring old forest patch.*

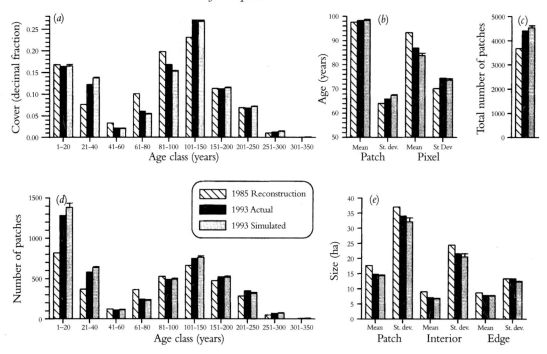

strips of old forest surrounded by regenerating strips. During the simulation period (1985–1993) the Forest removed some of the remnant patches of old forest between the strips. The author's method of reconstructing the pre-harvest age of these forests errs because the age of the old forests that were harvested becomes reconstructed to be the age of the adjoining regenerating strips, thus overestimating the amount of forest in this age class in 1985, and underestimating the age of older forests (i.e., 81–100, 201–250 in Table 11.3). For these ten indices (Table 11.3) the model is also likely to be valid, as the differences arise from the reconstruction, rather than the model itself.

Twelve indices are not validated, as their simulated 1993 values are simply too

Fig. 11.7. cont.

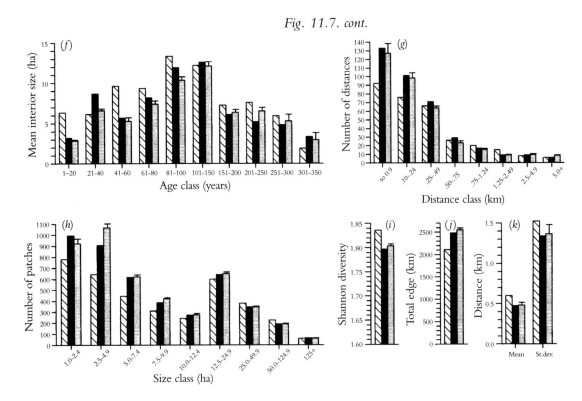

far from the actual 1993 value (Table 11.3). For these indices the predicted 1985–1993 change over- or under-estimates the actual 1985–1993 change by 50% or more. These include the standard deviation of mean patch age (Fig. 11.7(b)), mean pixel age (Fig. 11.7(b)), the standard deviation of mean patch size (Fig. 11.7(e)), the number of patches in the 2.5–4.9 ha and 7.5–9.9 ha size classes (Fig. 11.7(h)), and most of the nearest neighbor distance classes that are in the 0.25–0.49 km class or larger classes (Fig. 11.7(g)). An explanation for the lack of validation for these indices is lacking, and they should not be used until an explanation or remedy can be found. In summary, with the present level of development of LANDLOG, about 80% of the indices are valid for use, although ten should be interpreted with caution, as they overestimate actual change by 10–20% (Table 11.3). The model can still be improved, particularly by an improvement in the spread algorithm.

This short-term validation is all that is feasible, but short-term errors accumulate when the model is used for longer-term simulation (Kangas, 1997). Unfortunately, because management changes at 10-year intervals with Forest Plan revision, it will never be possible to use longer-term validation periods. Other approaches, such as systematic analysis of error propagation, may be needed (Mowrer, 1991).

Preliminary results from a long-term simulation

LANDLOG was used to simulate clearcut logging in the MBNF over the next century. The model begins with the current landscape as of 1997 (Fig. 11.4), and

Table 11.3. *Synopsis of validation results*

Actual 1993 value is within one standard deviation of simulated 1993 value	Simulated 1985–1993 change is within 20% of actual 1985–1993 change	1985 reconstruction is the source of the error in simulation
Mean patch age	St. dev. of pixel age	Cover: 21–40 years
Cover: 1–20 years	Cover: 61–80 years	Cover: 81–100 years
Cover: 41–60 years	Number of patches: 1–20 years	Cover: 201–250 years
Cover: 101–150 years	Number of patches: 61–80 years	Cover: 251–300 years
Cover: 151–200 years	Total number of patches	Number of patches: 21–40 years
Cover: 301–350 years	Mean interior size	Number of patches: 201–250 years
Number of patches: 41–60 years	Mean edge size	Interior size: 21–40 years
Number of patches: 81–100 years	Interior size: 1–20 years	Interior size: 61–80 years
Number of patches: 101–150 years	Mean patch size	Interior size: 81–100 years
Number of patches: 151–200 years	Total edge	Interior size: 201–250 years
Number of patches: 251–300 years		
Number of patches: 301–350 years		
St. dev. interior size	Indices not validated	
Interior size: 41–60 years	St. dev. patch age	
Interior size: 101–150 years	Mean pixel age	
Interior size: 151–200 years	St. dev. interior size	
Interior size: 251–300 years	St. dev. patch size	
Interior size: 301–350 years	Number of patches: 1.0–2.4 ha	
Number of patches: 5.0–7.4 ha	Number of patches: 2.5–4.9 ha	
Number of patches: 10.0–12.4 ha	Number of patches: 7.5–9.9 ha	
Number of patches: 12.5–24.9 ha	Number of distances: 0.25–0.49 km	
Number of patches: 25.0–49.9 ha	Number of distances: 0.50–0.74 km	

Table 11.3. (cont.)

Number of patches: 50.0–124.9 ha	Number of distances: 0.75–1.24 km
Number of patches: 125+ ha	Number of distances: 2.5–4.9 km
Shannon diversity of patch age classes	Number of distances: 5.0+ km
Mean nearest neighbor distance for old forests	
St. dev. nearest neighbor distance for old forests	
Number of distances: to 0.09 km	
Number of distances: 0.10–0.24 km	
Number of distances: 1.25–2.49 km	

Landscape structure indices are categorized by whether they are close to the actual value in 1993, whether the 1985 reconstruction is the source of errors in simulation, or whether they are not validated and not recommended for use.

continues the timber harvesting regime that followed the 1984 Forest Plan, which is the harvesting regime used in the validation. The purpose of the simulation is to determine the long-term effects on landscape structure of continuing the present harvesting policy.

The results of this experiment are presented in full elsewhere (Tinker and Baker, in press), but to illustrate the general trends some of the results of a single model run are presented here (Figs. 11.8: see color section, 11.9). While the model was run on the entire study area (Fig. 11.4), the trends in a smaller area in the south-central part of the study area are illustrated (Fig. 11.8). This landscape in 1997 contained numerous recent clearcuts (1–20 years old) in a matrix of older forest (note the blue patches, representing forest 201–300 years old). White areas in this figure are either considered unsuitable for timber harvesting or contain spruce-fir forests. By 2037, the model projects that much of the older forest will have been harvested, and the landscape will be dominated by smaller patches of young forest, particularly those in the 21–60 year age class. Harvesting after 2037 keeps the forest predominately young (< 100 years old) and with smaller patches throughout until the run ends in 2097 (Fig. 11.8).

A few of the landscape measures calculated for the whole study area and this single run of the model illustrate some of the major trends to be expected (Fig. 11.9), although these trends are very preliminary. Mean patch age declines after about 2017 at a nearly linear rate, and may be leveling off by about 2067 (Fig.

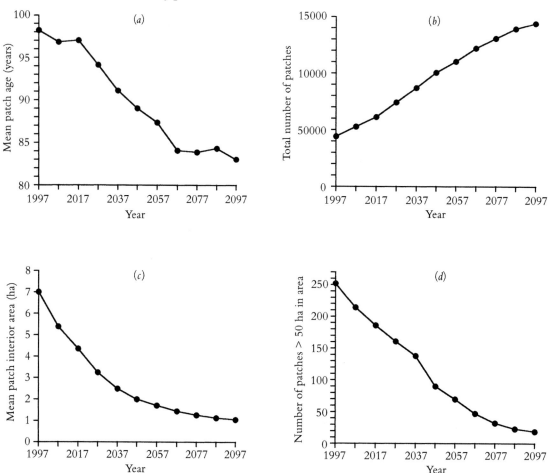

Fig. 11.9. Trends in landscape structure measures during a single run of LANDLOG for 100 years, continuing the timber harvesting policy practiced since 1984 on the MBNF. Measurements of landscape structure were made every 10 years. These trends are for the entire area of suitable lodgepole pine forest illustrated in Fig. 11.4. Measures are: (a) mean patch age, (b) total number of patches, (c) mean patch interior area, and (d) total number of patches > 50 ha in area.

11.9(a)). The total number of patches increases nearly linearly, with a slight hint of leveling off toward the end of the run (Fig. 11.9(b). The number of patches has about tripled after 100 years of simulation. The mean patch interior area declines in an apparently negative exponential pattern, appearing to level off toward the end of the run (Fig. 11.8(c)). By this time mean interior area is only about 1 ha, only about 15% of what it was in 1997. The total number of large patches (> 50 ha in area) declines nearly linearly to near zero by the end of the run (Fig. 11.8(d)).

Together, these trends (Fig. 11.9) and the pattern of change visible on a small part of the Forest (Fig. 11.8) suggest that present timber harvesting strategies will significantly transform the landscape over the next century if continued. The trends are toward increased fragmentation of large, old forest patches into small, young forest patches containing little or no interior habitat. These trends have previously been reported from other studies of forest fragmentation in the western United States (e.g., Reed *et al.*, 1996*a,b*) and from previous simulations of timber harvesting in the midwestern United States (e.g., Gustafson and Crow, 1996). However, this simulation is the first to use individual-stand age and cover type data as well as Forest Service suitability constraints with a validated model to show future trends expected under a certain management policy. Clearly, having this longer-term view, rather than just the snapshot view, allows us to see the long-term consequences of particular forest management policies before they are played out on the landscape in a way that limits future options. Landscape simulation of future consequences of proposed land management alternatives in landscapes subject to timber harvesting should become a standard part of environmental impact analysis.

References

Antonovski, M. Y., Ter-Mikaelian, M. T. and Furyaev, V. V. (1992). A spatial model of longterm forest fire dynamics and its applications to forests in western Siberia. In *A systems analysis of the global boreal forest*, ed. H. H. Shugart, R. Leemans and G. B. Bonan, pp. 373–403. Cambridge: Cambridge University Press.

Baker, W. L. (1989). Landscape ecology and nature reserve design in the Boundary Waters Canoe Area, Minnesota. *Ecology*, **70**, 23–35.

Baker, W. L. (1992*a*). The landscape ecology of large disturbances in the design and management of nature reserves. *Landscape Ecology*, **7**, 181–94.

Baker, W. L. (1992*b*). Effects of settlement and fire suppression on landscape structure. *Ecology*, **73**, 1879–87.

Baker, W. L. (1993). Spatially heterogeneous multi-scale response of landscapes to fire suppression. *Oikos*, **66**, 66–71.

Baker, W. L. (1994). Restoration of landscape structure altered by fire suppression. *Conservation Biology*, **8**, 763–9.

Baker, W. L. (1995). Longterm response of disturbance landscapes to human intervention and global change. *Landscape Ecology*, **10**, 143–59.

Baker, W. L. and Cai, Y. (1992). The r.le programs for multiscale analysis of landscape structure using the GRASS geographical information system. *Landscape Ecology*, **7**, 291–302.

Baker, W. L., Egbert, S. L. and Frazier, G. F. (1991). A spatial model for studying the effects of climatic change on the structure of landscapes subject to large disturbances. *Ecological Modelling*, **56**, 109–25.

Barrows, J. S. (1951). Fire behavior in northern Rocky Mountain forests. USDA Forest Service Station Paper 29, Northern Rocky Mountain Forest and Range Experiment Station, Missoula, Montana. 103 pp.

Baskent, E. Z. (1997). Assessment of structural dynamics in forest landscape management. *Canadian Journal of Forest Research*, **27**, 1675–84.

Baughman, R. G. (1981). Why windspeeds increase on high mountain slopes at night. USDA Forest Service Research Paper INT-276, Intermountain Forest and Range Experiment Station, Ogden, Utah. 6 pp.

Bettinger, P., Sessions, J. and Boston, K. (1997). Using Tabu search to schedule timber harvests subject to spatial wildlife goals for big game. *Ecological Modelling*, **94**, 111–23.

Boose, E. R., Foster, D. R. and Fluet, M. (1994). Hurricane impacts to tropical and temperate forest landscapes. *Ecological Monographs*, **64**, 369–400.

Boychuk, D. and Perera, A. H. (1995). FLAP-X 1.00 user's guide. Forest Fragmentation and Biodiversity Project, Report No. 20, Ontario Forest Research Institute, Sault Ste. Marie, Ontario, Canada. 36 pp.

Boychuk, D. and Perera, A. H. (1997). Modeling temporal variability of boreal landscape age-classes under different fire disturbance regimes and spatial scales. *Canadian Journal of Forest Research*, **27**, 1083–94.

Boychuk, D., Perera, A. H., Ter-Mikaelian, M. T., Martell, D. L. and Li, C. (1997). Modelling the effect of spatial scale and correlated fire disturbances on forest age distribution. *Ecological Modelling*, **95**, 145–64.

Burgan, R. E. and Rothermel, R. C. (1984). BEHAVE: Fire behavior prediction and fuel modeling system. USDA Forest Service General Technical Report INT-167.

CACI. (1987). *SIMSCRIPT II.5 programming language*. La Jolla, CA: CACI Products Co.

Canham, C. D. and Loucks, O. L. (1984). Catastrophic windthrow in the presettlement forests of Wisconsin. *Ecology*, **65**, 803–9.

Clark, T. L., Jenkins, M. A., Coen, J. and Packham, D. (1996). A coupled atmosphere-fire model: convective feedback on fire-line dynamics. *Journal of Applied Meteorology*, **35**, 875–901.

Clark, W. C., Jones, D. D. and Holling, C. S. (1977). Patches, movements and population dynamics in ecological systems: A terrestrial perspective. In *Spatial pattern in plankton communities*, ed. J. H. Steel, pp. 385–432. New York: Plenum Press.

Clarke, K. C., Brass, J. A. and Riggan, P. J. (1994). A cellular automaton model of wildfire propagation and extinction. *Photogrammetric Engineering and Remote Sensing*, **60**, 1355–67.

Costanza, R. (1989). Model goodness of fit: a multiple resolution procedure. *Ecological Modelling*, **47**, 199–215.

Crow, T. R. and Gustafson, E. J. (1997). Ecosystem management: managing natural resources in time and space. In *Creating a forestry for the 21st century: The science of ecosystem management*, ed. K. A. Kohm and J. F. Franklin, pp. 215–28. Washington, DC: Island Press.

Dale, V. H., O'Neill, R. V., Pedlowski, M. and Southworth, F. (1993b). Causes and effects of land-use change in central Rondônia, Brazil. *Photogrammetric Engineering and Remote Sensing*, **59**, 997–1005.

Dale, V. H., O'Neill, R. V., Southworth, F. and Pedlowski, M. (1994). Modeling effects of land management in the Brazilian Amazonian settlement of Rondônia. *Conservation Biology*, **8**, 196–206.

Dale, V. H., Southworth, F., O'Neill, R. V., Rosen, A. and Frohn, R. (1993a). Simulating spatial patterns of land-use change in Rondônia, Brazil. *Lectures on Mathematics in the Life Sciences*, **23**, 29–55.

ESRI. (1994). Understanding GIS: The ARC/INFO method. Environmental Systems Research Institute, Redlands, California.

Forman, R. T. T. (1995). *Land mosaics: The ecology of landscapes and regions*. Cambridge: Cambridge University Press. 632 pp.

Foster, D. R. and Boose, E. R. (1992). Patterns of forest damage resulting from catastrophic wind in central New England, USA. *Journal of Ecology*, **80**, 79–98.

Fowler, P. M. and Asleson, D. O. (1984). The location of lightning-caused wildland fires, northern Idaho. *Physical Geography*, **5**, 240–52.

Franklin, J. (1989). Toward a new forestry. *American Forests*, **95**(11/12), 37–44.

Franklin, J. F. and Forman, R. T. T. (1987). Creating landscape patterns by forest cutting: ecological consequences and principles. *Landscape Ecology*, **1**, 5–18.

Fuquay, D. M., Baughman, R. G., Taylor, A. R. and Hawe, R. G. (1967). Characteristics of seven lightning discharges that caused forest fires. *Journal of Geophysical Research*, **72**, 6371–3.

Gardner, R. H., Hargrove, W. W., Turner, M. G. and Romme, W. H. (1996). Climate change, disturbances and landscape dynamics, In *Global change and Terrestrial Ecosystems* ed. B. H. Walker and W. L. Steffen, pp. 149–72. IGBP Book Series No. 2, Cambridge, UK: Cambridge University Press.

Gilruth, P. T., Marsh, S. E. and Itami, R. (1995). A dynamic spatial model of shifting cultivation in the highlands of Guinea, West Africa. *Ecological Modelling*, **79**, 179–97.

Godron, M. and Forman, R. T. T. (1983). Landscape modification and changing ecological characteristics. In *Disturbance and ecosystems: Components of response*, ed. H. A. Mooney and M. Godron, pp. 12–28. New York: Springer-Verlag.

Gustafson, E. J. (1996). Expanding the scale of forest management: Allocating timber harvests in time and space. *Forest Ecology and Management*, **87**, 27–39.

Gustafson, E. J. and Crow, T. R. (1994). Modeling the effects of forest harvesting on landscape structure and the spatial distribution of cowbird brood parasitism. *Landscape Ecology*, **9**, 237–48.

Gustafson, E. J. and Crow, T. R. (1996). Simulating the effects of alternative forest management strategies on landscape structure. *Journal of Environmental Management*, **46**, 77–94.

Hansen, A. J., Garman, S. L., Marks, B. and Urban, D. L. (1993). An approach for managing vertebrate diversity across multiple-use landscapes. *Ecological Applications*, **3**, 481–96.

Hastings, J. R. and Turner, R. M. (1965). *The changing mile: An ecological study of vegetation change with time in the lower mile of an arid and semiarid region*. Tucson, AZ: University of Arizona Press. 317 pp.

He, H. S. and Mladenoff, D. J. (1999). Dynamics of fire disturbance and succession on a heterogeneous forest landscape: A spatially explicit and stochastic simulation approach. *Ecology*, 80: 81–99.

Heinselman, M. L. (1973). Fire in the virgin forests of the Boundary Waters Canoe Area, Minnesota. *Quaternary Research*, **3**, 329–82.

Heinselman, M. L. (1996). *The boundary waters wilderness ecosystem*. Minneapolis: University of Minnesota Press. 336 pp.

Hof, J. G. and Joyce, L. A. (1992). Spatial optimization for wildlife and timber in managed forest ecosystems. *Forest Science*, **38**, 489–508.

Kangas, A. S. (1997). On the prediction bias and variance in long-term growth projections. *Forest Ecology and Management*, **96**, 207–16.

Keane, R. E., Morgan, P. and Running, S. W. (1996). FIRE-BGC – a mechanistic ecological process model for simulating fire succession on coniferous forest landscapes of the northern Rocky Mountains. USDA Forest Service Research Paper INT-RP-484, Intermountain Research Station, Ogden, Utah. 122 pp.

Kessell, S. R. (1979). *Gradient modelling: Resource and fire management.* New York: Springer-Verlag.

Li, C., Ter-Mikaelian, M. and Perera, A. (1996). Ontario fire regime model: its background, rationale, development and use. Forest Fragmentation and Biodiversity Project, Report No. 25, Ontario Forest Research Institute, Sault Ste. Marie, Ontario, Canada. 42 pp.

Li, C., Ter-Mikaelian, M. and Perera, A. (1997). Temporal fire disturbance patterns on a forest landscape. *Ecological Modelling,* **99,** 137–50.

Li, H., Franklin, J. F., Swanson, F. J. and Spies, T. A. (1993). Developing alternative forest cutting patterns: A simulation approach. *Landscape Ecology,* **8,** 63–75.

Lieberman, D., Lieberman, M., Peralta, R. and Hartshorn, G. S. (1985). Mortality patterns and stand turnover rates in a wet tropical forest in Costa Rica. *Journal of Ecology,* **73,** 915–24.

Loehle, C. (1983). Evaluation of theories and calculation tools in ecology. *Ecological Modelling,* **19,** 239–47.

McGarigal, K. and Marks, B. J. (1995). FRAGSTATS: spatial pattern analysis program for quantifying landscape structure. USDA Forest Service General Technical Report PNW-GTR-351, Pacific Northwest Research Station, Portland, OR. 122 pp.

Medley, K. E., Pickett, S. T. A. and McDonnell, M. J. (1995). Forest-landscape structure along an urban-to-rural gradient. *Professional Geographer,* **47,** 159–68.

Meisner, B. N., Chase, R. A., McCutchan, M. H., Mees, R., Benoit, J. W., Ly, B., Albright, D., Strauss, D. and Ferrryman, T. (1994). A lightning fire ignition assessment model. *Proceedings of the Conference on Fire and Forest Meteorology,* **12,** 172–8.

Mertens, B. and Lambin, E. F. (1997). Spatial modelling of deforestation in southern Cameroon. *Applied Geography,* **17,** 143–62.

Mowrer, H. T. (1991). Estimating components of propagated variance in growth simulation model projections. *Canadian Journal of Forest Research,* **21,** 379–86.

Ohtsuki, T. and Keyes, T. (1986). Biased percolation: Forest fires with wind. *Journal of Physics A: Mathematics and General,* **19,** L281–7.

Overbay, J. C. (1992). Ecosystem management. In *Proceedings of the national workshop: Taking an ecological approach to management,* ed. P. E. Avers, pp. 3–15. [Salt Lake City, Utah–April 27–30, 1992]. USDA Forest Service, Watershed and Air Management, WO-WSA-3, Washington, DC.

Price, C. and Rind, D. (1994). Possible implcations of global climate change on global lightning distributions and frequencies. *Journal of Geophysical Research,* **99**(D5), 10 823–31.

Ratz, A. (1995). Long-term spatial patterns created by fire: a model oriented towards boreal forests. *International Journal of Wildland Fire,* **5,** 25–34.

Reed, R. A., Johnson-Barnard, J. and Baker, W. L. (1996*a*). Fragmentation of a forested Rocky Mountain landscape, 1950–1993. *Biological Conservation,* **75,** 267–77.

Reed, R. A., Johnson-Barnard, J. and Baker, W. L. (1996*b*). Contribution of roads to forest fragmentation in the Rocky Mountains. *Conservation Biology,* **10,** 1098–106.

Roberts, D. W. (1996*a*). Modelling forest dynamics with vital attributes and fuzzy systems theory. *Ecological Modelling,* **90,** 161–73.

Roberts, D. W. (1996b). Landscape vegetation modelling with vital attributes and fuzzy systems theory. *Ecological Modelling*, **90**, 175–84.

Rykiel, E. J., Jr. (1996). Testing ecological models: The meaning of validation. *Ecological Modelling*, **90**, 229–44.

Salwasser, H. (1991). New perspectives for sustaining diversity in U.S. National Forest ecosystems. *Conservation Biology*, **5**, 567–9.

Sargent, R. G. (1988). A tutorial on validation and verification of simulation models. In *Proceedings of the 1988 Winter Simulation Conference*, ed. A. Thesan, H. Grant and W. D. Kelton, pp. 33–9. Society for Computer Simulation, Washington, DC, USA.

Sessions, J., Reeves, G., Johnson, K. N. and Burnett, K. (1997). Implementing spatial planning in watersheds. In *Creating a forestry for the 21st century: the science of ecosystem management*, ed. K. A. Kohm and J. F. Franklin, pp. 271–9. Washington, DC: Island Press.

Sharpe, D. M., Guntenspergen, G. R., Dunn, C. P., Leitner, L. A. and Stearns, F. (1987). Vegetation dynamics in a southern Wisconsin agricultural landscape. In *Landscape heterogeneity and disturbance*, ed. M. G. Turner, pp. 137–55. New York: Springer-Verlag.

Stone, T. A., Brown, I. F. and Woodwell, G. M. (1991). Estimation, by remote sensing, of deforestation in central Rondinia, Brazil. *Forest Ecology and Management*, **38**, 291–304.

Tinker, D. B. and Baker, W. L. (1998). Using the LANDLOG model to analyze the fragmentation of a Wyoming forest by a century of clearcutting, In *Forest fragmentation in the Southern Rocky Mountains*, ed. R. L. Knight, F. W. Smith, S. W. Buskirk, W. H. Romme and W.L. Baker, Boulder, CO: University Press of Colorado.

Turner, M. G., Gardner, R. H., Dale, V. H. and O'Neill, R. V. (1989a). Predicting the spread of disturbance across heterogenous landscapes. *Oikos*, **55**, 121–9.

Turner, M. G., Costanza, R. and Sklar, F. H. (1989b). Methods to evaluate the performance of spatial simulation models. *Ecological Modelling*, **48**, 1–18.

Turner, M. G., Romme, W. H., Gardner, R. H., O'Neill, R. V. and Kratz, T. K. (1993). A revised concept of landscape equilibrium: Disturbance and stability on scaled landscapes. *Landscape Ecology*, **8**, 213–27.

Turner, M. G., Wear, D. N. and Flamm, R. O. (1996). Land ownership and land-cover change in the southern Appalachian Highlands and the Olympic peninsula. *Ecological Applications*, **6**, 1150–72.

USA–CERL (1994). *GRASS 4.1 user's manual*. Champaign, IL: US Army Corps of Engineers Construction Engineering Research Laboratory.

USDA Forest Service (1985). Land and resource management plan: Medicine Bow National Forest and Thunder Basin National Grassland. USDA Forest Service, Laramie, WY.

Vaillancourt, D. A. (1995). Structural and microclimatic edge effects associated with clearcutting in a Rocky Mountain forest. MA Thesis, University of Wyoming, Laramie, WY.

Vasconcelos, M. J. and Guertin, D. P. (1992). FIREMAP-simulation of fire growth with a geographic information system. *International Journal of Wildland Fire*, **2**, 87–96.

Walker, H. and Leone, J. M. Jr. (1996). The effect of elevation data representation on nocturnal drainage wind simulations. In *GIS and environmental modeling: Progress and research issues*, ed. M. F. Goodchild, L. T. Steyaert, B. O. Parks, C. Johnston, D. Maidment, M. Crane and S. Glendinning, pp. 105–10. Fort Collins, CO: GIS World Books.

Wallin, D. O., Swanson, F. J. and Marks, B. (1994). Landscape pattern response to changes in pattern generation rules: Land-use legacies in forestry. *Ecological Applications*, **4**, 569–80.

Wallin, D. O., Swanson, F. J., Marks, B., Cissel, J. H. and Kertis, J. (1996). Comparison of managed and pre-settlement landscape dynamics in forests of the Pacific Northwest, USA. *Forest Ecology and Management*, **85**, 291–309.

Wilkie, D. S. and Finn, J. T. (1988). A spatial model of land use and forest regeneration in the Ituri forest of northeastern Zaire. *Ecological Modelling*, **41**, 307–23.

Wu, Y., Sklar, F. H., Gopu, K. and Rutchey, K. (1996). Fire simulations in the Everglades landscape using parallel programming. *Ecological Modelling*, **93**, 113–24.

Xie, Y. (1996). A generalized model for cellular urban dynamics. *Geographical Analysis*, **28**, 350–73.

Xu, J. and Lathrop, R. G. Jr. (1994). Geographic information system (GIS) based wildfire spread simulation. *Proceedings of the Conference on Fire and Forest Meteorology*, **12**, 477–84.

Zhou, G. and Liebhold, A. M. (1995). Forecasting the spatial dynamics of gypsy moth outbreaks using cellular transition models. *Landscape Ecology*, **10**, 177–89.

12

HARVEST: linking timber harvesting strategies to landscape patterns

Eric J. Gustafson and Thomas R. Crow

Introduction and rationale

Providing a balance among the various benefits and values derived from forest lands has always been a challenge for managers. Determining this balance will be an even greater challenge in the future as increasing human populations consume greater amounts of natural resources from a decreasing land base. Obviously, not all multiple uses are compatible. Past attempts to reduce conflict have resulted in separate land allocations such as natural areas, developed recreation sites, non-motorized and semi-primitive areas, research natural areas, botanical areas, and so on. Most often these designations are made piecemeal, without a comprehensive spatial plan, resulting in *de facto* zoning of land use. Such an approach works only when there is a large land base available to make designations, and only when a small portion of this land base has already been designated.

Multiple use and sustained yield remain the guiding principles for managing many forest lands in the United States. In the case of national forests, these management principles have been codified into law as the Multiple-Use Sustained-Yield Act of 1960. In this Act, multiple use is defined as managing "the national forests so that they are utilized in the combination that will best meet the needs of the American people". Likewise, sustained yield in the context of this legislation refers to "the achievement and maintenance in perpetuity of a high-level annual or regular periodic output of the various renewable resources of the national forests". Although the emphasis is on utilization and outputs, the concepts of multiple use and sustained yield apply to a broad spectrum of benefits and values that are derived from forest land.

Dealing with the complex problem of integrating commodity production with other values and benefits requires a more comprehensive and spatial approach than has been traditionally applied to managing forest ecosystems (Brown and MacLeod, 1996). By taking a landscape perspective, combined with improved analytical tools, forest managers add consideration of both space and time to the multiple use and sustained yield mandates (Crow and Gustafson, 1997a,b). In this chapter, a management tool is described that combines remote sensing and geographic information systems (GIS) in a computer model that allows planners and managers to examine the long-term ecological consequences of management decisions and to compare the impacts of alternative management approaches in both a spatial and temporal context. The tool is a timber harvest allocation model (HARVEST) that generates landscape patterns with spatial attributes resulting from the initial landscape conditions and potential timber harvest activities. The model is simplistic in that it does not attempt to optimize timber production or quality, nor is it useful to predict the specific locations of future harvest activity, because it ignores many considerations such as visual objectives and road access. Instead, the model stochastically mimics the allocation of stands for harvest by forest planners, using the constraints imposed by the standards and guidelines, and management boundaries. Modeling this process allows experimentation to link variation in broad management strategies with the resulting pattern of forest openings and age class structure. HARVEST was designed to operate with minimal data input requirements, and readily simulates management on large areas ($>10^6$ ha). Therefore, it is a strategic, not a tactical, planning tool.

HARVEST was developed as part of a research project to compare the landscape patterns that would result under different management plans for the Hoosier National Forest (Indiana, USA). The initial goal was to reduce the harvesting standards and guidelines of each management plan to a set of rules that could be applied to the landscape. A simulation model approach was adopted that allowed flexible input of parameters related to the standards and guidelines for timber harvest, and incorporated spatial information (in the form of GIS maps) about the boundaries of management areas where various management goals were assigned. With HARVEST, the object is not to find a scheduling solution (i.e., determining the order in which individual stands should be harvested), but to assess the spatial pattern consequences of general management strategies.

The conceptual basis for simulation of cutting patterns at landscape scales can be traced back at least to Franklin and Forman's (1987) paper in *Landscape Ecology*. Other similar pattern-generation models include LSPA (Li et al., 1993), CASCADE (Wallin et al., 1994, 1996), and the DISPATCH model of Baker (1995; this volume) as modified to simulate disturbance by timber harvest. Harvest scheduling programs, e.g., FORPLAN (Johnson and Rose, 1986), SNAP (Sessions and Sessions, 1991), STEPPS (Arthaud and Rose, 1996) have much greater data requirements, and were not designed for this type of strategic landscape pattern assessment.

Description of the model

HARVEST is a cell-based (raster) model designed to simulate harvest methods that produce openings greater than one cell in size. It was designed to simulate even-age silvicultural methods such as clearcutting, shelterwood and seed-tree, and the uneven-age group selection method. The group selection algorithm can also be used to simulate patch cutting, in which several small clearcuts are dispersed throughout a stand. HARVEST is not able to simulate other uneven-age harvest systems such as single-tree selection or variable retention (Franklin *et al.*, 1997) unless the cell size is smaller than the size of a tree crown.

HARVEST simulates one time step per model run. The length of time represented by the model run is input by the user. Although this feature requires the user to initiate a run at each time-step of a simulation of long time periods, it allows for modification of the management parameters at each step. The inputs to the model for multiple time steps are typically recorded in a batch file, so that real-time intervention is not required. This also allows replicates of simulations to be easily produced.

A number of simplifying assumptions were made in the development of HARVEST to reduce input data requirements, and enable it to quickly simulate harvest activity over a relatively large area. The first is that harvest allocations within timber production zones typically take a spatially random distribution over the period represented by the time-step of the model run. However, this assumption does not nullify the spatial constraints most important in management planning: harvest allocations are constrained by the locations of existing stands that are older than the rotation length, and by the boundaries of management zones. It is important to note that spatially clustered harvests can readily be produced by HARVEST when timber production zones are delineated to force clustering. For example, HARVEST simulations reported in Gustafson (1996) demonstrate the spatial effects of varying degrees of clustering harvests. It is only within timber production zones that HARVEST distributes harvests randomly. The spatially random assumption is based on an analysis of stands reaching rotation age, and past harvest allocations. Using nearest neighbor analysis (Davis, 1986) on ten subsets of Hoosier National Forest (HNF; located in southern Indiana, USA) stand maps (mean size of subsets = 3366 ha, s.d. = 1062 ha), the observed mean nearest-neighbor distance between stands of similar age was compared to the distance expected if stands were randomly distributed, and a z-statistic was computed. The null hypothesis that stands are randomly distributed could not be rejected at the 95% confidence level for eight of the ten subsets (see Gustafson and Crow, 1996).

HARVEST also ignores specific forest types, with the exception of a single, secondary generic class (e.g., conifer), for which the user can (optionally) define a different size distribution for harvests. This feature was incorporated into HARVEST to allow larger harvest units on conifer plantations. Stands of forest types that will not be harvested at all should be excluded from the input map presented to HARVEST. If some forest types will not be harvested in proportion to their abundance within the timber land base, then harvest of various forest types would need to be simulated independently, and the results mosaicked using a GIS. HARVEST uses age as a surrogate for merchantability, and ignores stocking density and size class. Access and operability are assumed to be uniform across the land base. Significantly large areas known to be inoperable should be excluded from the timber land base for the simulations. Remember that HARVEST was designed to allow comparison of the impacts of broad management strategies on forest spatial pattern over large areas, and not for more detailed, stand-specific decisions.

HARVEST uses a number of parameters that are commonly specified in the standards and guidelines of management plans. These include harvest size distributions, total area harvested, rotation length (understood by HARVEST to mean the minimum age of stands that can be harvested), silvicultural method (even-age or group selection) and the width of buffers that must be left around harvests. An important capability of HARVEST is the ability to allocate harvests only in portions of the landscape that are designated for harvest. Equally important is the ability to apply different management strategies to different portions of the landscape (management areas). The primary output is a forest age map that includes the location(s) of canopy-removing harvest activity. The age of even-age regeneration can be used as a surrogate for a number of forest structure characteristics including canopy closure and seral stage. Successional change (from one forest type to another) is not modeled by HARVEST.

HARVEST was written in FORTRAN 77, and was originally developed using ERDAS Toolkit routines to run within the ERDAS v. 7.4$^+$ GIS environment. The ERDAS version is still available as Version 3.2. Version 4.1 is independent of ERDAS, allowing use of data exported from other raster GIS systems. Utility programs are available at the HARVEST Web site (URL given below) to convert map files in various text formats to the ERDAS 7.4 format, and to convert the output ERDAS files back into text format. These utilities can read and write data values in 1-, 2-, 3- and 4-digit integer formats, and also 8- and 16-bit ASCII characters. Input data must be in fixed column format. ARC/INFO users may use the command GRIDIMAGE to convert ARC grids into ERDAS 7.4 GIS files. However, ARC does not produce a full-length final record for ERDAS files, and so the files are not directly compatible with HARVEST. For ARC users, there is also available a utility that can be used to convert ARC-generated ERDAS files to a format that can be read by HARVEST. ARC/INFO users can use IMAGEGRID to convert HARVEST output files into ARC grid files.

Input data requirements for HARVEST are minimal: a stand age map, a stand ID map, and an optional forest type map if it is necessary to distinguish between two forest types (e.g., deciduous and conifer) that have different size distributions of harvest units. Timber harvest allocations are made by HARVEST using the stand age map, where grid-cell values reflect the age (in years) of the forest in that cell, and areas that are not to be harvested have a value of zero. HARVEST takes this GIS age map as input, and produces a new age map incorporating harvest allocations, where harvested cells take a value of 1 and unharvested non-zero cells increase in age by the time-step specified by the user. HARVEST also requires a map that contains a stand identification number for each forested cell. HARVEST records harvest information for each stand harvested (described below), which can be spatially linked to the landscape through this stand ID layer. This map is typically produced by passing the stand age map through a GIS clumping procedure so that contiguous cells of the same age are assigned the same stand ID number. Alternatively, an existing stand ID map can be used, provided that each stand has a unique ID value. The third (optional) input map is a land cover map that (minimally) contains the second forest type (e.g. conifer).

The parameters that are controlled by the user are listed in Table 12.1, with brief descriptions of their meaning and use. HARVEST allows control of the size distribution of harvest openings, the total area of forest to be harvested, and the rotation length (by specifying the minimum age on the input age map where harvests may be allocated). HARVEST generates a normal distribution of harvest sizes with a user-specified mean and standard deviation, and the user may truncate either tail of the distribution if desired. HARVEST allows the user to specify a different size distribution of harvests for up to one additional specific land cover type (for example, conifer). Should the user wish to harvest entire stands regardless of their size, HARVEST provides an option to constrain harvests by stand boundaries. The user can enter a very large mean size value, and each harvest will terminate when the stand becomes completely harvested. This option is also useful if management activity is to be constrained by existing stand boundaries. The model also allows buffers to be left between harvests, and between harvests and non-forested habitats. There are few parameters affecting the behavior of the model that are not under user control. Exceptions are: (i) the user cannot control the value assigned to harvested cells in the output age map (always a "1"), and (ii) the rules used to determine the type of forest found in the focal stand are fixed (described below). The random number generator is part of the source code, and can be examined or modified by interested users.

An ASCII file containing a record for each stand is produced on the first run of HARVEST, in which is recorded a user-specified integer code ("treatment code") for each stand harvested during that model run. This "treatment file" can be used on subsequent runs to control how HARVEST allocates harvests in stands that were treated during a previous run, and to record allocations made during the current run. For example, the user may specify minimum and maximum "treatment code" values, and HARVEST will not allocate harvests in stands with "treatment codes" outside that range. The user can also use this file to force HARVEST to revisit group selection stands at the appropriate time, and/or to prevent additional harvests in partially harvested stands. This file is the link between successive model runs, and represents institutional memory of previous management activity.

The two algorithms built into the model for determining the spatial dispersion of allocations are a random dispersion and a group selection dispersion. Under both algorithms, HARVEST selects initial harvest locations randomly within the timber management zones, checking first to ensure that the forest is old enough to meet rotation length requirements. Under the random dispersion algorithm, this suitable cell becomes the focal cell around which a harvest allocation will be made. Under the group selection algorithm, the stand in which this cell is located becomes the focal stand in which group openings will be allocated.

Group selection is implemented by HARVEST such that a proportion of a group-selected stand is cut during each entry. The number of cells (n) harvested

Table 12.1. *Parameters used by HARVEST to simulate alternative management strategies*

Parameter	Valid range	Data type	Description
Time step	≥1 (years)	Integer	Number of years represented by a model run. This value is added to the age of each (non-zero) cell in the input age map prior to harvest allocations.
Mean harvest size of primary forest type (PFT)	≥ 1 (cells)	Real	Specifies the mean value of the distribution of harvest sizes for the primary forest type.
Standard deviation (PFT)	≥ 0.0 (cells)	Real	Controls the width of the distribution of harvest sizes for the PFT.
Minimum size (PFT)	≥ 1 (cells)	Integer	Specifies the minimum allowable harvest size for the PFT. Enables user to truncate the left tail of the size distribution.
Maximum size (PFT)	≥ Minimum size (cells)	Integer	Specifies the maximum allowable harvest size for the PFT. Enables user to truncate the right tail of the size distribution.
Mean harvest size of secondary forest type (SFT)	≥ 1 (cells)	Real	Specifies the mean value of the distribution of harvest sizes for the secondary forest type. (optional)
Standard deviation (SFT)	≥ 0.0 (cells)	Real	Controls the width of the distribution of harvest sizes for the SFT. (optional)
Minimum size (SFT)	≥ 1 (cells)	Integer	Specifies the minimum allowable harvest size for the SFT. Enables user to truncate the left tail of the size distribution. (optional)
Maximum size (SFT)	≥ Minimum size (cells)	Integer	Specifies the maximum allowable harvest size for the SFT. Enables user to truncate the right tail of the size distribution. (optional)
Harvest mode:	"Group", "non-group"	Character	Group: small openings scattered within randomly selected stands; Non-group: harvests allocated at random locations.
Proportion (p) of a stand cut under group-selection	$0.0 < p < 1.0$	Real	How much of each group-selection stand will be allocated to small openings during the current model run.

Table 12.1. *cont.*

Total area to be harvested	≥ 1 (cells)	Integer	Total number of cells to be harvested. The model run is terminated when this number is reached.
Rotation length	≥ Time step (years)	Integer	Minimum age-value for cells to be harvested. Must include the time step that was added to ages at initialization.
Maximum age	< maximum age in map (years)	Integer	Maximum age-value for cells to be harvested. Must reflect the time step that was added to ages at initialization.
Width of buffers	0–20 (cells)	Integer	Distance (in cells) that must be left between harvested cells and any cells with value of 1 (prior harvests) or 0 (excluded areas).
Stay-within-stand option	"Y", "N"	Character	A switch to force HARVEST to stay within the stand boundaries for each allocation.
Treatment code	1–255	Integer	User-specified integer to flag all stands harvested during the current model run. For group-selection re-entries, this value is used to identify stands to be re-entered.
Minimum and maximum treatment codes	0–255	Integer	HARVEST will not allocate harvests in stands with "treatment codes" outside the range specified here.
Work array size	5–169 (cells)	Odd integer	Specifies the size of the work arrays. Used to improve performance when the harvest size distribution is small.
Random number seed	$0-(2^{31}-1)$	Integer	If a zero is entered, the random number generator is seeded using the system clock.

in a stand during each entry is calculated by HARVEST as a user-specified proportion (p) of the size of the stand (A):

$$n = (A \times p)$$

Selection of new stands for group selection is achieved with the "Generate" option of HARVEST, in which stands are randomly selected from those stands with an age greater than the prescribed rotation length, and small openings (groups of trees) within those stands are then randomly placed, with at least the user-specified distance between openings. Re-entry into previously group selected stands is achieved with the "Lookup" option, in which HARVEST allocates groups in all stands with a specific "treatment code" value stored in the "treatment file". During re-entry, groups are allocated on previously unharvested cells within the stand. The stands are allocated in numerical order by stand ID. The user must ensure that re-entries occur by invoking the "Lookup" option and specifying the same "treatment code" used to initiate the group-selected stands that are to be re-entered in the current model run. Because the group-selection algorithm disperses openings throughout a stand, it could be used to mimic single tree selection if the cell size is smaller than a tree crown. However, the ecological significance of such openings may be different than those produced when using HARVEST at the broader scale(s) for which it was designed.

Portions of the land base can be excluded from timber harvest by presenting HARVEST with an age map of only areas where harvest is allowed. Independent runs of HARVEST on different portions of a larger map may be used to simulate different management strategies on different portions (management areas) of the landscape. These management areas can later be mosaicked with the rest of the land base (using a GIS) to produce a map that characterizes the entire land base. A detailed example of the mechanics of implementing HARVEST is given in Gustafson (in press).

HARVEST has modest runtime memory requirements (approximately 390K), even for very large areas because the input maps are not loaded into memory. HARVEST uses an algorithm to access portions of the maps for processing. A log file is produced for each model run, recording all the input parameters and a summary of the runtime results.

HARVEST is relatively simple to use. The digital input maps can be derived from spatial data commonly maintained by land managers (i.e., stand maps with associated inventory information). The model parameters are relatively intuitive, and can often be derived from management planning documents. The (non-graphical) model interface requests parameter input from the user, and the model requires little technical skill to install and run. The user documentation is modeled after the ERDAS 7.5 user manual. HARVEST is useful as both a planning/management tool and a research tool. The model was developed to make it useful to strategic planners who wish to get coarse-filter answers to broad questions about potential alternatives, without the need for extensive training and technical support. Its initial development was spurred by research questions related to the landscape pattern consequences of forest management alternatives on the Hoosier National Forest. This dual role is consistent with the concept of management plans as working hypotheses (Levins, 1966; Baskerville, 1997). HARVEST allows exploration of the spatial consequences of management alternatives, and the results can be linked with other models relating spatial pattern to specific ecological processes (e.g., habitat suitability related to the pattern of fragmentation).

Algorithms

Simulations are implemented by HARVEST following the general algorithm shown in Fig. 12.1. After the files are open and the parameters input, HARVEST enters either the group-selection mode or the non-group mode, depending on the user's choice. Harvests are allocated until the specified number of cells have been harvested. The model run is terminated after the "treatment file" has been updated and all files closed. Because HARVEST was developed to run in an MS-DOS environment, memory limitations prohibited loading the input maps into memory. Harvest allocations are made by loading a portion of the map into a work array, modifying the cells harvested, and then writing the work array back into the map. The size of this array is controlled by the user, allowing faster execution speed when the expected size of harvest units is small (e.g., group selection).

Non-group harvest allocations

A randomly selected cell on the input map is examined to see if: (i) the cell is forested, (ii) the forest on the cell is older than the rotation length, and (iii) previous harvest activity has made the stand in which the cell is located unsuitable for harvest. If the cell is not suitable for harvest, new cells are chosen until a suitable cell is found. If HARVEST cannot find a suitable cell, it will notify the user and terminate. This suitable cell functions as a focal cell around which other cells will be harvested. HARVEST examines the land cover of the cells in a 17×17 cell neighborhood to determine the relative abundance of the primary and secondary generic forest type (e.g., conifer). The secondary type harvest size distribution is used if the abundance of the secondary type is >33%. A harvest size is then randomly selected from the appropriate harvest size distribution. If the size is outside the user-specified minimum and maximum, a new size is randomly generated.

The portions of the age map and stand ID map surrounding the focal cell are loaded into two work arrays. The data in these work arrays provide information used to determine whether individual cells meet the criteria for harvest as described below. Two logical arrays having the same dimensions as the work arrays are used to track the allocation of individual cells to the current harvest. The initial state of each array element is .FALSE. except for the center element (focal cell). Cells are added to the harvest in concentric rings (squares) around the focal cell. The algorithm uses a local Boolean expand operator, as described by James (1988, Chapter 6.) The two logical arrays are alternatively passed and returned as arguments to a subroutine that expands the current harvest by one additional concentric ring. Before a cell is actually harvested (set to .TRUE.) a check is made to verify that the cell is: (i) older than the rotation age, (ii) within the same stand as the focal cell (if the user has specified that harvests must stay within stand boundaries), and (iii) not too close (i.e., beyond the buffer distance) to another harvested cell or excluded area (zero valued cell). Thus harvests are usually square in shape, although

Fig. 12.1. Flow chart outlining the algorithm used by HARVEST to simulate timber harvest allocations. Input parameters are given in Table 12.1. The model run terminates (DONE?) when the user-specified number of cells have been harvested. See text for more details of the algorithms used in specific processes.

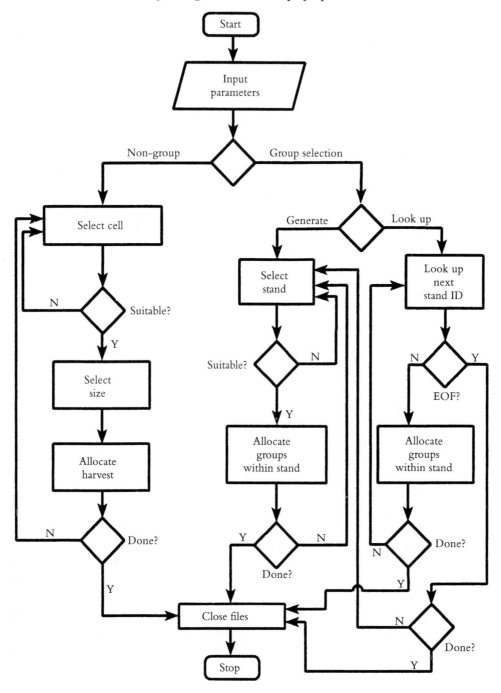

they do wrap around obstacles (e.g., cells within a buffer zone, or stand boundaries), and can take the shape of stands when large harvests are constrained to stay within stand boundaries. This allocation algorithm can only allocate contiguous cells. If the selected harvest size cannot be achieved because not enough suitable cells are contiguous, or the stand is too small, the allocation is made using the available cells. However, if the selected harvest size cannot be achieved because the work array is too small, HARVEST notifies the user, and records this event in the log file. The harvested cells are recorded by mapping all .TRUE. cells in the final logical array to a value of 1 in the age map work array. This array is then written back into the age map file, and the current "treatment code" is stored in the stand record for the stand of the focal cell in the "treatment file". The model continues to select new focal cells, iterating the allocation process until the total number of cells to be harvested is reached.

Group-selection harvest allocations

Group selection is simulated in two distinct modes: "Generate" (where group selection is first initiated in stands) and "Lookup" (where previously initiated stands are re-entered to allocate additional groups.) The "Generate" mode is implemented so that stands are selected randomly from the input age map, and then small openings (groups) are randomly placed within the stand boundaries. Stands are selected by randomly selecting cells using the procedure just outlined for non-group harvests. The cell is checked to determine if it meets the suitability requirements as outlined above. If it does, the stand containing the selected cell becomes the focal stand, and HARVEST loads the portion of the map containing the stand into the work arrays, and calculates the size of the stand. The total number of cells to be harvested within the stand is calculated as a proportion (input by the user) of the size of the stand. Each group (small opening within the stand) is allocated using the approach described for non-group harvests, except that all harvest allocations (groups) must be located within the focal stand.. The size of a group is chosen from the size distribution input by the user. (Forest types are not considered under group selection.) A random cell is chosen from within the stand, and it is checked to determine that it meets the age and buffer distance requirements. If it does, then the cell becomes a focal cell, and additional cells are allocated using the procedures outlined for non-group harvests. Groups are allocated until the proportion of cells to be harvested is reached. The age map work array is updated by the logical arrays and written back into the age map file, and the stand record in the "treatment file" is updated with the current "treatment code". The model continues to select new focal stands, iterating the allocation process until the total number of cells to be harvested is reached.

Group selection usually requires that stands be re-entered at intervals, to harvest additional groups. HARVEST provides the "Lookup" option to re-enter previously group-selected stands. In this case, HARVEST does not randomly select stands, but examines the "treatment file" to identify stands with a specific "treatment code" (input by the user). For each stand having the specific "treatment code," HARVEST finds the location of that stand, loads it into the work array, and allocates new groups on previously unharvested cells within that stand. If the total number of cells to be harvested has not been reached after all the stands to be re-entered are harvested, HARVEST enters the "Generate" mode to locate additional stands to be harvested by group selection (Fig. 12.1). The user is responsible to invoke the "Lookup" option at the appropriate time step(s), and to indicate the appropriate "treatment code" to properly simulate re-entries into the stands.

Model behavior and testing

HARVEST has been tested and used extensively by the authors since 1993, and it is well established that the model functions as it was designed. The behavior of the model has been demonstrated by testing the sensitivity of the results to variation in the main parameters (Gustafson and Crow, 1994). Mean harvest size was varied between 1 and 100 ha, in 10-ha increments, with a standard deviation of 10% of the mean; and total area harvested per decade was varied between 0 and 8% of the forest area within 23 593-ha forested landscapes, in 1% increments. We assumed that canopy closure occurred 20 years after harvest, and the amount of forest interior and forest edge habitat was calculated for each scenario using a GIS (Fig. 12.2). Group selection produces amounts of forest interior comparable to that produced in non-group mode when the number of stands harvested is held constant, because the same number of stands have canopy openings. However, when the number of cells harvested is held constant, group selection requires many more stands to reach the target, and fragmentation is usually higher (Gustafson and Crow, 1994, 1996). Group selection invariably produces more edge because of the higher perimeter-area ratios of smaller openings. Increasing the width of buffers that must be left around harvests reduces the amount of interior habitat when harvest levels are high, because the buffers serve to reduce clustering of harvest openings. Our studies have consistently shown that amounts of interior increase as harvests are clustered (Gustafson and Crow, 1996; Gustafson, 1996). However, the reduction of interior by increased buffer widths is negligible compared to the effect of dispersing harvests throughout a landscape at each time step (Gustafson and Crow, 1996; Gustafson, 1998b; Crow and Gustafson, 1997b). Forest fragmentation levels are most sensitive to the spatial restriction of timber harvests, even if these restrictions are temporary and their locations move across the landscape over time (Gustafson and Crow, 1996; Gustafson, 1996). Our experience has also shown that the variability in measures of forest fragmentation produced by replicates of model runs is generally quite low, and three replicates are usually adequate for robust results.

To verify that HARVEST produces patterns that mimic those produced by timber management, the past two decades of timber cutting were simulated on three study areas on the HNF (size range 34 053–49 515 ha). The ages of stands in 1968 and 1978 were reconstructed by subtracting 20 and 10 years, respectively, from each stand age in the 1988 stand age map. Stands with a calculated value >1

Fig. 12.2. *Response surface showing (a) the area of forest interior (>210 m from an edge or forest stand <50 years of age), and (b) showing the total length of forest edge (forest adjacent to an opening or forest stand <50 years of age) as a function of mean clearcut size and total area of forest harvested per decade. Simulations conducted using the landscape on a portion of the Pleasant Run Unit of the Hoosier National Forest.*

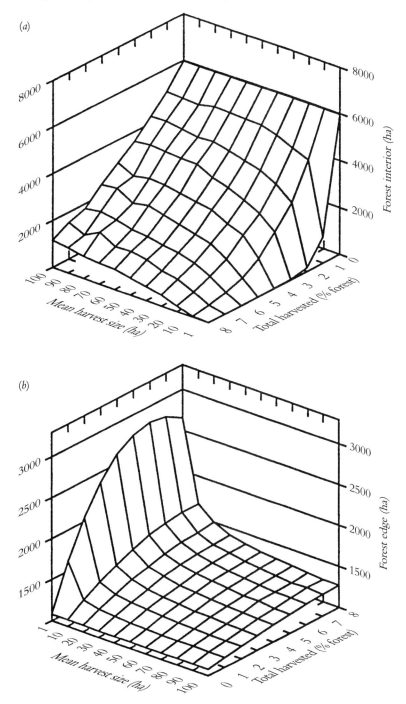

Fig. 12.3. Comparison of amounts of forest interior predicted by HARVEST and actual amounts estimated on three study areas (Lost River (LRIV), Pleasant Run (PRUN) and Tell City (TELL): see Gustafson and Crow, 1996 for more details) on the HNF. Error bars show 1 standard deviation based on three replicates. The possible range of values on each study area is shown by stippling, and the deviation from the actual amount as a percentage of the range is shown for each comparison.

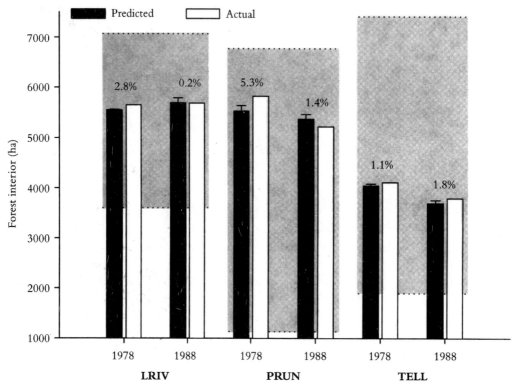

were likely regenerated during the interval, and it was assumed that these stands were closed-canopy forest prior to regeneration. The mean size and total area of stands actually regenerated in each decade were determined by analyzing the size distribution and total area of stands that had their year of origin within each decade. These values were used as parameters to simulate harvest during the two decades since 1968, and the amount of forest interior and edge predicted by HARVEST compared to the amounts derived from the 1978 and 1988 age maps. A forest opening was defined as a harvested area <20 years of age, and the amount of forest interior (forest >210 m from a harvest opening or forest edge) and linear forest edge on each study area calculated for each decade. The amount of forest interior on a managed landscape is the result of the spatial distribution of forested lands, harvests, and areas reserved from harvesting. The possible range of forest interior on each study area given the current configuration of non-forested lands and HNF ownership is shown by stippling in Fig. 12.3. HARVEST predicted the amount of forest interior within 3% of the range possible on each landscape except PRUN in 1978, where the mean prediction departed from the actual by 5.3% (Fig. 12.3). Predicted amounts of forest edge deviated little from the actual amounts (not shown). This is not surprising since edge is related more to the size of harvests than their spatial location, and HARVEST is able to closely match the size distribution of the actual harvests. These results suggest that the simple rules used by HARVEST can mimic patterns produced by forest planners who typically allocate harvest units under complex constraints. The prudent user should verify that the assumptions of HARVEST do not diverge appreciably from practices to be implemented on the planning unit(s).

Applications of HARVEST

Most applications of HARVEST have been made on the Hoosier National Forest in conjunction with their forest planning. The 1985 Forest Plan for the HNF specified primarily clearcutting across most of the Forest. Due to public opposition to this Plan, an Amended Forest Plan in 1991 specified primarily group selection (removal of small groups of trees), limited to a much smaller portion of the Forest. Using stand information and Landsat TM imagery from 1988 as initial conditions, we simulated the effects of implementing each plan over a 150-year period (Gustafson and Crow, 1996). Assessments were made in terms of the amount of forest edge and the amount of interior forest produced by each over the simulation period. Despite the 60% decrease in timber production in the 1991 Plan compared to the 1985 Plan, the treatments proposed in the 1991 Plan produced almost as much forest edge due to the reliance on small harvest units with large perimeter-to-area ratios. Further, the restriction of harvesting to a more limited area in the 1991 Plan had a greater effect on landscape pattern than did differences in harvest intensity between the two Plans. If the management goal is to reduce forest fragmentation, our simulations suggest that the most effective strategy is to establish areas managed to maintain a continuous canopy along with areas of intensive harvesting, rather than reducing cutting intensity across the entire forest (Gustafson and Crow, 1996).

Our prior work has assumed that canopies are not opened by harvest practices on the private land interspersed throughout the HNF. Currently, most timber is harvested on private land using a diameter-limit harvest method (cutting every tree with a DBH >16–18″), and less than 2% of privately-owned timber is removed by clearcutting (Jack Nelson, personal communication). However, approximately 30% of diameter-limit harvests remove most of the canopy, leaving only sapling-size (<4″) trees (Glen Durham, personal communication) To determine the possible effects on forest fragmentation of timber harvesting on private land, canopy-removing harvest activities were simulated on 32% of private land (over a five-decade period), and those results linked with simulations of the two HNF Forest Plans reported elsewhere (Gustafson and Crow, 1996). Private lands tend to be on the margins of the contiguous forested blocks in this area, and our objective was to determine whether private timber cutting has the potential to significantly increase forest fragmentation. Our simulation of cutting on 32% of private land may represent a worse-case scenario. Six scenarios were simulated using a 3 by 2 factorial: three harvest regimes on the HNF (1985 Plan, 1991 Plan and no harvest) and two regimes on private land (harvest and no harvest). Three replicates of each scenario were produced. Parameters defining the size distribution of harvest openings on private land were derived from a sample of open-canopy harvests on private land using aerial photographs (1:40 000). Parameters for harvests on the HNF were derived from published Forest Plans (USDA Forest Service, 1985, 1991).

A portion (34 053 ha) of the Pleasant Run Unit of the HNF was used as a study area, in which 32.8% of the area is under private ownership (Fig. 12.4). Although the stand size distribution on private land is not known, conversations with local county foresters suggest that stand sizes are similar to those on the HNF. Because a stand age map of privately owned land was not available, a stand age map was generated with stand size and age distributions similar to those found on adjacent HNF land. This was done by simulating past harvest activity on private land since settlement (120 years), beginning with a homogeneous forest. The simulation of past activity was implemented so that the total area of stands produced each decade was varied to produce a stand age distribution similar to that found on the adjacent HNF land within the study area. Although today there are differences between private and public forest management, stands now reaching maturity were initiated when the (now) public land was privately owned, so it is not unreasonable to assume that such stands would have similar size and age distributions even though some are now under public ownership.

To evaluate the simulation results, a forest opening was defined as a harvested area <20 years of age, and calculated the amount of forest interior (forest >210 m from a harvest opening or forest edge) and linear forest edge at each time step for each scenario. The variation in forest interior over time was a consequence of: (i) a recent timber cutting moratorium on the study area that caused relatively high levels of interior in decades 0–2 due to regeneration, and (ii) the establishment of an equilibrium pattern under each harvest regime (Fig. 12.5(a)). Levels of forest interior and edge approach an equilibrium at decade 5 that persists for ten more decades (Gustafson and Crow, 1996). Two striking features are seen in the spatial distribution of forest interior: (i) cutting on private land has minimal impact on forest interior because it primarily occurs on the margins of contiguous forested blocks (compare Fig. 12.6(a) and (c) and (b) and (d)), and (ii) locating areas reserved from canopy-opening cutting within large contiguous blocks of forest can produce significant amounts of continuous canopy (compare Fig. 12.6(c) and (e)). Differences in the amount of cutting on private land also had very little impact on linear forest edge (Fig. 12.5(b)). Harvesting on the HNF under the 1991 Plan was primarily by group selection, with amounts of edge comparable to that produced by the 1985 Plan, even though the total area cut was much less.

Additional insights into strategies for minimizing forest fragmentation were

Fig. 12.4. *Location and map of the study area showing the distribution of public and private land. HNF refers to land owned by the Hoosier National Forest, COE to land owned by the Corps of Engineers, and the remainder is privately owned. HNF land within the "reserved" polygons were withdrawn from timber production under the 1985 Land and Resource Plan and the 1991 Amended Plan.*

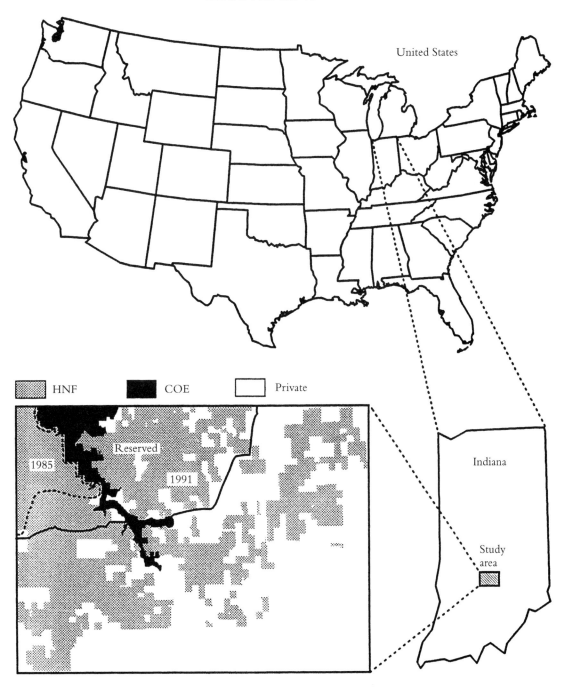

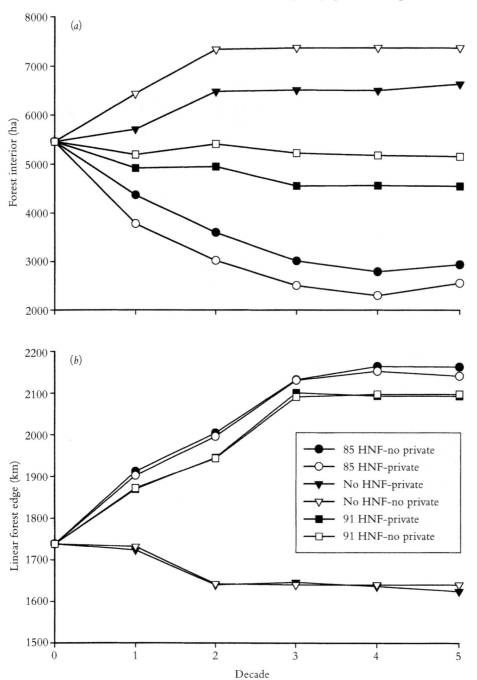

Fig. 12.5. Changes in the amount of (a) *forest interior habitat (forest >210 m from an edge or forest stand <20 years of age)* and (b) *linear forest edge over time resulting from simulation of timber harvest and no-harvest scenarios on the HNF and on private land. 85 HNF refers to harvests simulated under the 1985 Plan, 91 HNF refers to the 1991 Amended Plan, and No HNF represents no harvest openings produced on HNF lands. Private refers to harvest openings produced on 32% of privately owned land during the five decades simulated, and No Private refers to no harvest openings produced on private lands.*

Fig. 12.6. Maps of forest interior (forest >210 m from an edge or forest stand <20 years of age) at decade 5 under alternative timber harvest scenarios: (a) harvesting as specified in the 1985 Plan on the HNF, no harvest openings on private land; (b) no harvest openings on the HNF, harvest on private land; (c) harvesting as specified in the 1985 Plan on the HNF, harvest on private land; (d) no harvest openings on the HNF, no harvest openings on private land; (e) harvesting as specified in the 1991 Plan on the HNF, harvest on private land. Openings located on COE and Reserved lands (see Fig. 12.4) represent non-forest land cover that existed at the beginning of the simulation (1988) that were assumed to persist throughout the simulated period.

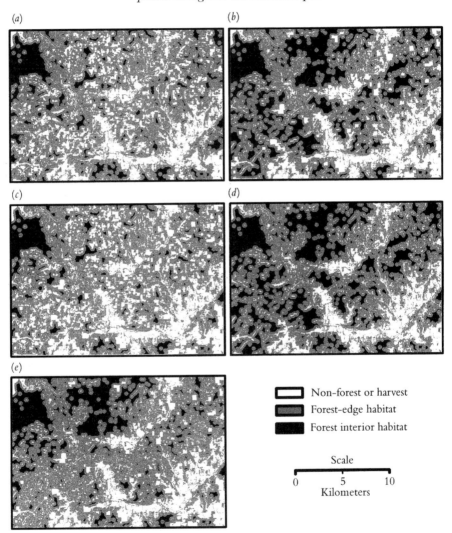

Fig. 12.7. *Stand age distributions after 210 years of timber harvests under the* (a) *1985 HNF Forest Plan, and* (b) *1991 HNF Forest Plan. Adapted from Gustafson and Crow (1996).*

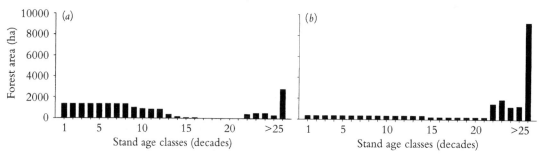

gained by applying HARVEST in both a temporal and spatial domain (Gustafson, 1996, 1998a). Again, based on simulations with real landscapes on the HNF, harvest strategies that produced the greatest clustering in time and space provided the greatest reductions in forest fragmentation when considered over the entire forest. This approach, called "dynamic zoning" (Gustafson, 1996), allowed increased timber production while reducing forest fragmentation. It illustrates the importance of thinking along scales of time and space when considering harvest allocations, and more generally, when contemplating alternative landscape designs for multiple uses. It is also a good example of the investigation of the spatial consequences of novel management paradigms. With the aid of HARVEST, the ability to explore the spatial and temporal elements of the multiple-use concept has been added.

Changes in age-class structure as related to alternative harvest strategies can also be studied with HARVEST. Again, using HNF simulations as examples, striking age-class distributions result when a static management strategy is maintained over long periods (Fig. 12.7). In the case of the 1985 Plan, even distributions of young age-classes occur up to rotation age. In this case, two-thirds of the timber base was to be managed with a rotation length of 80 years. The harvest strategy proposed in the 1991 Plan resulted in a preponderance of young and old stands, with a dramatic reduction in mid-age stands (Fig. 12.7) when simulated over 150 years. The premise that management applications can be held constant over long periods is obviously questionable. However, the ability to identify age-class distributions and potential gaps in the distributions that are produced by various management alternatives has great value for forest planners and managers.

Linking pattern with process

While predicting changes in landscape structure using spatially explicit models is relatively straightforward, understanding the impacts of these changes on ecological processes is more difficult. In an attempt to understand the relation between landscape pattern and ecological process, Liu *et al.* (1994) linked a spatially explicit model with an object-oriented model to simulate the population dynamics and extinction probability of Bachman's Sparrow (*Aimophila aestivalis*) in landscapes with different amounts and distributions of mature pine. Schulz and Joyce (1992) investigated the spatial application of a pine marten (*Martes americana*) habitat model. Hastings (1990) provided a more general discussion about incorporating spatial heterogeneity in population models. We have assessed the potential consequences of alternative patterns produced by HARVEST for a generalized neotropical migrant forest bird using a GIS model predicting the spatial distribution of the relative vulnerability of forest birds to brood parasitism by brown-headed cowbirds (Gustafson and Crow, 1994). While most models that relate ecological process with landscape pattern are simplistic – they generally deal with a single species, or at best, a guild of species – they serve as useful prototypes for future work in understanding the relation between land management, biological process, and ecological function.

Future utility for HARVEST is envisioned as a generator of forest patterns that can be used as input to models of population dynamics and other ecosystem processes. The premise of landscape ecology is that there is a strong relationship between the spatial pattern of ecosystem elements and ecological processes. The coupling of ecological process models to patterns expected under specific management alternatives provides a link that is desperately needed by management planners. HARVEST also holds promise as a tool for the investigation of the effects of scale (resolution and extent) on the representation and quantification of spatial pattern as it is related to forest ecosystems. These issues remain unresolved, but are critical to the practical implementation of ecosystem management (Wiens, 1997; Gustafson, 1998*b*).

Significance of approach

HARVEST is a timber harvest allocation model designed for efficiency and flexibility. The model simulates the impacts of even-age harvest (clearcut and shelterwood) and group selection on landscape pattern through time. Realism in the model is provided through application to existing landscapes by using classified Landsat TM imagery and digitized maps of stand ages as model inputs. Further, the rules controlling the allocation and size of harvest units can be based on the actual standards and guidelines developed in forest management plans. Other input variables include the rotation age (specified as a minimum age for harvest) and the percent of forested area to be harvested per unit time.

The utility of HARVEST is enhanced by using digital maps for both input to and output from the model. HARVEST is designed to be used in conjunction with a grid-cell geographic information system (GIS), with routines for direct input and output of ERDAS files along with support for moving files from other raster-based GIS programs. Timber harvest allocations are made by HARVEST using a digital stand map, where grid-cell values represent the age of the forest in that cell. Using the stand age map as input, HARVEST then produces a new stand age map incorporating harvest allocations. The output from the model can be integrated with a landscape map produced from other sources (e.g., satellite imagery). While landscape patterns incorporated in these maps can be quantified using any number of landscape metrics, comparisons among management alternatives can be expedited by simple visual comparisons of simulated landscape patterns (e.g., Fig. 12.6.). This spatially explicit tool, when applied through time, allows for considering both the spatial (where?) and temporal (when?) components of resource management questions (Crow and Gustafson, 1997a,b). Because the model requires little technical skill to install and run, and because it has minimal data requirements, it provides a strategic modeling tool that can be useful to planners seeking to assess general and perhaps novel management alternatives. Consideration of the most creative and novel alternatives are discouraged when their evaluation is costly and time-consuming. HARVEST represents a tool that that may encourage exploration of creative and risky options at the beginning of the planning process. It is at this stage that such options have the greatest likelihood of becoming viable. It is the large-scale decisions about the spatial zoning of timber management activities that have the most profound impact on spatial pattern and ecological processes, and the ability to simulate large-scale strategic options is critical for effective and efficient land management decisions.

Availability of HARVEST

HARVEST is available without cost on the Internet. Detailed user documentation, the software and source code can be downloaded from the North Central Forest Experiment Station World Wide Web site (http://www.ncfes.umn.edu) under "Research Products". Information about how to contact the authors is kept current at the Web site. A Windows version of HARVEST will be available on this web site by summer of 1999.

Acknowledgments

The development of HARVEST was funded by the North Central Forest Experiment Station of the US Forest Service, and by the Hoosier National Forest. Critical comments by W. L. Baker and an anonymous reviewer improved the manuscript.

References

Arthaud, G. J. and Rose, D. W. (1996). A methodology for estimating production possibility frontiers for wildlife habitat and timber value at the landscape level. *Canadian Journal of Forest Research*, **26**, 2191–200.

Baker, W. L. (1995). Longterm response of disturbance landscapes to human intervention and global change. *Landscape Ecology*, **10**, 143–59.

Baskerville, G. L. (1997). Advocacy, science, policy, and life in the real world.

Conservation Ecology, [online] **1**(1), 9. Available from the Internet. URL: http://www.consecol.org/vol1/iss1/art9

Brown, J. R. and MacLeod, N. D. (1996). Integrating ecology into natural resource management policy. *Environmental Management*, **20**, 289–96.

Crow, T. R. and Gustafson, E. J. (1997a). Concepts and methods of ecosystem management lessons from landscape ecology. In *Ecosystem management, applications for sustainable forest and wildlife resources*, ed. M. S. Boyce and A. Haney, pp. 54–67. New Haven, CT: Yale University Press.

Crow, T. R. and Gustafson, E. J. (1997b). Ecosystem management: managing natural resources in time and space. In *Creating a forestry for the 21st century: The science of ecosystem management*, ed. K. A Kohm and J. F. Franklin, pp. 424–50. Washington, DC: Island Press.

Davis, J. C. (1986). *Statistics and data analysis in geology*. New York: John Wiley.

Franklin, J. F. and Forman, R. T. T. (1987). Creating landscape patterns by forest cutting: ecological consequences and principles. *Landscape Ecology*, **1**, 5–18.

Franklin, J. F., Berg, D. R., Thornburgh, D. A. and Tappeiner, J. C. (1997). Alternative silvicultural approaches to timber harvesting: variable retention harvest systems. In *Creating a forestry for the 21st century: The science of ecosystem management*, ed. K. A. Kohm and J. F. Franklin, pp. 111–39. Washington, DC: Island Press.

Gustafson, E. J. (in press). HARVEST: A timber harvest allocation model for simulating management alternatives. In *Landscape ecological analysis: Issues and applications*, ed. J. M. Klopatek and R. H. Gardner, New York: Springer-Verlag.

Gustafson, E. J. (1998a). Clustering timber harvests and the effect of dynamic forest management policy on forest fragmentation. *Ecosystems*, **1** 489–92.

Gustafson, E. J. (1998b). Quantifying landscape spatial pattern: what is the state of the art? *Ecosystems*, **1**, 143–56.

Gustafson, E. J. (1996). Expanding the scale of forest management: allocating timber harvests in time and space. *Forest Ecology and Management*, **87**, 27–39.

Gustafson, E. J. and Crow, T. R. (1994). Modeling the effects of forest harvesting on landscape structure and the spatial distribution of cowbird brood parasitism. *Landscape Ecology*, **9**, 237–48.

Gustafson, E. J. and Crow, T. R. (1996). Simulating the effects of alternative forest management strategies on landscape structure. *Journal of Environmental Management*, **46**, 77–94.

Hastings, A. (1990). Spatial heterogeneity and ecological models. *Ecology*, **71**, 426–8.

James, M. (1988). *Pattern recognition*. New York: John Wiley.

Johnson, K. N. and Rose, D. W. (1986). *FORPLAN version 2: an overview*. Washington, DC: USDA Forest Service Land Management Planning System Section.

Levins, R. (1966). Strategy of model building in population biology. *American Scientist*, **54**, 421–31.

Li, H, Franklin, J. F., Swanson, F. J. and Spies, T. A. (1993). Developing alternative forest cutting patterns: A simulation approach. *Landscape Ecology*, **8**, 63–75.

Liu. J., Cubbage, F. W. and Pulliam, H. R. (1994). Ecological and economic effects of forest landscape structure and rotation length: simulation studies using ECOLECON. *Ecological Economics*, **10**, 249–63.

Schulz, T. T. and Joyce, L. A. (1992). A spatial application of a marten habitat model. *Wildlife Society Bulletin*, **20**, 74–83.

Sessions, J. and Sessions, J. B. (1991). Tactical harvest planning. In *Proceedings of the 1991 SAF National Convention*, August 4–7, 1991, San Francisco, California.

USDA Forest Service. (1985). *Land and Resource Management Plan, Hoosier National Forest.* Bedford, IN: USDA Forest Service, Eastern Region, Hoosier National Forest.

USDA Forest Service. (1991). *Plan Amendment, Land and Resource Management Plan, Hoosier National Forest.* Bedford, IN: USDA Forest Service, Eastern Region, Hoosier National Forest.

Wallin, D. O., Swanson, F. J. and Marks, B. (1994). Landscape pattern and response to changes in the pattern-generation rules: Land-use legacies in forestry. *Ecological Applications*, **4**, 569–80.

Wallin, D. O., Swanson, F. J., Marks, B., Kertis, J. and Cissel, J. (1996). Comparison of managed and pre-settlement landscape dynamics in forests of the Pacific Northwest, U.S.A. *Forest Ecology and Management*, **85**, 291–310.

Wiens, J. A. (1997). Scientific responsibility and responsible ecology. *Conservation Ecology*, [online] **1**(1), 16. Available from the Internet. URL: http://www.consecol.org/vol1/iss1/art16.

13

Progress and future directions in spatial modeling of forest landscapes

William L. Baker and David J. Mladenoff

Introduction

Today, we are confronted by new questions underlying natural resource disputes and forest management throughout the world. How will our future forest landscapes look? What products, services, and opportunities will they afford for people? How well will the biological diversity and functioning of forest ecosystems be perpetuated? Increasingly, these questions demand a long-term outlook on large land areas, and a spatially sophisticated framework.

Now, we are asked not just how much old growth forest will there be, but will old growth forest patches in 2050 AD be sufficiently connected to other patches to insure movement and viability of metapopulations of area-sensitive forest species? This kind of question is difficult to answer with chronosequences, experiments, stand-level plot studies, and other techniques that have been the basis of research fundamental to forest management in the past. There is little doubt that spatial models of forest landscapes will have to play an increasing role in addressing these questions.

Forest landscape ecological models (FLMs) have matured over the last decade from simple, checkerboard-scale, abstract, game-like models to complex models of large landscapes with feedbacks, spatial interactions, and linkages to other models. Here, the achievements and findings evident in the diversity of presentations in this book will be reviewed, present shortcomings will be outlined, and a little about future directions of research will be speculated.

Themes in forest ecology represented in models

The themes that are the subject of our modeling efforts are those that produce structure in forest landscapes from the scale of individual trees to that of entire landscapes and regions. Models with roots in the gap or individual-tree modeling traditions are rich with detail about tree-to-tree interactions and the resulting successional process (Caspersen *et al.*, Chapter 2; Liu *et al.*, Chapter 3; Urban *et al.*,

Chapter 4). The LANDIS and VAFS/LANDSIM models use vital attributes or life history traits in modeling succession with natural disturbances (Roberts and Betz, Chapter 5; Mladenoff and He, Chapter 6). Natural disturbance itself is another significant theme, with most models focused on fires. The intricate details of spread of individual fires are modeled mechanistically in *FARSITE* (Finney, Chapter 8), while the longer-term behavior of fires and fire effects on landscapes are the foci of one version of DISPATCH (Baker, Chapter 11) and SAFE FORESTS (Sessions *et al.*, Chapter 9). The spectrum of approaches to fire modeling in landscapes is reviewed by Gardner *et al.* (Chapter 7). Perhaps, because of their central importance globally, timber harvesting and deforestation are a focus of models oriented toward human disturbances (Liu *et al.*, Chapter 3; ZELSTAGE of Urban *et al.*, Chapter 4; Dale and Pearson, Chapter 10; Baker, Chapter 11; Gustafson and Crow, Chapter 12).

Treatment of forest ecology in landscape models

What are the essential processes and structures in forest landscapes that must be modeled? The answer to this question is still evolving, in part from our modeling efforts. In a related area of expanding research, on metapopulations, it was a modeling study (Levins, 1970), rather than an empirical study, that stimulated much of the subsequent empirical and modeling research (e.g., McCullough, 1996; Hanski & Gilpin, 1997). The stimulus for FLMs, in contrast, has followed empirical research and natural resource controversies that have shifted the focus from the stand level to the landscape ecology of forests (e.g., Harris, 1984; Thomas *et al.*, 1990; Laurance and Bierregaard, 1997). Some of the major processes and structures that appear essential to model can perhaps now be identified, yet our image of the forest and our understanding of its complexity continue to change.

Processes and structures of at least four scales now appear to be of interest in FLMs (Table 13.1). At the patch level are the many well-known processes that generate spatial and vertical structure within a forest stand, based largely on the response of individuals. These processes have been the subject of much research and modeling effort (e.g., Botkin *et al.*, 1972; Oliver and Larson, 1990). Processes and structures at the landscape scale are less well studied, but well represented in our models. Regional and global processes may influence landscape dynamics as well, but these also have received less attention.

Patch processes in landscape models

A forest stand is a patch of forest that is internally relatively homogeneous. This somewhat arbitrary forest unit has been the subject of much previous empirical research. In FLMs a stand is represented as either a single pixel or as a group of pixels or a polygon. Indeed, in grid-based FLMs the concept of a forest stand may be unnecessary, and processes may center around individual pixels that represent repeating square units of fixed size within a forest.

Table 13.1. *Major processes and structures affecting tree populations in FLMs*

Scale	Processes	Structures
Patch	Within-patch dispersal	Within-patch mosaic of seed densities
	Regeneration	Within-patch mosaic of trees of different ages/sizes
	Growth	Variation in vertical/spatial array of tree sizes
	Mortality	Spatial and vertical mosaic of dead and downed wood
	Competition	Modifies the regeneration and growth processes
	Within-patch herbivory	Modifies the regeneration and growth processes
	Succession	Successional stages, communities
	Small natural disturbances	Within-patch mosaic of successional stages/tree groups
	Boundary/edge differentiation	Edge versus interior environments
Landscape	Patch-to-patch dispersal	Patch-to-patch mosaic of seed densities
	Metapopulation dynamics	Patch-to-patch mosaic of sub-populations
	Herbivore movements	Patch-to-patch variation in herbivory effects
	Large natural disturbances	Disturbance patches
Region	Fluctuation in regional species pool	Regional seed and pollen rain
	Herbivore migration	Spatial variation in herbivory effects
Global	Neotropical migrant declines	Decline in bird-disseminated seeds
	Climatic change	Variation in regeneration, growth, and mortality processes

Models differ in the within-stand tree-to-tree interactions and vertical forest structure that is simulated. In gap models and derivatives, the stand has vertical structure and interactions that control light and moisture regimes (Caspersen *et al.*, Chapter 2; Urban *et al.*, Chapter 4). In FACET, for example, the stand consists of a grid of cells potentially occupied by individual trees, in which shading by adjoining trees and overstory trees affects light availability monitored at 1 m vertical increments (Urban *et al.*, Chapter 4). Moreover, the soil has multiple layers in which soil water is maintained and in which roots grow. Overstory trees influence soil water through interception and modification of transpiration rates. Light, water, nutrients, and temperature influence tree establishment, growth, and mortality. Competition between trees thus arises from effects on light and moisture in the canopy and, potentially, below ground. In LANDIS and VAFS/LANDSIM, only tree species age classes are simulated, and individual trees, vertical layers in the canopy, and soil are not simulated. When cells or polygons approach tree size, then these models may track individual tree locations. Species are ranked by shade tolerance, and can reproduce in their own shade or the shade of less tolerant species, as an approximation of vertical shading interactions (Roberts and Betz, Chapter 5; Mladenoff and He, Chapter 6). In FORMOSAIC, the landscape is divided into grid-cells, but the location and birth, death, and growth of each tree within a grid-cell is tracked. Tree size and neighborhood pressure influence individual tree growth, using a demographic approach, so vertical canopy layers, shading, and moisture-driven functions are not used (Liu et al., Chapter 3). In other FLMs, there is no vertical or horizontal structure within the stand or pixel, and within-stand processes operate only on aggregate variables. For example, the age of each forest pixel increases (MetaFor of Urban et al., Chapter 4; Baker, Chapter 11; Gustafson and Crow, Chapter 12) or carbon accumulates (Dale and Pearson, Chapter 10).

Within-stand processes also affect natural disturbance in several models. Within-stand fire and fuel-buildup processes are reviewed in Gardner et al. (Chapter 7). A species-specific function adds leaf litter, branches, foliage, and whole stems to time-lag fuel moisture classes in relation to tree demography in FACET (Urban et al., Chapter 4). More simply, rates of fuel accumulation and decomposition vary with land type and cell age in LANDIS (Mladenoff and He, Chapter 6) and VAFS/LANDSIM (Roberts and Betz, Chapter 5), while probability of fire is conditioned on soil moisture and time since fire in MetaFor (Urban et al., Chapter 4). Fires initiate probabilistically, based on mean fire interval (Urban et al., Chapter 4; Roberts and Betz, Chapter 5; Mladenoff and He, Chapter 6). Fire intensity depends upon fuel load only (Mladenoff and He, Chapter 6) or also includes moisture (Urban et al., Chapter 4). In *FARSITE*, fire spread rate and intensity are calculated from the Rothermel equations, which use physical fuel properties, moisture content, wind speed, and slope (Finney, Chapter 8), and a simplified version of this approach is used in SAFE FORESTS (Sessions et al., Chapter 9). Damage within a stand depends upon fire intensity and species ability to tolerate fire (Urban et al., Chapter 4; Roberts and Betz, Chapter 5; Mladenoff and He, Chapter 6) or on crown scorch height (Finney, Chapter 8) and established fire effects models (Sessions et al., Chapter 9). FORMOSAIC explicitly and spatially models the effects of pigs on sapling recruitment in tropical rainforest and the effects of windthrow as a mortality agent (Liu et al., Chapter 3). Some models do not contain large-scale disturbance components at the present time (e.g., Caspersen et al., Chapter 2), and some disturbances (e.g., disease outbreak) have not been modeled.

The within-stand conditions that lead to increased probability of disturbance by timber harvesting and tenant farmers are also modeled in LANDLOG, FORMOSAIC, and DELTA. In LANDLOG and HARVEST the within-stand contribution to the susceptibility of a stand to harvesting depends only upon stand age, which increases as the model runs (Baker, Chapter 11; Gustafson and Crow, Chapter 12). In FORMOSAIC trees must reach a certain diameter before they are eligible for harvesting, which then occurs as selective logging, with an associated, larger impact zone around the tree (Liu *et al.*, Chapter 3). In DELTA, the carbon content of the forest recovers linearly following cultivation (Dale and Pearson, Chapter 10). SAFE FORESTS uses an optimization approach to allocate harvest and other silvicultural activities to reach forest structure and harvest goals subject to watershed constraints (Sessions *et al.*, Chapter 9).

The environment of a stand plays a significant role in modifying within-stand tree growth, regeneration, and mortality, as well as the probability of natural and human disturbances. In FACET, temperature and incoming radiation, adjusted for elevation, slope, and aspect, alter within-stand rates of tree growth, regeneration, and mortality in part directly, but also through effects on soil water balance (Urban *et al.*, Chapter 4). In the present landscape implementation of SORTIE the focus is on the effects of spatial variation in soil moisture on probability of sapling mortality (Caspersen *et al.*, Chapter 2). In FORMOSAIC, growth functions for individual trees are in part a function of slope and elevation (Liu *et al.*, Chapter 3). In LANDIS and VAFS/LANDSIM individual pixels or polygons are classified into landtypes or habitat types, based on environment, and the types influence species regeneration, fire characteristics, and fuel accumulation within the stand (Roberts and Betz, Chapter 5; Mladenoff and He, Chapter 6). Fuel moisture is a function of terrain variables in *FARSITE* (Finney, Chapter 8), and custom fuel models are modified by terrain variables in SAFE FORESTS (Sessions *et al.*, Chapter 9). Soil suitability for agriculture and the carbon content of the original vegetation type influence colonists' choice of lots to clear in the DELTA model (Dale and Pearson, Chapter 10). In LANDLOG, the suitability of a particular pixel for timber harvesting is determined by a combination of elevation, soils, slope, and proximity to riparian areas (Baker, Chapter 11). In MetaFor, elevation, slope, and aspect influence temperature and moisture indices that constrain species-specific establishment probabilities (Urban *et al.*, Chapter 4). The environment is thus influential, but is limited to a static function of topography and is not spatially interactive in most models. For example, while the soil varies both vertically and spatially inside stands in FACET (Urban *et al.*, Chapter 4), there is not yet simulation of horizontal hydrologic interactions, such as runoff or subsurface flow, or spatial flows of organic matter or nutrients.

FLMs have seldom to date been used to address the effects of forest fragmentation, so it perhaps is unsurprising that they commonly do not directly model some patch-level phenomena that are a consequence of fragmentation. For example, the edge environment of a patch often contains a different micro-environment from the interior (Murcia 1995), leading to different rates of birth, growth, and death. This can be modeled using FACET (Urban *et al.*, Chapter 4), by preventing tree growth in the cells representing the opening, since the light regime is three-dimensional. The FORMOSAIC application presented here specifically addresses the effects of adjoining oil palm plantations on dynamics in rainforest fragments differing in size, but does not include micro-environmental edge effects (Liu *et al.*, Chapter 3). Other models also do not presently have this capability, perhaps largely because they have been developed for use in continuous forests. Other potential patch-level effects on tree populations include changes in pollinator abundance, changes in herbivory as exemplified in the effects of pigs on rainforest (Liu et al., Chapter 3), and an altered disturbance regime inside the fragment. Application of FLMs to fragmented forest landscapes may require further development of patch-level phenomena.

Landscape-level influences and interactions

A primary process that links patches in most present tree-based FLMs is seed dispersal, which may be a distance function, a neighborhood function, or not be spatially determined. In the landscape implementation of SORTIE, seedling density is a function of the diameter of the source tree and is modeled as a radially symmetric cubic function of distance from the source tree (Caspersen et al., Chapter 2). In LANDIS species can disperse with high probability a certain "effective distance", but dispersal declines exponentially beyond that distance (Mladenoff and He, Chapter 6). In VAFS/LANDSIM species can regenerate in a stand only if there are sexually mature individuals in the stand itself or in one of its immediately adjacent neighbors (Roberts and Betz, Chapter 5). In FACET, seedling establishment is not currently spatially linked, but is simply a function of the environment within the stand, while in MetaFor the abundance of neighboring cells occupied by a species also influences its probability of establishment (Urban et al., Chapter 4). In FORMOSAIC, the oil palm plantations surrounding a rainforest patch do not, but a surrounding species-rich forest does, provide seeds into the rainforest patch (Liu et al., Chapter 3).

In the DELTA model, the movement of tenant farmers among lots is somewhat analogous to seed dispersal in tree models. However, the probability of abandonment of a lot is a function of time on a lot, and the probability of choosing another lot is a function of lot size, soil quality, distance to market along roads, and current carbon storage, although other factors can be added (Dale et al., 1993). In many ways this is a more complex movement component than is present in tree-based FLM models to date, but it is essential to the success of DELTA in replicating the spatial dynamics of deforestation (Dale and Pearson, Chapter 10).

In addition to the movement and dispersal of organisms, another spatial linkage in FLMs is the spread of natural disturbances, most commonly fire, and in two instances also wind (Liu et al., Chapter 3; Mladenoff and He, Chapter 6). In LANDIS, fire spreads to susceptible neighbor pixels once ignited, but is more likely to spread in a wind direction determined at ignition; fire size is constrained by mean, maximum, and minimum sizes specified as inputs (Mladenoff and He, Chapter 6). In VAFS/LANDSIM, fire can spread to neighboring polygons based on their mean fire interval, but fire size is also constrained by inputs (Roberts and Betz, Chapter 5). Similarly, in SAFE FORESTS, fire spreads to neighboring polygons, with fire size constrained by inputs (Sessions et al., Chapter 9). In MetaFor, fires spread probabilistically to neighbors based on their soil moisture status and time-since-fire, also constrained by a specified maximum fire size (Urban et al.,

Chapter 4). In one version of DISPATCH, fires spread probabilistically to neighbors based only on time since the last fire (Baker, Chapter 11). In the most complex fire-spread model, *FARSITE*, fire-spread is modeled as a vector-process influenced by physical fuel properties, moisture conditions, wind speed, and topography (Finney, Chapter 8). Other models do not simulate fire spread.

Some models also spread human disturbances. In LANDLOG, once a timber harvesting operation is initiated in a pixel, it attempts to spread to produce an approximately rectangular harvest unit with some random shape modifications as spread completes, but spatial constraints and adjoining stands typically lead to irregularly-shaped units (Baker, Chapter 11). HARVEST uses a similar approach, and also includes clear cutting, shelterwood, and seed-tree, as well as group-selection silvicultural approaches (Gustafson and Crow, Chapter 12). Deforestation operates on individual lots in DELTA, rather than spreading in grid-cell space (Dale and Pearson, Chapter 10). ZELSTAGE includes the algorithms from CASCADE (Wallin *et al.*, 1994) that model dispersed-patch and aggregated timber harvesting strategies (Urban *et al.*, Chapter 4). ZELSTAGE can also do hierarchically nested management in which within-stand prescriptions, such as thinning, can be distributed spatially across the landscape. Harvesting has also been incorporated into the most recent version of LANDIS (Gustafson *et al.*, unpublished data).

Some FLMs approach having the structure of models of metapopulations, or a set of sub-populations weakly linked by dispersal (Hanski and Gilpin, 1997), yet the themes of metapopulation modeling (viability, extinction) have not been themes of FLMs. The essential features of metapopulation models are within-patch birth, growth, and death processes (including catastrophes), typically modified by environmental conditions in the patch, coupled with dispersal between patches. Perhaps the reason that FLMs have not been applied in a metapopulation sense is that the focus to date has been upon dominant trees that produce pattern on the landscape scale. These trees typically are not distributed in isolated sub-populations weakly linked by dispersal, as are, perhaps, some rarer trees or other plants in forests. As a consequence, FLMs have not had a focus upon the within-stand small population processes (e.g., demographic stochasticity, genetic deterioration) that may lead to sub-population extinction (Wilcove, 1986). The most demographically explicit models (SORTIE; Caspersen *et al.*, Chapter 2; FORMOSAIC; Liu *et al.*, Chapter 3) may be most suitable for this use in the future.

As mentioned earlier, FLMs have not commonly been used to address problems associated with forest fragmentation, and this may explain the absent or incipient attention to landscape-scale effects on population processes. Animal studies have emphasized the role of corridors, barriers to movement (e.g., roads), and the resistance of the matrix to movement as factors influencing small populations in patches (Forman, 1995). Some of these processes, as well as the within-patch fragmentation processes mentioned earlier, do also affect plant populations, and might in the future be useful additions to FLMs when applied to fragmentation problems. FORMOSAIC does use the movement of pigs and seeds to analyze adjacency, one significant aspect of forest fragmentation (Liu *et al.*, Chapter 3).

FLMs do not presently include spatially interactive land surface processes. Wind direction may be specified for the spread of an individual fire at the onset (Mladenoff and He, Chapter 6; Finney, Chapter 8), but wind direction modifications by topography are not presently tractable in FLMs. Runoff and subsurface flow processes that spatially redistribute precipitation in watersheds also are not included. However, these are comparatively subtle additions to models that already account for the primary topographic effects on the temperature and moisture regimes important to tree populations.

Regional and global influences in landscape models

Regional processes and structures can have effects on local dynamics, even at the patch level, through secondary effects. Regional declines in forest abundance may influence local bird populations (Askins et al., 1987). In the case of bird-disseminated plant propagules, there may be subsequent changes in dispersal rates. Similar inter-scale interactions may also come from the global scale. For example, declines in Neotropical migratory birds due to forest loss in their wintering grounds may mean declines in these birds in temperate forest landscapes where they play roles in seed dispersal and in regulating insect abundance (Hagan and Johnston, 1992). It could perhaps be argued that these kinds of interactions from regional and global scales are less significant than are the basic environmental and disturbance processes that produce most of the pattern in our landscapes. Indeed, one of the difficulties of modeling these kinds of effects is that it may require decades or even centuries for their impact to become significant. However, 500 or 1000 years of forest dynamics are now routinely being simulated, and at this temporal scale the relevance of regional and global processes is potentially significant.

Along a similar vein, few of our models now have in place a mechanism for linkage to large-scale exogenous influences, such as global climate change, yet here the potential effects are well known. Gap models have been used to analyze the potential response of forests to global climate change (e.g., Solomon, 1986), but the ramifications for entire forested landscapes have not been effectively explored using FLMs. Simple scenarios for the response of disturbance landscapes to global change have been explored (Baker, 1995), but there remains considerable potential for using more complex models containing tree populations to explore the landscape implications of global change, perhaps through direct linkage to global climate models (GCMs).

Interactions among scales and world views in FLMs

Interactions between the patch, landscape, regional, and global scales (Table 13.1) are known, as mentioned above, from empirical research, yet our models reflect differing emphases about the relative importance of these scales. To a large extent, this reflects the development and genealogy of FLMs. Model development in landscape ecology has been done by individuals with training in various fields of ecology and forestry, but not landscape ecology explicitly, because it is such a new field (Mladenoff and Baker, Chapter 1). In part, this means the field is a diverse collection of researchers with experience at a range of spatial scales, with a similarly diverse set of attempts to develop FLMs. The field contains both new models built from the ground up to address larger scales, and approaches that use existing, fine-scale models as building blocks.

These differences in development and genealogy may also reflect differences in world views underlying our modeling approaches and emphases. Is the most significant source of pattern in landscapes individual trees or larger-scale forces, such as natural disturbance? Models that include individual trees or age/size-classes of trees emphasize the generation of pattern at the landscape scale from within-stand processes modified by landscape-scale environmental variation (Caspersen et al., Chapter 2; Liu et al., Chapter 3; Urban et al., Chapter 4; Roberts and Betz, Chapter 5; Mladenoff and He, Chapter 6). The landscape version of SORTIE seeks to employ models to evaluate how large scale patterns of the distribution and abundance of species emerge from small scale processes (Caspersen et al., Chapter 2). Proponents of this perspective may even suggest that the multiple scales that influence forest landscapes (Table 13.1) can all essentially emerge from models of individuals:

> Individual-based models link all of these separate levels in the ecological hierarchy. The responses of individuals to their local environment is based on physiological and behavioral responses. The aggregation of all individuals of a species produces the population dynamics of that species. The aggregation of all individuals of many species interacting with each other and with their environment produces community dynamics. Ecosystem dynamics result from the aggregation of individual-environment interactions into large-scale material and energy fluxes.
> (Huston et al., 1988, p. 690)

While there is reason to be enthusiastic about the ability of individual-based models to capture landscape-scale patterns, other modelers may argue that there are important processes, such as large-scale disturbance, global change, and the behavior of the global economic system that impose considerable structure on the fate of individual trees in forests, and that do not primarily emerge from the behavior of individuals. For example, two-way interactions between pathogens and forest development or landscape patterning suggest that some within-patch population processes and structures are in part controlled by landscape-scale processes and structures (Castello et al., 1995). Similarly, there is increasing evidence that local climate is in part a reflection of land surface structures and processes (e.g., Copeland et al., 1996), so that the rates of tree natality, growth, and mortality in a landscape may in part be an indirect reflection of landscape structure or regional forest abundance. Certainly, large-scale economic and social forces are constraining the fate of patches of tropical rainforest (Dale and Pearson, Chapter 10) and individual high-value trees in temperate forest landscapes. However, it may be difficult

Table 13.2. *Analogies between world views and modeling frameworks*

World view	Models and authors
Behavioralist	SORTIE: Caspersen et al., Chapter 2
	ZELIG version FACET: Urban et al., Chapter 4
	DELTA: Dale and Pearson, Chapter 10
Structurationist	FORMOSAIC: Liu et al., Chapter 3
	METAFOR: Urban et al., Chapter 4
	ZELSTAGE: Urban et al., Chapter 4
	VAFS/LANDSIM: Roberts and Betz, Chapter 5
	LANDIS: Mladenoff and He, Chapter 6
	FARSITE: Finney, Chapter 8
	SAFE FORESTS: Sessions et al., Chapter 9
Structuralist	LANDLOG: Baker, Chapter 11
	HARVEST: Gustafson and Crow, Chapter 12

Models are placed where their primary emphasis is at the present time.

to determine whether individual trees are controlling landscape processes or landscape processes are controlling individual trees (Castello et al., 1995), so an interactive view is also reasonable.

These differences in emphasis recall differences in world view that underlie how people explain the functioning of social systems, which can be broadly painted as behavioralist, structuralist, and structurationist (e.g., Zimmerer, 1991). Behavioralists attribute primacy to individuals (agency) and emphasize the power of the individual relative to structural constraints, which are often treated as simply context. Structuralists tend to emphasize that the behavior of individuals is so constrained by large-scale political and economic structures that there is little point in focusing on individuals as agents of change. Structurationists, in contrast, emphasize the mutual dependence of structure and agency. Giddens (quoted from Zimmerer, 1991), a chief proponent of structuration, suggests that "the structural properties of social systems are both the medium and the outcome of practices that constitute these systems" (Giddens, 1979, p. 69).

There is, then, an analogy between social theory and the theory or world views underlying our modeling emphases (Table 13.2). Individuals in human social systems are analogous to individual trees in a forest landscape. In our modeling approaches, there appears to be a convergence toward an inclusion of processes that represent both structure and agency and their interactions. Even in models that focus on large-scale disturbance (ostensibly a "structural" focus), such as *FARSITE*, there is considerable mechanistic influence at the stand level (local physical fuel properties), as well as influences on the local level from regional weather (e.g., influencing local fuel moisture and wind speeds). This structurationist world view, blending processes and structures at several scales and including their interactions, may be becoming the norm in forest landscape ecological models because of the recognition that landscape dynamics derive from the interaction of processes and structures at scales ranging from the individual tree to the patch, region, and even globe (Table 13.1).

Capabilities, limits, and needs in the forest landscape modeling process

Landscape size, resolution, and scaling

In the last decade computer capabilities have increased by more than an order of magnitude, and this is reflected in the scale and resolution of present models. Grid-based models now commonly work with extents close to, or greater than 1000 rows × 1000 columns (e.g., Baker, Chapter 11). Distributing a model over many workstations enables thousands of detailed gap-level plots to be simulated in a few hours (Urban *et al.*, Chapter 4). The individual-tree-based models of Caspersen *et al.* (Chapter 2) and Liu *et al.* (Chapter 3) can now be run effectively at the scale of hundreds of thousands of individual trees. Timber harvesting models are now feasible at the scale of entire National Forests (Baker, Chapter 11; Gustafson and Crow, Chapter 12). Complex polygon-based models with hundreds of polygons, ten species, and 30 habitat types can be run in minutes (Roberts and Betz, Chapter 5). DELTA can simulate 3000 lots on a 300 000 ha land area for 50 years in a few minutes (Dale *et al.* 1993). *FARSITE* can simulate a large landscape fire in complex topography in a few minutes (Finney, Chapter 8). Clearly, spatial modeling of forest landscapes has reached the level at which complex forest processes can be simulated in reasonable times on large land areas.

Model design, modularity, and modeling languages

There is potentially considerable advantage to creating forest landscape models or components of these models that are generic and modular in design (Reynolds & Acock, 1997; Sequeira *et al.*, 1997). For example, LANDIS (Mladenoff and He, Chapter 6), FORMOSAIC (Liu *et al.*, Chapter 3), and *FARSITE* (Finney, Chapter 8) use a relatively newer object-oriented programming design and the C++ language to compartmentalize or encapsulate the various program modules. In this way the internal duties of a module are separate from the internal dynamics of other modules, and the interaction of those modules (Mladenoff and He, Chapter 6). This approach can allow for easier program modification and additions, without broadly affecting other portions of the model code. Such a design may lead to a collection of modules or a toolbox approach, which can be selected from and joined, depending on need. Cross-platform compatibility (Windows or Unix) is also maintained as a part of this philosophy. However, an integrated modeling approach that addresses a range of scales and processes would be needed to facilitate this kind of generic development, and may be less likely to occur if it depends on individual investigator-driven research. More recent development in computer languages, such as Java, carry modularity and generic, cross-platform compatibility

Table 13.3. *Levels of coupling of GIS and FLMs during model runs*

No linkage to GIS	Liu *et al.*, Chapter 3;
GIS preprocessing of input data or GIS display of final output maps, but no use of GIS during model runs	Caspersen *et al.*, Chapter 2; Urban *et al.*, Chapter 4; Roberts and Betz, Chapter 5; Mladenoff and He, Chapter 6; Finney, Chapter 8; Sessions *et al.*, Chapter 9; Dale and Pearson, Chapter 10; Gustafson and Crow, Chapter 12
Files transferred from model to GIS during model runs; some GIS functions used during model runs	Baker, Chapter 11;
GIS and model share common files and memory and use a common interface	None
Model embedded in GIS as one system	None

further. Java was developed for Internet applications that need to operate on all computing platforms transparently. These developments in programming languages may further encourage a generic landscape modeling toolbox.

Use of GIS software, functions, and capabilities

Although geographical information systems (GIS), software packages for manipulating map data, contain programs potentially useful in FLMs, these models do not always use GIS. FLMs and GIS can be coupled at several levels of integration (Nyerges, 1993; Fedra, 1993), but present FLMs either do not use GIS or are only loosely coupled with GIS (Table 13.3). For some models there may be no particular value in linkage with GIS. Where use of GIS is advantageous, a loose coupling requires the least development, but as larger land areas with finer resolution are simulated, there may be significant time constraints from file transfer operations. For example, with LANDLOG (Baker, Chapter 11), the file transfer operations consume more than half of the processing time when simulating a 894 column × 1209 row area.

Most FLMs coupled with GIS currently use the GIS only for display or for a few data processing functions. Tighter integration with GIS may be advantageous if these models require more spatial sophistication, using more complex GIS functions, or require more frequent interaction with the GIS. For this integration to be possible, however, the GIS software itself must be reasonably open, so that embedded models can directly use GIS functions and data structures. For example, in the popular ARC/INFO GIS software (ESRI 1997) it was necessary to write model operations using a macro-language prior to Version 7.2, when an application-programming interface, that allows GIS functions to be directly called from standard programming languages, became available.

Validation of forest landscape models

Many of our landscape models are not yet validated for use as predictive models, simply reflecting their stage of development, or their intended purpose. However, validation is not a singular matter. Rykiel (1996) suggests that it is necessary to carefully state what the validation criteria will be based upon the purpose of the model, its desired performance, and the context for its use. He distinguishes (i) operational validation, where model output is tested relative to desired performance standards, (ii) conceptual validation, where theories or assumptions and model logic are evaluated, and (iii) data validation, where the quality of the input data are determined to meet a specified standard. Rykiel reviews Sargents' (1988) explanation of suitable validation testing procedures.

Testing approaches from previous ecosystem modeling research are useful starting points for landscape models, but may require further development in some instances. Sensitivity analysis and error analysis can lead to model refinement and effective data collection by identifying the needed precision of data collection and most important model parameters (Gardner *et al.*, 1981). Predictive spatial models may attempt to replicate cell-by-cell patterns or may only aim to predict aggregate measures of landscape structure. Turner *et al.*, (1989) and Costanza (1989) review model goodness-of-fit evaluation procedures for these two approaches. Loehle (1997) suggests that a hypothesis testing framework, rather than goodness-of-fit testing, will lead to better models. A neutral-model approach has been adapted for spatial modeling (Gardner *et al.*, 1987) and Henebry (1995) suggests an autocorrelation-based approach that can be used in a hypothesis-testing framework. Error analysis in a spatial framework may also require new approaches, such as an analysis of the contribution of interpolated input data to output model error (Phillips and Marks, 1996).

Applications of forest landscape dynamic models

Richard Hobbs has recently provided a strong challenge to landscape ecology:

> I suggest that the products of landscape ecology (i.e., theory, methodology etc.) are best assessed, not on their intrinsic interest or popularity in the scientific literature, but on the impact they have on the planning and management of real landscapes . . . I suggest that in its present condition, landscape ecology has surprisingly little to offer those wishing to plan and manage the landscapes of the future.
> (Hobbs, 1997, p. 6)

Can this challenge be met? Do our models offer something useful to those wishing to plan and manage future landscapes?

Our models appear to be almost universally firmly linked to the real world, and to have considerable value for planning and management, although some of the potential is just over the horizon at the present time. Dale and Pearson (Chapter 10) address a global problem, tropical deforestation, using a rich empirical database on the landscape and on human behavior, and with important model outputs (carbon output, forest loss). Baker (Chapter 11) and Gustafson and Crow (Chapter 12) use US government data to project the potential consequences of continuing present timber harvesting practices on a National Forest into the future, a forest planning need globally. ZELSTAGE is designed to analyze the effects of forest management and natural disturbance on landscapes in the Cascade and Coast ranges in Oregon (Urban et al., Chapter 4). Liu et al. (Chapter 3) examine the impacts of exotic pigs and timber harvesting on a tropical rainforest fragment. The LANDIS model has been used to examine the long-term consequences of landscape recovery from a century of human use (Mladenoff and He, Chapter 6).

VAFS/LANDSIM has been used to analyze the long-term impacts of lengthening the fire return interval on vegetation and landscapes in Bryce Canyon National Park, Utah (Roberts and Betz, Chapter 5). The *FARSITE* model can be used for predicting the potential pattern of fire spread as a fire is burning, or can be used to examine potential fires that might burn given certain fuel management options, both of which are very useful for fire managers (Finney, Chapter 8). SAFE FORESTS (Sessions et al., Chapter 9) is focused on the joint management of fire, late-successional forests, and timber harvesting in the Sierra Nevada Mountains. Caspersen et al. (Chapter 2) have, in the landscape version of SORTIE, established a strong empirical link to tree population dynamics and the physical landscape in their study area, and now are in a position to explore applied problems. ZELIG version FACET has the potential to address many problems, as evidenced by the application of gap models over the last two decades, but to do so now in a spatially informative manner.

Future directions

Where will forest landscape modeling be in the near future, and what are the potential directions of development that appear most fruitful? First, technological developments will surely lead to faster computers, making simulation of large, satellite-image-scaled areas and more complex models more feasible, even as the volume of satellite data leapfrogs this development by an order of magnitude. Second, some progress toward modular and generic models (e.g., Mladenoff and He, Chapter 6) may make it possible to reach the stage where development is through incremental improvements in process algorithms rather than new model construction. Third, further development of socioeconomic drivers for landscape models (e.g., Dale and Pearson, Chapter 10) will be needed to make our models more relevant to planning and management. Addition of meaningful output variables (e.g., fraction of old growth forest, volume of timber) will also increase the utility of our models (e.g., Sessions et al., Chapter 9). Fourth, a pluralism of emphases, from individual-based to regional/global models will continue to be useful for addressing problems at multiple scales, with meta-modeling (Urban *et al.*, Chapter 4) used when linkage is needed. Fifth, models more deeply embedded in GIS will become more feasible and more desirable due to greater use of GIS functions. Sixth, increasing attention to development of spatial algorithms will be needed, particularly to overcome the constraints in both grid-based and vector-based approaches. Finney's (Chapter 8) creative integration of vector-algorithms and raster-data is an example. Diverse efforts to develop process concepts and efficient algorithms is healthy, as is a pluralism of emphases. Seventh, it would behoove us to move our models into a phase of testing and validation that will shore up current developments and lead to refinements in process concepts and algorithms. Along with new testing and validation techniques, the field is now maturing to a level where detailed model comparisons can be made. Such comparisons, on a single landscape or dataset, should be a priority for research. This would give us valuable information concerning the appropriate model and scale to be used for specific questions, as well as help in evaluating comparative model designs and algorithms.

Lastly, linkage of landscape models to global climate models, and other process models, may be achievable, with potential benefits to both modeling efforts. It would be useful to reach the level of development of global climate models, where alternative models and empirical data are each contributing to refinement of models and data that are capable of helping us choose the future that Hobbs (1997) suggests we are not yet helping to shape:

> Landscape ecologists must decide whether they wish to participate in the process of shaping future landscapes, or simply act as passive recorders of changes in landscape patterns.
>
> (Hobbs, 1997, p. 7)

References

Askins, R. A., Philbrick, M. J. and Sugeno, D. S. (1987). Relationship between the regional abundance of forest and the composition of forest bird communities. *Biological Conservation*, **39**, 129–52.

Baker, W. L. (1995). Longterm response of disturbance landscapes to human intervention and global change. *Landscape Ecology*, **10**, 143–59.

Botkin, D. B., Janak, J. F. and Wallis, J. R. (1972). Some ecological consequences of a computer model of forest growth. *Journal of Ecology*, **60**, 849–72.

Castello, J. D., Leopold, D. J. and Smallidge, P. J. (1995). Pathogens, patterns, and processes in forest ecosystems. *BioScience*, **45**, 16–24.

Copeland, J. H., Pielke, R. A. and Kittel, T. G. F. (1996). Potential climatic impacts of vegetation change: a regional modeling study. *Journal of Geophysical Research*, **101**(D3), 7409–18.

Costanza, R. (1989). Model goodness of fit: a multiple resolution procedure. *Ecological Modelling*, **47**, 199–215.

Dale, V. H., Southworth, F., O'Neill, R. V., Rosen, A. and Frohn, R. (1993).

Simulating spatial patterns of land-use change in Rondônia, Brazil. *Lectures on Mathematics in the Life Sciences*, **23**, 29–55.

ESRI (1997). *Understanding GIS: the ARC/INFO method*. Redlands, CA: Environmental Systems Research Institute.

Fedra, K. (1993). GIS and environmental modeling. In *Environmental modeling with GIS*, ed. M. F. Goodchild, B. O. Parks and L. T. Steyaert, pp. 35–50, New York: Oxford University Press.

Forman, R. T. T. (1995). *Land mosaics: the ecology of landscape and regions*. Cambridge: Cambridge University Press.

Gardner, R. H., Milne, B. T., Turner, M. G. and O'Neill, R. V. (1987). Neutral models for analysis of broad-scale landscape pattern. *Landscape Ecology*, **1**, 19–28.

Gardner, R. H., O'Neill, R. V., Mankin, J. B. and Carney, J. H. (1981). A comparison of sensitivity analysis and error analysis based on a stream ecosystem model. *Ecological Modelling*, **12**, 173–90.

Giddens, A. (1979) *Central problems in social theory: action, structure, and contradiction in social analysis*. Berkeley, CA: University of California Press.

Hagan, J. M. III and Johnston, D. W. (eds.) (1992). *Ecology and conservation of Neotropical migrant landbirds*. Washington, DC: Smithsonian Institution Press.

Hanski, I. A. and Gilpin, M. E., eds. (1997). *Metapopulation biology: ecology, genetics, and evolution*. London: Academic Press.

Harris, L. D. (1984). *The fragmented forest*. Chicago, IL: The University of Chicago Press.

Henebry, G. M. (1995). Spatial model error analysis using autocorrelation indices. *Ecological Modelling*, **82**, 75–91.

Hobbs, R. J. (1997) Future landscapes and the future of landscape ecology. *Landscape and Urban Planning*, **37**, 1–9.

Huston, M., DeAngelis, D. and Post, W. (1988). New computer models unify ecological theory. *BioScience*, **38**, 682–91.

Laurance, W. F. and Bierregaard, R. O., Jr., eds. (1997). *Tropical forest remnants: ecology, management, and conservation of fragmented communities*. Chicago, IL: The University of Chicago Press.

Levins, R. (1970). Extinction. *Lecture Notes in Mathematical Life Sciences*, **2**, 75–107.

Loehle, C. (1997). A hypothesis testing framework for evaluating ecosystem model performance. *Ecological Modelling*, **97**, 153–65.

McCullough, D. R., ed. (1996). *Metapopulations and wildlife conservation*. Washington, DC: Island Press.

Murcia, C. (1995). Edge effects in fragmented forests: implications for conservation. *Trends in Ecology and Evolution*, **10**, 58–62.

Nyerges, T. L. (1993). Understanding the scope of GIS: its relationship to environmental modeling. In *Environmental modeling with GIS*, ed. M. F. Goodchild, B. O. Parks and L. T. Steyaert, pp. 75–93, New York: Oxford University Press.

Oliver, C. D. and Larson, B. C. (1990). *Forest stand dynamics*. New York: McGraw-Hill.

Phillips, D. L. and Marks, D. G. (1996). Spatial uncertainty analysis: propagation of interpolation errors in spatially distributed models. *Ecological Modelling*, **91**, 213–29.

Reynolds, J. F. and Acock, B. (1997). Modularity and genericness in plant and ecosystem models. *Ecological Modelling*, **94**, 7–16.

Rykiel, E. J., Jr. (1996). Testing ecological models: the meaning of validation. *Ecological Modelling*, **40**, 229–44.

Sargents, R. G. (1988). A tutorial on validation and verification of simulation models. *Proceedings of the Winter Simulation Conference*, **1988**, 33–9.

Sequeira, R. A., Olson, R. L. and McKinion, J. M. (1997). Implementing generic, object-oriented models in biology. *Ecological Modelling*, **94**, 17–31.

Solomon, A. M. (1986). Transient response of forest to CO_2-induced climate change: simulation modeling experiments in eastern North America. *Oecologia*, **68**, 567–79.

Thomas, J. W., Forsman, E. D., Lint, J. B. et al. (1990). *A conservation strategy for the northern spotted owl: A report of the Interagency Scientific Committee to address the conservation of the northern spotted owl*. USDA Forest Service, USDI Bureau of Land Management, US Fish and Wildlife Service, and USDI National Park Service, Portland, Oregon.

Turner, M. G., Costanza, R. and Sklar, F. H. (1989). Methods to evaluate the performance of spatial simulation models. *Ecological Modelling*, **48**, 1–18.

Wallin, D. O., Swanson, F. J. and Marks, B. J. (1994). Landscape pattern response to changes in pattern generation rules: land-use legacies in forestry. *Ecological Applications*, **4**, 569–80.

Wilcove, D. S. (1987). From fragmentation to extinction. *Natural Areas Journal*, **7**, 23–9.

Zimmerer, K. S. (1991). Wetland production and smallholder persistence: agricultural change in a highland Peruvian region. *Annals of the Association of American Geographers*, **81**, 443–63.

INDEX

Abies
  balsamea 158
  concolor 113
Acer
  rubrum 18, 21–5, 28–34, 36
  saccharum 22–5, 28–34, 36, 143–53, 157–8
age-classes 82, 99, 106, 126, 137–8, 328, 335
aspect 82–3, 193

basal area 28
Brazil 256–72

calibration 21, 142
California 84, 220–52
canopy
  leaf area 74, 79, 90
carbon release 259–72
Clements, Frederick 2–3
competition 15–16, 20, 47
Connecticut 21–38
cover type 82, 86
cutting
  see forest management

deforestation 257, 259–72
demography, tree 76, 83–4
  see also recruitment, mortality, growth
dispersal 18, 20, 38, 45, 47, 82–3, 100, 105, 129–30, 134–5, 338
distributed queuing 80
  see also parallel processing
disturbance 3, 6, 8
  fire 7, 37, 74, 77–9, 84, 92, 102–6, 129–31, 138–40, 143–51, 154–8, 163–80, 187–203, 217–19, 231–52, 279, 282, 284, 287, 334, 336, 338
    see also fuels
  gap-phase 19
  regime 277–8
  tolerance 102

wind 7, 49, 55, 125–31, 140–1, 143–4, 154–8, 338
drought 23–4, 35, 76–7
  tolerance 35

ecosystem management 42
elevation 58, 82–3, 86, 193, 337
  digital elevation model 82, 138
establishment
  see recruitment
evapotranspiration 76

Fagus grandifolia 18, 22, 28–34, 36
farmers 263–4
fire
  see disturbance, fire
forest fragmentation, 9, 42, 256–72, 288, 299, 301–3, 323–9, 337, 339
forest management 87, 288–303, 309–30
Fraxinus americana 18, 21–5, 28–34, 36
fuels 78–9, 84, 104–5, 129, 139–40, 169, 192–4, 225–7
fuzzy sets 100, 106

GIS
  data sets 32, 220–52, 263, 344
  linkage to 109, 121, 141, 261–2, 284, 289, 312, 329, 344
Gleason, H. 2–3, 15
global effects 340
gradient
  analysis 14–15
  moisture 14, 16, 20–5, 29–32, 34–7
  models 34
grid-cell format 45, 73, 82, 88, 126, 128, 189, 282, 312, 334, 343, 347
growth
  age increases 106, 136
  radial increment 18, 22–4, 77, 90

habitat type 103
  see also landtype
harvesting
  see forest management
herbivory by pigs 44, 49–52, 55–63
humidity 192
hydrology 76

Idaho 201
intermediate disturbance hypothesis 61

Kessell, S. 7

Landsat TM 32, 91, 154, 262, 265–6, 270
landscape ecology
  application of 8
  history of 1
landscape structure
  measures 106, 115–20, 261, 294–303, 324–8
  software 128, 141
landtype 128–30, 138
land-use change 258–72
late-successional forest
  see old-growth forest
leaf area
  see canopy, leaf area
life-history traits 27
  see also models, vital-attributes
light 17–19, 22, 32, 74–5, 83, 194, 335
litter 74
logging
  see timber harvesting

Malaysia 44–5
mechanisms 14, 16, 186, 202
metapopulations 339
models
  by acronym
    BEHAVE 172–4, 187, 231
    DELTA 8, 259–72, 280, 336–8, 343
    DISPATCH 8, 167, 169, 176–7, 279, 283–303, 310, 334, 338, 346
    EMBYR 169, 174–5
    FARSITE 8, 173, 174, 187–203, 279, 334, 336–8, 342, 343, 346
    FIRE-BGC 169, 173, 279
    FIREMAP 169, 172
    FM 97.3 73–81
    FORPLAN 4, 6, 217–18, 280, 288, 310
    FOREST 4–5
    FORET 5, 126
    FORMOSAIC 8, 44–63, 336–9, 343, 346
    HARVEST 8, 309–30, 338, 346
    JABOWA 4–5, 72, 126
    LANDIS 8, 125–59, 334–8, 343, 346
    LANDLOG 289–303, 336–8
    METAFOR 8, 81–5, 336–8
    MOSAIC 85–8
    SAFE FORESTS 8, 220–52, 334, 336–8, 346
    SORTIE 5, 17–38, 337–8, 341, 346
    VAFS/LANDSIM 8, 99–122, 126–7, 334–8, 346
    ZELIG version facet 8, 43, 72–81, 335–8, 346
    ZELSTAGE 87–91, 334, 338, 346
  cellular automata 8, 81–4, 92, 283
  distributional 7
  exploratory 170–1
  gap 4–5, 8, 43, 58, 71, 73–81, 86, 92
  growth and yield 6
  individual-tree 8, 14, 16–17, 38, 43–4, 48, 48, 59, 70, 333, 341
    Markov 3–4
    matrix 3–4
    meta-models 71–3, 81, 94–5
    neutral 94
    percolation 8
    physical 171–4
    point 70–2
    probabilistic 174–6
    resource competition 15
    rule-based 4
    semi-Markov 85–8, 92
    spatially explicit 7, 43–4, 99–100, 126, 330
    stage-structured 87–93
    statistical 176–7
    theoretical 168, 279
    transition 3–4, 5, 85, 88
    vital-attributes 3, 4, 99–103, 110–11, 127
      see also life-history traits
moisture
  see gradient, moisture
  see soil moisture
Montana 201
mortality 18–19, 22–6, 47, 77–8, 82–3, 90, 107, 136

National Forest
  Eldorado 216–50
  Hoosier 323–8
  Medicine Bow 288–303
National Park
  Bryce Canyon 112–21
  Sequoia 84
natural disturbance
  see disturbance
niche 15
nutrients
  nitrogen 74

object-oriented 45, 128, 131–4, 343
old-growth forest 210–14, 220–52
optimization 215–20
Oregon 87, 90

parallel processing 72, 281
  see also distributed queuing
patches 86, 277
Pinus
  banksiana 156–8
  ponderosa 113–14
  resinosa 156
  strobus 18, 22, 28–34, 37, 157–8
Populus tremuloides 143, 157–8
plantations 45, 49, 63

precipitation 75, 83
Pseudotsuga menziesii 113

Quercus rubra 18, 22, 28–34, 37, 143–53, 156–8

radiation
   see light
raster
   see grid-cell format
recruitment 19, 25–6, 47, 76, 82–3, 100, 129, 136, 147–9, 152–3
regeneration
   see recruitment
regional effects 340
roads 257–72

saplings 19, 22–3
scale
   multiple 44–6, 53, 70–1, 80, 110–11, 128–33, 215–16, 334–5
   of processes 9, 70–1, 80, 110–11, 128–33, 334–5
   spatial 46, 53, 70, 215–16
   spatial and temporal 2, 133
seedlings 19
sensitivity analysis 52, 110, 143–53, 320, 345
shade-tolerance 16, 35, 74, 101, 129, 336
shading 75, 77, 335, 336
Sierra Nevada Mountains 84, 210, 220–52
sites 135
slope 58, 82–3, 193, 337
soil
   fertility 77, 337
   type 36–7, 73, 75, 337
   moisture 14, 29–32, 34, 36–7, 74–8, 82–3, 335
species richness 56–62

stochastic 44, 142, 219, 259
succession 129, 134
   history of models of 2–9
   individualistic response to 2
   organismic concept of 2
   species trends during 28–34, 119, 143, 157–8

temperature 77, 82–3
timber harvesting 48, 53, 88–91, 235–6, 242–7, 280–1, 288–303, 309–30, 334
topography 20, 76, 82–3
tropical rainforest 45, 87, 257–72
Tsuga canadensis 18, 21–5, 28–35, 37–8, 157–8

uncertainty analysis 52
Utah 112–21

validation 32–4, 52, 93–4, 121, 142, 283, 290–9, 320–3, 345
vector format 109, 129, 190–2, 347
vegetative reproduction 101
   see also recruitment
verification
   see validation
vital attributes
   see models, vital-attributes
   see also life-history traits

water balance 76
Wilderness
   Selway Bitterroot 201
wind
   see disturbance, wind
Wisconsin 149, 154–8
world views 342